Mycoses in
AIDS Patients

Mycoses in AIDS Patients

Edited by

Hugo Vanden Bossche
Janssen Research Foundation
Beerse, Belgium

Donald W. R. Mackenzie
Central Public Health Laboratory
London, United Kingdom

Geert Cauwenbergh and Jan Van Cutsem
Janssen Research Foundation
Beerse, Belgium

Edouard Drouhet and Bertrand Dupont
Institut Pasteur
Paris, France

PLENUM PRESS • NEW YORK AND LONDON

Library of Congress Cataloging-in-Publication Data

Symposium on Topics in Mycology on Mycoses in AIDS Patients (1989 :
Paris, France)
    Mycoses in AIDS patients / edited by Hugo Vanden Bossche ... [et
al.].
        p.   cm.
    "Proceedings of the Third Symposium on Topics in Mycology on
Mycoses in AIDS Patients, held November 20-23, 1989, in Paris,
France"--T.p. verso.
    Includes bibliographical references.
    Includes index.

    1. Mycoses--Congresses.   2. AIDS (Disease)--Complications and
sequelae--Congresses.   3. Opportunistic infections--Congresses.
I. Vanden Bossche, H.   II. Title.
    [DNLM: 1. Acquired Immunodeficiency Syndrome--complication.
2. Mycoses--complication--congresses.   WD 308 S98967m 1989]
RC117.A2S93   1989
616.97'92015--dc20
DNLM/DLC
for Library of Congress   90-14250
                                                              CIP

Proceedings of the Third Symposium on Topics in Mycology on Mycoses
in AIDS Patients, held November 20-23, 1989, in Paris, France

ISBN-13: 978-1-4612-7912-9   e-ISBN-13: 978-1-4613-0655-9
DOI: 10.1007/978-1-4613-0655-9

© 1990 Plenum Press, New York
Softcover reprint of the hardcover 1st edition 1990
A Division of Plenum Publishing Corporation
233 Spring Street, New York, N.Y. 10013

PREFACE

The World Health Organization estimates that at least five million people worldwide are infected with human immunodeficiency virus (HIV). Of these about 100,000 are in Asia and Oceania, 500,000 in Europe, 2 million in the Americas and 2.5 million in Africa (Mann, 1989). The acquired immunodeficiency syndrome is characterized by a derangement in cell-mediated immunity leading to opportunistic infections with for example *Mycobacterium* spp., *Candida* spp., *Cryptococcus neoformans*, *Pneumocystis carinii*, *Toxoplasma gondii* and *Cryptosporidium*.

The third symposium on "Topics in Mycology" brought together 265 experts from 32 countries to discuss the epidemiology, immmunological and pathogenetic aspects of AIDS and its opportunistic infections in general and fungal infections in particular.

Pneumocystis carinii pneumonia is by far the commonest opportunistic infection in AIDS patients. The nature and classification of *P. carinii* is still controversial. In search for its true taxonomic affinities an introductory paper formulates a number of key questions. Candidosis is another frequent opportunistic infection. A number of papers discuss the possibility that selective pressures may operate on *Candida albicans* within the AIDS population and influence its nature: this might have an impact on prophylaxis and curative and/or suppressive therapy.

AIDS has become the leading predisposing factor for cryptococcosis. Other infectious complications in AIDS patients include histoplasmosis and coccidioidomycosis. In contrast, paracoccidioidomycosis and blastomycosis are rare, as are aspergillosis and sporotrichosis. However, an unusual mycosis, penicilliosis (caused by *Penicillium marneffei*) has been proposed as indicative of AIDS. HIV infection and/or treatment may induce the appearance of other rare mycoses such as trichosporonosis, saccharomycosis and fusariosis. A great number of scientists of high international standing discuss the epidemiological, immmunological and clinical aspects of these mycoses and generate a number of questions that should stimulate further research.

These Proceedings, which are comprised of the papers presented at this symposium, contain a wealth of information on AIDS-indicative mycosis, dermatomycosis and rare mycoses encountered in some HIV positive patients, and provide up-to-date information on fungal models in immunocompromised animals, on pharmacokinetics and on the mode of action of antifungals of use in immunocompromised patients. The actual status and perspectives in the treatment of mycoses in patients with AIDS are highlighted.

The Symposium was organized in the "Institut Pasteur" (Paris), an institute that, since its founding in 1887, has contributed to the discovery of several pathogenic agents, a tradition renewed with the discovery of two viruses: HIV-1 and HIV-2.

The Symposium was sponsored jointly by the Janssen Research Foundation, the International Society for Human and Animal Mycology and the Mycology Unit of the Pasteur Institute.

The Editors

Mann, J., 1989, Global AIDS into the 1990's, <u>World Health</u>, October 1989: 6.

ACKNOWLEDGEMENTS

The organizers are grateful to the Pasteur Institute (Paris), the International Society for Human and Animal Mycology and the Janssen Research Foundation for providing the opportunity to organize this symposium.

Thanks are due to chairmen, speakers, authors and all participants and especially to those who presented posters and contributed so much to the discussions.

We have been greatly assisted in our organizational duties by Mrs. M. Verbaandert, Mrs. H. Dergent, Mrs. A. Siegers, and Mrs. C. Volkerick.

Special thanks are due to Mrs. H. Dergent for retyping the manuscripts.

Hugo Vanden Bossche
Donald Mackenzie
Geert Cauwenbergh
Jan Van Cutsem
Edouard Drouhet
Bertrand Dupont

CONTENTS

Introductory Papers

Epidemiology of AIDS and Its Opportunistic Infections 3
    F.W. Chandler

Immunologic and Pathogenic Aspects of HIV Infections:
    Current Hypothesis 13
    D. Dormont

Mycoses in AIDS Patients. An Overview 27
    E. Drouhet and B. Dupont

Pneumocystis carinii: A Nomadic Taxon 55
    D.W.R. Mackenzie

Candida and Candidosis in AIDS Patients

Epidemiology of *Candida* Infections in AIDS 67
    F.C. Odds, J. Schmid and D.R. Soll

Candidemia in Patients with Acquired Immunodeficiency Syndrome   75
    F.E. Chu, M. Carrow, A. Blevins and D. Armstrong

Immunological Aspects of Candidosis in AIDS Patients 83
    D.W. Warnock

Controversial Aspects of Candidiasis in the Acquired
    Immunodeficiency Syndrome 93
    J.D. Sobel

Cryptococcosis

Ecology of *Cryptococcus neoformans* and Prevalence of Its Two
    Varieties in AIDS and Non-AIDS Associated Cryptococcosis   103
    K.J. Kwon-Chung, A. Varma and D.H. Howard

Clinical Aspects of Cryptococcosis in Patients with AIDS   115
    J. Vandepitte

Immunological Aspects of Cryptococcosis in AIDS Patients   123
    J. Müller

Dermatomycoses and Rare Mycoses in AIDS Patients

Dermatophytes and *Pityrosporum* in AIDS Patients   135
    Ch. De Vroey and M. Song

Clinical Aspects of Dermatomycoses in AIDS Patients   141
    R.J. Hay

Unusual Mycoses in AIDS Patients   147
    M.A. Viviani and A.M. Tortorano

Mechanism of Pathogenesis- Why Are Certain Mycoses Rare in
    AIDS Patients ?   155
    H.P.R. Seeliger and K. Tintelnot

Dimorphic Fungi

Histoplasma in AIDS Patients   163
    M.G. Rinaldi

Coccidioides immitis in AIDS Patients   171
    J.N. Galgiani, N.M. Ampel, C.L. Dols and D.G. Fish

The Major Endemic Mycoses in the Setting of AIDS:
    Clinical Manifestations   179
    J.R. Graybill, P.K. Sharkey, P. Johnson and S. Nightingale

Immunological Aspects of Dimorphic Fungi in AIDS   191
    J.W. Murphy

Chemotherapy

Current Status and Perspectives of Antifungal Therapy   199
    J.E. Bennett

Fungal Models in Immunocompromised Animals   207
    J. Van Cutsem

Mode of Action of Antifungals of Use in Immunocompromised
  Patients. Focus on *Candida glabrata* and *Histoplasma*
  capsulatum 223
  H. Vanden Bossche, P. Marichal, J. Gorrens, D. Bellens,
  M.-C. Coene, W. Lauwers, L. Le Jeune, H. Moereels and
  P.A.J. Janssen

Pharmacokinetics of Antifungals 245
  G. Cauwenbergh and J. Heykants

Skin Candidosis in AIDS Patients. Effects of Ketoconazole
  and Itraconazole. Focus on Tissue Levels 255
  H.C. Korting

The Modern Revolution in Antifungal Drug Therapy 265
  J.R. Graybill

Treatment of Candidosis in AIDS Patients 279
  G. Just, D. Steinheimer, M. Schnellbach, C. Böttinger,
  E.B. Helm and W. Stille

Cryptococcal Meningitis in AIDS Patients- A Pilot Study
  of Fluconazole Therapy in 52 Patients 287
  B. Dupont, I. Hilmarsdottir, A. Datry, M. Gentilini,
  P. Dellamonica, E. Bernard, S. Lefort, J. Frottier,
  P. Choutet, J.L. Vilde and the French study group
  on fluconazole in cryptococcal meningitis in AIDS
  patients

Oral Itraconazole Therapy of Cryptococcal Meningitis and
  Cryptococcosis in Patients with AIDS 305
  D.W. Denning, R.M. Tucker, J.S. Hostetler, S. Gill and
  D.A. Stevens

Treatment of Dermatomycosis in AIDS Patients 325
  H. Degreef

Index.. 329

Introductory Papers

EPIDEMIOLOGY OF AIDS AND ITS OPPORTUNISTIC INFECTIONS

Francis W. Chandler

Department of Pathology, BF-230, Medical College of Georgia
Augusta, Georgia 30912, USA

The acquired immunodeficiency syndrome (AIDS), now known to be caused
by a novel retrovirus known as the human immunodeficiency virus (HIV)
(Barre-Sinoussi et al., 1983; Gallo et al., 1984), is characterized by a
profound derangement in cell-mediated immunity leading to multiple
opportunistic infections and unusual neoplasms. AIDS was first recognized
in the spring of 1981 when *Pneumocystis carinii* pneumonia (PCP) and
Kaposi's sarcoma (KS) appeared in previously healthy homosexual men in
California and New York (CDC, 1981a,b). It soon became apparent that other
opportunistic diseases were present in addition to PCP and KS, and that
groups other than homosexual men were also at risk including intravenous
(IV) drug users, recipients of blood and blood products, and heterosexual
contacts of persons at increased risk (CDC, 1981c; CDC, 1982a). By autumn
of 1981, the CDC had created a task force to conduct epidemiologic,
clinical and laboratory investigations on AIDS. For purposes of national
surveillance, a case definition of AIDS was developed in 1982 that required
the diagnosis of one or more of certain diseases at least moderately
predictive of a defect in cell-mediated immunity in persons without
underlying conditions known to cause immunodeficiency, but did not
generally require laboratory confirmation of either immunodeficiency or
infection with HIV (CDC, 1982b). Indicator diseases used in the initial CDC
case definition, before HIV was identified as the causative agent of AIDS,
are listed in Table 1. This original surveillance definition was
subsequently modified in 1985 to include other severe manifestations such
as disseminated histoplasmosis capsulati, bronchial or pulmonary
candidiasis, chronic isosporiasis, and certain non-Hodgkin's lymphomas as
knowledge about HIV infection increased (CDC, 1985). In 1987, the CDC case
definition was revised and expanded for persons with laboratory evidence of
HIV infection (e.g., positive HIV-antibody test or positive culture for
HIV) to include a broader spectrum of diseases found in persons with HIV
infection such as extrapulmonary tuberculosis, disseminated
coccidioidomycosis, certain bacterial infections, HIV encephalopathy, and
HIV wasting syndrome, and the presumptive diagnosis of selected diseases
(see Table 2) (CDC, 1987).

Mycoses in AIDS Patients, Edited by
H. Vanden Bossche *et al.,* Plenum Press, New York, 1990

Table 1. Diseases used in the initial (1982) CDC case definition of AIDS,
before HIV was identified, and considered at least moderately
indicative of underlying immunodeficiency

A. Viral infections (noncongenital)

Chronic (>1 month) mucocutaneous herpes simplex infection
Histologically evident cytomegalovirus infection of an organ other than
liver or lymph node
Progressive multifocal leukoencephalopathy caused by papovavirus (JC virus)

B. Bacterial infections

Disseminated *Mycobacterium avium* complex or *M. kansasii* infection

---

C. Fungal infections

Candidal esophagitis
Cryptococcal meningitis or disseminated infection

D. Protozoan and helminthic infections

Pneumocystis carinii pneumonia
Toxoplasma gondii encephalitis or disseminated infection (noncongenital)
Chronic (>1 month) *Cryptosporidium* enteritis
Strongyloidiasis (pneumonitis, encephalitis, or disseminated infection
beyond the gastrointestinal tract)

E. Neoplasms

Kaposi's sarcoma in a person <60 years of age
Primary lymphoma limited to the central nervous system

As of January 1989, 139,886 AIDS cases had been reported from 142 of
177 countries or territories throughout the world participating in the
Global Programme on AIDS of the World Health Organization (WHO)[*],[**]
(Berkelman et al., 1989) . The USA had the most reported cases followed by
France, Uganda, and Brazil. According to WHO, which coordinates worldwide
surveillance for AIDS, the present number of AIDS cases is markedly
underestimated; underreporting ranges from 10% to 80% by country. WHO
estimates that the actual number of AIDS cases worldwide is approximately
377,000. It is now well-recognized that AIDS is only part of the clinical
spectrum of HIV infection. Infection by HIV invariably causes a variety of
milder clinical manifestations that precede AIDS, and these in turn are
usually preceded by a long asymptomatic incubation period. As many as 5
million persons are estimated to be infected with HIV worldwide[**]
(Berkelman et al., 1989) .

---

[*] Heyward, W.L., AIDS Program, Center for Infectious Diseases, Centers for
  Disease Control, Public Health Service, U.S. Department of Health and
  Human Services, Atlanta, Georgia, Personal communication.
[**] World Health Organization, January 31, 1989, Update, AIDS cases reported
  to the Global Programme on AIDS.

As of January 1, 1989, a total of 82,764 AIDS cases had been reported to CDC from the USA (CDC, 1989c). The number of AIDS cases reported each year continues to increase; however, the rate of increase has steadily declined, except in 1987, when the expansion of the CDC case definition resulted in an abrupt increase in reported cases. Of the AIDS cases reported from the USA, 90% were in males ≥13 years of age, and the mean age at the time of diagnosis was 37 years. Groups at increased risk for AIDS, and the percentages of the total number of patients in each group, included those with histories of homosexual/bisexual contact without IV-drug use (68%), IV-drug users without homosexual/bisexual contact (17%), and those with both homosexual/bisexual contact and IV-drug use (8%). Another 2% had histories of blood transfusion, 1% had hemophilia or other coagulation disorder, 1% had heterosexual contact with sex partners at increased risk for or known to be infected with HIV, 1% were born in countries with predominantly heterosexual transmission of HIV (CDC, 1988b), and 3% had undetermined means of infection.

Eight percent of the AIDS cases from the USA have been reported among women ≥13 years of age. Fifty-two percent had histories of IV-drug use, 18% had sex partners with histories of IV-drug use, 7% had sex partners otherwise at increased risk for or known to be infected with HIV, 11% had histories of blood transfusion, 4% were born in countries with predominantly heterosexual transmission of HIV (CDC, 1988b), and 8% had undetermined means of infection.

The remaining 2% of AIDS cases reported to CDC from the USA have been <13 years of age; 55% were male. Eighty-two percent of pediatric cases were <5 years of age at diagnosis, and 40% were <1 year of age. Seventy-eight percent had mothers at increased risk for or known to be infected with HIV, 13% had received a blood transfusion, 6% had received blood products used to treat hemophilia, and 4% had undetermined means of exposure to HIV.

In the USA, the cumulative incidence of AIDS cases is disproportionately higher in blacks (3.2 to 1) and Hispanics (2.8 to 1) when compared with whites (CDC, 1989b; Selik et al., 1988). The higher rate of IV-drug use among black and Hispanic groups resulting in greater risk of HIV exposure appears to be a contributing factor to this racial disproportion.

Among all AIDS patients from the USA, 59% are reported to have died. However, the median survival time is about one year, and by 7 years after diagnosis, at least 80% have died; the actual case-fatality rate is higher because of incomplete reporting of deaths. AIDS incidence continues to be highest in the most populous metropolitan areas in the USA, and the states of California, Florida, New Jersey, New York, and Texas account for 65% of all cases.

Projections of the number of AIDS cases that will be diagnosed and reported to the CDC in the future have been made using mathematical and statistical models (CDC, 1989a; Morgan and Curran, 1986). Using an extrapolation model, the CDC estimated in May 1988 that 365,000 cumulative AIDS cases will have been diagnosed in the USA by the end of 1992 (CDC, 1989a). Of these 365,000 cases, 263,000 cumulative deaths are predicted by the end of 1992. Projections also indicate that a total of 172,000 AIDS patients in the USA will require medical care during 1992 at a cost expected to range from $ 5 - 13 billion. The above figures are probably underestimates of HIV-related morbidity, since many clinical manifestations of HIV infection are not diagnosed or reported to CDC, even with the revised and expanded 1987 case definition. It has been estimated by mortality studies that only 70% to 90% of HIV-related deaths are identified through USA surveillance of AIDS (Buehler et al., 1989)

In Europe and the eastern Mediterranean region, 19,094 AIDS cases had been reported from 45 countries as of January 1989. Most of these cases

occurred in homosexual men or in persons using IV drugs. Countries
reporting the most cases included France (5,655), Italy (3,008), the
Federal Republic of Germany (2,779), Spain (2,165), and the United Kingdom
(1,982). However, the highest rates for AIDS (number of cumulative cases
per million population) were reported from Switzerland (66.5), France
(65.3), and Denmark (51.4) (WHO, 1988).

In the Americas (excluding the United States), 13,595 cases of AIDS
had been reported as of January 1989. Leading countries reporting AIDS
cases were Brazil (4,709), Canada (2,196), Haiti (1,661), and Mexico
(1,642). In the Caribbean, the absolute numbers of reported AIDS cases are
miniscule when compared to other regions, but the incidence rates are
reported to be among the highest in the world (Berkelman et al., 1989). In
some Caribbean countries, the number of heterosexually transmitted AIDS
cases is greater than the total number of cases acquired by homosexual
transmission and IV drug use (Berkelman et al., 1989; Pape et al., 1986);
the male:female ratio among AIDS cases is reported to range from 4:1 to
1:1, depending on the country.

Fifty-two African countries had reported a total of 21,213 AIDS cases
at the beginning of 1989, with 7 countries reporting more than 1,000 cases:
Uganda (5,508), Tanzania (3,055), Kenya (2,732), Malawi (2,586), Burundi
(1,408), Zambia (1,296), and Congo (1,250). The male:female ratio among
AIDS cases reported from Africa was approximately 1:1, and the number of
cases attributed to heterosexual transmission exceeded 90% in some regions
(Berkelman et al., 1989). However, in Africa, AIDS patients have clinical
features and immunologic derangements similar to those reported in patients
from the USA and Europe.

As of January 1989, 1,460 AIDS cases had been reported from 18 of 35
countries in Asia and the Pacific basin. The majority of cases (1,168) were
reported from Australia. On this continent, the epidemiology of HIV
infection parallels that of the United States and western Europe.

Worldwide, HIV infection via sexual contact, blood and body fluids,
and mother to child continue to be the predominant modes of transmission.
Berkelman et al. 1989, emphasized that the differences in the epidemiologic
patterns of HIV infection and AIDS in specific countries or regions are
primarily due to differences in the proportions of the modes of
transmission. They further pointed out that all areas of the world have
shown large rates of increase of AIDS, and that the areas only differ
according to the time in which HIV infection was introduced.

Although active surveillance for AIDS did not begin until 1982, the
epidemic of AIDS appears to have arisen in the USA, central Africa, and
Haiti at about the same time in the late 1970's. Retrospective studies have
revealed that persons fitting the 1982 CDC case definition for AIDS were
present in New York City as early as 1977 (Bigger, 1988). These findings
were followed by similar reports of persons fitting the CDC case definition
in Haiti (Pape et al., 1983) as well as central Africa (Clumeck et al.,
1984). The oldest stored blood sample that has been found to be positive
for antibodies to HIV-1 by western blot assay was collected in 1959 in
Zaire (Nahmias et al., 1986) The oldest clinical specimen from which HIV-1
has been isolated was a serum sample taken in 1976 also from a patient in
Zaire (Getchell et al., 1987). The geographic origin of HIV-1 remains
unknown.

A new AIDS virus, HIV-2, was discovered in 1985 (Clavel et al., 1987;
Horsburg and Holmberg, 1988). This virus, which has been found
predominantly in west Africans, can produce clinical manifestations
indistinguishable from those caused by HIV-1 (Clavel et al., 1987; CDC,
1988a; CDC, 1989b). Like HIV-1, sexual contact is the most important route
of transmission of HIV-2, and it may have a long incubation period.

Isolated cases of HIV-2 infection have been reported in several countries, including France (Brücker et al., 1987;), Italy (Ferroni et al., 1987), Brazil (Veronesi et al., 1987), the Federal Republic of Germany (Marquart et al., 1988) and the USA (CDC, 1989b). However, unlike HIV-1, HIV-2 is generally restricted to west Africa at present.

OPPORTUNISTIC INFECTIONS

Opportunistic diseases are the predominant causes of morbidity and mortality in AIDS, and their clinical and pathologic features are well described (Gold, 1985; Gold, 1988; Gal et al., 1988; Selik et al., 1987). Although PCP is by far the most commonly reported opportunistic infection in AIDS patients, 24 other infectious and neoplastic diseases used as indicators of AIDS in the 1987 revision of the CDC case definition are also encountered (CDC, 1987). The percentages of AIDS cases reported with various AIDS-indicative diseases among cases diagnosed since the 1987 revision of the CDC case definition and reported through 1988 in the USA are given in Table 2. Retrospective reviews of autopsy reports and medical records have demonstrated that the cumulative incidence of some opportunistic diseases in AIDS patients is much greater than the frequencies reported to CDC based on surveillance data (Reichert et al., 1983; Welch et al., 1984). Estimates of the cumulative incidence based on surveillance alone are generally lower than the true values because most diseases that occur in AIDS patients after the patients have initially been reported are not reported to CDC.

Protozoa most often encountered in AIDS are *P. carinii*, *Toxoplasma gondii*, *Cryptosporidium* spp., and Isospora *belli*. Based on clinical and autopsy findings, approximately 85% of AIDS patients develop PCP during their illness (Mills, 1986). AIDS patients may also develop neurologic manifestations because of toxoplasmic encephalitis, and *T. gondii* is the most common cause of central nervous system (CNS) mass lesions in AIDS. The prevalence of toxoplasmic encephalitis in AIDS patients has been estimated to be as high as 25% in San Francisco and 12% in New York City (Wong et al., 1984). The clinical course of cryptosporidiosis caused by *Cryptosporidium* spp. in AIDS is characterized by severe, protracted , watery diarrhea. The coccidian *Isospora belli* also causes chronic diarrhea in AIDS. It is usually restricted to the epithelial cells of the small intestine, but disseminated extraintestinal isosporiasis in an AIDS patient has recently been reported (Restrepo et al., 1987).

Although AIDS or HIV-infected patients are susceptible to all the mycoses, their profound cellular immunodeficiency makes them extraordinarily susceptible to certain opportunistic fungi of normally low pathogenicity (Chandler, 1985; Holmberg and Meyer, 1986). The most frequently encountered agents are *Candida albicans*, *Cryptococcus neoformans*, and *Histoplasma capsulatum* var. *capsulatum*. The percentage of AIDS patients with mycoses caused by these agents based on 3,170 cases reported to CDC between May 1983 and June 1984 included candidal oropharyngitis (41.8%), candidal esophagitis (9.4%), cryptococcal meningitis or disseminated infection (5.9%), and disseminated histoplasmosis capsulati (0.51%) (Chandler, 1985). With the revised and expanded 1987 AIDS case definition, there have been slight to moderate increases in the reported frequencies of occurrence of these mycoses in AIDS patients (see Table 2). However, these percentages are still underestimates because follow up information is seldom available or obtained on opportunistic infections that occur after the initial report of an AIDS case to CDC.

Retrospective surveys have revealed that 58% to 81% of AIDS or HIV-infected patients develop one or more fungal infections at some time during their illness (Reichert et al., 1983; Welch et al., 1984) . Oropharyngeal

Table 2. Percentages of AIDS cases reported with various AIDS indicative
diseases among cases diagnosed since the 1987 revision of the CDC
case definition and reported through 1988 in the USA

AIDS-Indicative Disease	Percent* (N=28,920)
Pneumocystis carinii pneumonia: definitive diagnosis	45
presumptive diagnosis	11
HIV wasting syndrome	14
Kaposi's sarcoma: definitive diagnosis	8.2
presumptive diagnosis	1.6
Candidiasis of esophagus: definitive diagnosis	6.7
presumptive diagnosis	5.2
HIV encephalopathy (dementia)	5.8
Cryptococcosis, extrapulmonary	5.7
Herpes simplex, causing esophagitis, pneumonitis, or chronic mucocutaneous ulcers	3.2
Candidiasis of bronchi, trachea, or lungs	3.0
Cytomegalovirus disease other than retinitis	3.0
Mycobacterium avium complex or *M. kansasii* disease, disseminated or extrapulmonary: definitive diagnosis	2.5
presumptive diagnosis	0.2
Lymphoma, immunoblastic or equivalent (other than brain)	1.7
Cryptosporidiosis, causing chronic diarrhea	1.6
Mycobacterium tuberculosis, extrapulmonary disease	
definitive diagnosis	1.6
presumptive diagnosis	0.3
Cytomegalovirus retinitis: presumptive or definitive	1.4
Toxoplasmosis of CNS: definitive diagnosis	1.2
presumptive diagnosis	2.5
Mycobacterial disease caused by other or unidentified species, disseminated or extrapulmonary:	
definitive diagnosis	0.9
presumptive diagnosis	0.3
Histoplasmosis capsulati, disseminated or extrapulmonary	0.9
Lymphoma, Burkitt's or equivalent (other than in brain)	0.7
Progressive multifocal leukoencephalopathy	0.6
Salmonella septicemia, recurrent (counted separately in adults only)	0.5
Lymphoma, primary in brain	0.4
Bacterial infections, multiple or recurrent, serious pyogenic (counted in children only)	0.3
Coccidioidomycosis, disseminated or extrapulmonary	0.2
Isosporiasis, causing chronic diarrhea	0.2
Lymphoid interstitial pneumonia (in children only)	
definitive diagnosis	0.1
presumptive diagnosis	0.1

*The sum of percentages exceeds 100% because some cases had >1 disease
reported.

candidiasis (thrush) occurs in most AIDS patients, but candidal esophagitis
is less common; disseminated candidiasis is rare, even at autopsy. The
incidence of cryptococcosis in AIDS patients ranges from 6% to 29%
depending on the patient group selected for study (Pitchenik et al., 1983;
Vandepitte et al., 1983; Kovacks et al., 1985; Zuger et al., 1986).
Although candidiasis, cryptococcosis, histoplasmosis capsulati, and
coccidioidomycosis are the mycotic indicators in the current (1987) CDC
case definition for AIDS, a review of the literature and our laboratory

Table 3. AIDS-associated mycoses, listed in alphabetical order, and their etiologic agents

1. Aspergillosis *Aspergillus fumigatus*	9. Hyalohyphomycosis *Fusarium proliferatum*
2. Blastomycosis *Blastomyces dermatitidis*	*Pseudallescheria boydii* 10. Malasseziasis
3. Candidiasis	*Malassezia furfur*

1. Aspergillosis
   Aspergillus fumigatus
2. Blastomycosis
   Blastomyces dermatitidis
3. Candidiasis
   Candida albicans
   C. krusei
   C. parapsilosis
   C. pseudotropicalis
   C. tropicalis
4. Coccidioidomycosis
   Coccidioides immitis
5. Cryptococcosis
   Cryptococcus neoformans
6. Dermatophytosis
   Trichophyton rubrum
7. Histoplasmosis capsulati
   Histoplasma capsulatum var.
   capsulatum
8. Histoplasmosis duboisii
   Histoplasma capsulatum var.
   duboisii

9. Hyalohyphomycosis
   Fusarium proliferatum
   Pseudallescheria boydii
10. Malasseziasis
    Malassezia furfur
11. Paracoccidioidomycosis
    Paracoccidioides brasiliensis
12. Penicilliosis
    Penicillium marneffei
13. Phaeohyphomycosis
    Rhinocladiella atrovirens
14. Sporotrichosis
    Sporothrix schenckii
15. Torulopsosis
    Torulopsis candida
    Torulopsis glabrata (*Candida glabrata*)
16. Zygomycosis
    Absidia corymbifera
    Cunninghamella bertholletiae

records has revealed other AIDS-associated mycoses. These mycoses and their etiologic agents are listed in Table 3.

Disseminated mycobacteriosis caused by *Mycobacterium avium* complex, a ubiquitous soil saprophyte, is by far the most common bacterial infection in AIDS patients. Microbiologic and pathologic studies indicate that at least 50% of patients develop this infection during their illness (Gold, 1988; Armstrong et al., 1985).

The most common viral infections in AIDS are caused by cytomegalovirus (CMV), herpes simplex virus (HSV), and papovavirus (JC virus). At autopsy, almost all AIDS patients have histopathologically confirmed disseminated CMV infection. Severe, persistent, ulcerative perianal HSV infection was one of the first indicator diseases recognized in AIDS patients (Siegal et al., 1981). Herpetic esophagitis is also a common finding at autopsy, but unlike CMV, severe disseminated HSV infection is rare.

JC virus is the cause of progressive multifocal leukoencephalopathy, a relatively rare but severe demyelinating and necrotizing infection of the CNS in AIDS patients.

ADDENDUM

Update on cumulative total of AIDS cases worldwide:
As of July 1, 1989, 167,373 AIDS cases from 149 of 177 participating countries or territories had been reported to the Global Programme on AIDS of WHO. The actual cumulative number of AIDS cases worldwide is now estimated by WHO to be approximately 480,000 instead of 377,000 as had been estimated in January 1989. About 70% (112,839) of the global total has been reported from 43 countries in the Americas, and 85% of these cases have been reported from the USA. Africa has reported 30,064 cases and Europe 22,609 cases, approximately 18% and 13% of the world total, respectively.

ACKNOWLEDGMENT

The author wishes to thank Melita Posey for preparation of the manuscript.

REFERENCES

Armstrong, D., Gold, J.W.M., Dryjanski, J., Whimbey, E., Polsky, B.,
    Hawkins, C., Brown, A.E., Bernard, E., and Kiehn, T.E., 1985,
    Treatment of infections in patients with the acquired
    immunodeficiency syndrome, Ann. Intern. Med., 103:738.
Barre-Sinoussi, F., Chermann, J.C., Rey, F., Nugeyre, M.T., Chamaret, S.,
    Gruest, J., Dauguet, C., Axler-Blin, C., Vezinet,-Brun, F.,
    Rouzioux, C., Rozenbaum, W., and Montagnier, L., 1983, Isolation of
    a T-lymphotropic retrovirus from a patient at risk for the acquired
    immune deficiency syndrome (AIDS), Science, 220:868.
Berkelman, R.L., Heyward, W.L., Stehr-Green, J.K., and Curran, J.W., 1989,
    Epidemiology of human immunodeficiency virus infection and acquired
    immunodeficiency syndrome, Am. J. Med., 86:761.
Bigger, R.J., 1988, AIDS-related Kaposi's sarcoma in New York City in 1977
    [Letter], N. Engl. J. Med., 318:252.
Bruecker, G., Brun-Vezinet, F., Rosenheim, M., Rey, M.A., Katlama, C., and
    Gentilini, M., 1987, HIV-2 infection in two homosexual men in France
    [Letter], Lancet, 1:223.
Buehler, J., Berkelman, R.L., and Devine, O., 1989, Estimate of HIV-related
    deaths in young adult men, United States, 1986 [Abstract], V
    International Conference on AIDS. Montreal.
Centers for Disease Control, 1981a, Pneumocystis pneumonia--Los Angeles,
    MMWR, 30:250.
Centers for Disease Control, 1981b, Kaposi's sarcoma and Pneumocystis
    pneumonia among homosexual men--New York City and California, MMWR,
    30:305.
Centers for Disease Control, 1981c, Follow-up on Kaposi's sarcoma and
    Pneumocystis pneumonia, MMWR, 30:409.
Centers for Disease Control, 1982a, Update on Kaposi's sarcoma and
    opportunistic infections in previously healthy persons--United
    States, MMWR, 31:294.
Centers for Disease Control, 1982b, Update on acquired immunodeficiency
    syndrome (AIDS)--United States, MMWR, 31:507.
Centers for Disease Control, 1985, Revision of the case definition of
    acquired immunodeficiency syndrome for national reporting--United
    States, MMWR, 34:373.
Centers for Disease Control, 1987, Revision of the CDC surveillance case
    definition for acquired immunodeficiency syndrome, MMWR, 36:Suppl,
    1S.
Centers for Disease Control, 1988a, AIDS due to HIV-2 infection--New
    Jersey, MMWR, 37:33.
Centers for Disease Control, 1988b, Update: acquired immunodeficiency
    syndrome (AIDS)--worldwide, MMWR, 37:286 and 293.
Centers for Disease Control, 1989a, AIDS and human immunodeficiency virus
    infection in the United States: 1988 update, MMWR, 38:5-4.
Centers for Disease Control, 1989b, Update: HIV-2 infection--United States,
    MMWR, 38:572.
Centers for Disease Control, 1989c, Update: acquired immunodeficiency
    syndrome--United States, 1981-1988, MMWR, 38:229.
Chandler, F.W., 1985, Pathology of the mycoses in patients with the
    acquired immunodeficiency syndrome (AIDS), Curr. Topics Med. Mycol.,
    1:1.
Clavel, F., Mansinko, K., Chamaret, S., Guetard, D., Favier, V., Nina, J.,
    Santos-Ferreira, M.A., Champalimaud, J.L., and Montagnier, L., 1987,
    Human immunodeficiency virus type 2 infection associated with AIDS
    in west Africa, N. Engl. J. Med., 316:1180.
Clumeck, N., Sonnet, J., Taelman, H., Mascart-Lemone, F., De Bruyere, M.,
    Vandeperre, P., Dasnoy, J., Marcelis, L., Lamy, M., Jonas, C., et
    al., 1984, Acquired immunodeficiency in African patients, N. Engl.
    J. Med., 310:492.
Ferroni, P., Tagger, A., Lazzarin, A., and Moroni, M., 1987, HIV-1 and HIV-
    2 infections in Italian AIDS/ARC patients [Letter], Lancet, 1:869.

Gal, A.A., Koss, M.N., Hartman, B., et al, 1988, A review of pulmonary pathology in the acquired immune deficiency syndrome, *Surg. Pathol.*, 1:325.

Gallo, R.C., Salahuddin, S.Z., Popovic, M., Shearer, G.M., Kaplan, M., Haynes, B.F., Palker, T.J., Redfield, R., Oleske, J., Safai, B., et al., 1984, Frequent detection and isolation of cytopathic retroviruses (HTLV-III) from patients with AIDS and at risk for AIDS, *Science*, 224:500.

Getchell, J.P., Hicks, D.R., Srinivasan, A., Heath, J.L., York, D.A., Malonga, M., Forthal, D.N., Mann, J.M., and McCormick, J.B., 1987, Human immunodeficiency virus isolated from a serum sample collected in 1976 in central Africa, *J. Infect. Dis.*, 156:833.

Gold, J.W.M., 1985, Clinical spectrum of infections in patients with HTLV-III-associated diseases, *Cancer Res.*, 45(Suppl):4652S.

Gold, J.W.M., 1988, Infectious complications in patients with HIV infection, *AIDS*, 2:327.

Holmberg, K., and Meyer, R.D., 1986, Fungal infections in patients with AIDS-related complex, *Scand. J. Infect. Dis.*, 18:179.

Horsburgh, C.R., and Holmberg, S.D., 1988, The global distribution of human immunodeficiency virus type 2 (HIV-2) infection, *Transfusion*, 28:192.

Kovacs, J.A., Kovacs, A., Polis, M., Wright, W.C., Gill, V.J., Tuazon, C.U., Gelmann, E.P., Lane, H.C., Longfield, R., Overturf, G., et al., 1985, Cryptococcosis in the acquired immunodeficiency syndrome, *Ann. Intern. Med.*, 103:533.

Marquart, K.-H., Muller, H.A.G., and Brede, H.D., 1988, HIV-2 in West Germany [Letter], *AIDS*, 2:141.

Mills, J., *Pneumocystis carinii* and *Toxoplasma gondii* infections in patients with AIDS, *Rev. Infect. Dis.*, 8:1001.

Morgan, W.M., and Curran, J.W., 1986, Acquired immunodeficiency syndrome: current and future trends, *Public Health Rep.*, 101:459.

Nahmias, A.J., Weiss, J., Yao, X., Lee, F., Kodsi, R., Schanfield, M., Matthews, T., Bolognesi, D., Durack, D., Motulsky, A., et al., 1986, Evidence for human infection with an HTLV-III/LAV-like virus in central Africa, 1959 [Letter], *Lancet*, 1:1279.

Pape, J.W., Liautaud, B., and Thomas, F., 1986, Changing patterns of AIDS epidemiology [Abstract], *Clin. Res.*, 34:528A.

Pape, J.W., Liautaud, B., Thomas, F., Mathurin, J.R., St. Amand, M.M., Boncy, M., Pean, V., Pamphile, M., Laroche, A.C., and Johnson, W.S., Jr., 1983, Characteristics of the acquired immunodeficiency syndrome (AIDS) in Haiti, *N. Engl. J. Med.*, 309:945.

Pitchenik, A.E., Fischl, M.A., Dickinson, G.M., Becker, D.M., Fournier, A.M., O'Connell, M.T., Colton, R.M., and Spira, T.J., 1983, Opportunistic infections and Kaposi's sarcoma among Haitians: evidence of a new acquired immune deficiency state, *Ann. Intern. Med.*, 98:277.

Reichert, C.M., O'Leary, T.J., Levens, D.L., Simrell, C.R., and Macher, A.M., 1983, Autopsy pathology in the acquired immune deficiency syndrome, *Am. J. Pathol.*, 112:357.

Restrepo, C., Macher, A.M., and Radamy, E.H., 1987, Disseminated extraintestinal isosporiasis in a patient with acquired immune deficiency syndrome, *Am. J. Clin. Pathol.*, 87:536.

Selik, R.M., Starcher, E.T., and Curran, J.W., 1987, Opportunistic diseases reported in AIDS patients: frequencies, associations, and trends, *AIDS*, 1:175.

Selik, R.M., Castro, K.G., and Pappaioanou, M., 1988, Distribution of AIDS cases by racial/ethnic group and exposure category, United States, June 1, 1981 - July 4, 1988, *MMWR*, 37(SS-3):1.

Siegal, F.P., Lopez, C., Hammer, G.S., Brown, A.E., Kornfeld, S.J., Gold, J., Hassett, J., Hirschman, S.Z., Cunningham-Rundles, C., Adelsberg, B.R., Parham, D.M., Siegal, M., Cunningham-Rundles, S., and Armstrong, D., 1981, Severe acquired immunodeficiency in male homosexuals manifested by chronic perianal ulcerative herpes simplex lesions, *N. Engl. J. Med.*, 305:1439.

Vandepitte, J., Verwilghen, R., and Zachee, P., 1983, AIDS and cryptococcosis (Zaire 1977), Lancet, 1:925.

Veronesi, R., Mazza, C.C., Santos Ferreira, M.O., and Lourenco, M.H., 1987, HIV-2 in Brazil [Letter], Lancet, 2:402 (1987).

Welch, K., Finkbeiner, W., Alpers, C.E., Blumenfeld, W., Davis, R.L., Smuckler, E.A., and Beckstead, J.H., 1984, Autopsy findings in the acquired immune deficiency syndrome, JAMA, 252:1152.

Wong, B., Gold, J.W.M., Brown, A.E., Lange, M., Fried, R., Grieco, M., and Mildvan, D., 1984, Central nervous system toxoplasmosis in homosexual men and parenteral drug abusers, Ann. Intern. Med., 100:36.

World Health Organization, 1988, Global strategy for the prevention and control of AIDS--report by the Director General, WHO:1.

Zuger, A., Louie, E., Holzman, R.S., Simberkoff, M.S., Rahal, J.J., 1986, Cryptococcal disease in patients with acquired immunodeficiency syndrome. Diagnostic features and outcome of treatment, Ann. Intern. Med., 104:234.

IMMUNOLOGIC AND PATHOGENIC ASPECTS OF HIV INFECTIONS: CURRENT HYPOTHESIS

Dominique Dormont

Laboratoire de Neurovirologie, DPS/SPE/CRSSA, CEA, BP6,
92265 Fontenay aux Roses Cedex, France

INTRODUCTION

The acquired immunodeficiency syndrome (AIDS) (Gottlieb et al., 1981) is the major clinical expression of infection with the Human Immunodeficiency Virus (HIV) (Barre-Sinoussi et al., 1983) in man, and with the Simian Immunodeficiency Virus (SIV) in Macaques (Daniel et al., 1985). These viruses (HIV 1 and 2, and SIV $_{MAC}$) belong to the lentivirus family, which are retroviruses. Although theses viruses are highly cytopathogenic for numerous CD4+ cells *in vitro* (Klatzmann et al., 1984; Haseltine, 1988; Gallo, 1988), they induce *in vivo* a clinical picture for the main characteristics are a long dormant stage, and a slow decrease of CD4+ cells in blood and haematopoïetic systems. These differences between *in vitro* and *in vivo* behaviours might be related to several features, including, first, close relationships between latent infection and expression of both intracellular activation signals and cytokines, and, second, infection of monocyte/macrophages, in which lentivirus replication is essentially intracellular. Therefore, HIV and SIV *in vivo* pathogenicities may be related to the viral genetic mechanisms, to the nature of the target cells, to the activation level of the infected cells, and to the probability of these cells being activated by cytokines and/or intracellular signals.

HIV STRUCTURE

HIV is a viral particle 100 nm in diameter, with a density of 1.16 in sucrose gradient (Barre-Sinoussi et al., 1988). This particle is surrounded by an envelope which contains proteins embedded in a lipid membrane, which is organized externally into spikes. The envelope protein is a glycoprotein, derived from a heavily modified precursor by cleavage and carbohydratation. This 160 kd glycoprotein is divided into two different parts in the viral envelope: the carboxy portion (GP41) is stuck to the viral membrane, and the N-terminal part (GP120) is entirely outside the viral envelope; GP 120 is non covalently attached to the GP41. These two glyproteins are products of the *env* gene of HIV. Inside the envelope, one may identify the viral core, which is usually eccentric, with an either truncated or rounded shape; this core contains proteins and the viral genome. Main core proteins are P25, P18 and P13; the major core protein, P25, is a methionine rich molecule, which is, like P18 and P13, a *gag* gene product. Ultra thin sections of HIV-producing cells show budding particles on the cell membrane. Typical morphology of human lentiviruses is close to mature D particles and very different from type C virions (Barre-Sinoussi et al., 1983).

Mycoses in AIDS Patients, Edited by
H. Vanden Bossche *et al.,* Plenum Press, New York, 1990

HIV BIOLOGICAL CYCLE

The HIV biological cycle may be summarized as followed:

1) Interaction between GP120 (external envelope protein) and specific cellular receptor (CD4 molecule).
2) Fusion of the cell membrane with the viral envelope which may or may not be after internalization.
3) After fusion, the viral core enters the cytoplasm of the host cell.
4) Reverse transcription occurs: in the presence of a tRNA-lysine and P13 (the smallest viral protein), as initiation factors, the viral reverse transcriptase (RT) copies viral RNA in a single strand DNA; the result is a RNA-DNA hybrid. Then, RT associated RNAse H digests viral RNA, and DNA dependant - DNA polymerase synthesizes a double stranded DNA, which becomes circular; this DNA is the proviral DNA.
5) Proviral DNA migrates into the cell nucleus, and may or may not be integrated into the host genome; the precise mechanism of this HIV biological cycle phase is not yet known.
6) Proviral DNA may be latent for days, months or years in the infected cell, depending on the cellular status. Usually, in lymphocytes, the activation process is often followed by cell mitosis, and induction of viral expression: viral messengers RNAs are synthesized, and they may follow two different paths: 1) various splicings occur, in order to drive viral protein synthesis; 2) no splicing occurs, and these RNAs are incorporated in mature viral particles, as genetic information.
7) Spliced RNAs are submitted to transcription, and precursors of viral proteins are synthesized.
8) Structural proteins precursors are cleaved and glycosylated.
9) Assembly occurs near the cell membrane: viral genome stabilized by P13, is surrounded by core proteins and envelope glycoproteins; the final structure is completed during budding.

The consequence of viral replication in the host cell is a cytopathogenic effect: as it occurs in the CD4+ lymphocyte population, the host cells die; the mechanism of cell death may be relevant from the points of view of both cell energetic metabolism exhaustion, and specific viral biological properties. In other cell populations, cells may survive the HIV replication, and, thus, may act as a reservoir for HIV; these cells are usually long-lived cells, like macrophages, reticulo-endothelial system (RES) cells or cells from the central nervous system (CNS).

CELLULAR TROPISM OF HIV

Cellular tropism of HIV has been described as selective for CD4+ cells (Klatzmann et al., 1984). The external part of the GP120 interacts with the fourth domain of the CD4 molecule, between the binding site of two monoclonal antibodies, OKT4a and anti-Leu3a. These two antibodies are able to block cellular infection when preincubated with the susceptible cells before HIV infection. This selective cellular tropism of HIV may partially explain the slow decrease of CD4+ blood cell numbers observed the course of HIV infection. Nevertheless, HIV is not directly detectable in blood cells during such classical procedures as in situ hybridization, or with immunofluorescence. Only PCR permits identification of HIV proviral sequences in fresh blood cells. If one admits that less than a few percent of CD4+ lymphocytes are infected in patients' blood, one cannot explain HIV infection and a fatal disease by infection of only this very small number of cells. Other cells may be infectable, and may constitute a viral reservoir; these cells may release slowly small amounts of viral particles, which progressively kill CD4+ cells. These infections of non lymphocytic cells are probably the most important facts in AIDS pathogenetic mechanisms. These cells have been principally identified as monocytes-macrophages (MM) (Haseltine, 1988; Gallo, 1988; Gartner et al., 1986), Langerhans cells (LC) (Tschachler et al., 1986), dendritic cells (DC) (Gallo, 1988; Muller et al., 1986), and Central Nervous System (CNS) cells subpopulations (Gallo, 1988; Rosenberg and Fauci, 1989; Koenig et al. 1986). They are mainly CD4+ cells, but the level of CD4 expression on their cell membranes is variable, and one might hypothesise a non CD4 related infection route for some of them. Debates occur on the possible role of immune complexes in infections of macrophages, and the potential role of Fc receptors on cell membranes. Fc receptors might bind HIV-anti-HIV complexes, and the antigen-antibody complex could then be internalised, and infect the cell. Other infection mechanisms have been described *in vitro*, but not demonstrated *in vivo*. Infected cells may express GP120 on their membranes; these GP120 might interact with CD4 molecules located on the cell membrane of non-infected lymphocytes when this molecular interaction occurs, it could be followed by cell fusion, resulting in an infected syncytium.

HIV GENOME

Biological behaviour of HIV is also a consequence of the viral genes. The HIV genome is 9.1 kb long (Alizon et al., 1984; Wain-Hobson et al., 1985): its full sequence has now been determined (Alizon et al., 1984; Alizon et al., 1986; Wain-Hobson et al., 1985). As described in other retroviruses, three structural genes may be identifiable: *gag*, *pol*, and *env*. *Gag* gene encodes the nucleocapside proteins, *pol* gene encodes proteins which support all the functions of the reverse transcription (Reverse transcriptase, RNAse H, DNA polymerase, endonuclease, and integrase) and the viral protease (there is a common precursor to *gag* and *pol* polyproteins, which has to be cleaved by the viral aspartyl-protease) , and *env* gene encodes a precursor of glycoproteins which is eventually carbohydrated and cleaved into GP120 and GP41 (Haseltine, 1988). Other open reading frames should be listed in the HIV genome: *vif* (virion infectively factor), *nef* (negative factor), *tat* (transactivator), *rev* (regulator of viral expression), *vpu* (viral protein u, present only in HIV1), *vpx* (viral protein x, present only in HIV2 and SIV), *vpr* (viral protein r). These open reading frames correspond to regulatory genes. All the functions of the proteins encoded by these genes are not known; nevertheless, regulation mechanisms involving *vif*, *vpu*, *nef*, *tat*, and *rev* have been partly determined.

- *vif*: if a mutation occurs in *vif*, mutated virions cannot infect susceptible cells via the classical CD4 route. Infection is possible through the cell to cell infection route (cell fusion mechanism); thus, *vif* gene may participate in the control of the viral biological cycle after virus binding and before circularization (Haseltine, 1988; Fischer et al., 1987; Strebel et al., 1987).
- *vpu*: this protein is found specifically in HIV1. Terwilliger et al. (1989) first suggested that a defect in the *vpu* gene results in an increase of the replication rate, and that *vpu* may be a negative regulation gene (Haseltine, 1988; Matsuda et al., 1988). Others have noted a decrease of syncytia and cell death in "*vpu-*" mutant-infected cell culture, associated with an enhancement of viral particle numbers in the cell culture supernatant. These data suggest that *vpu* may help in viral particle excretion, and, therefore, increasing the spread of HIV (Strebel et al., 1988; Terwilliger et al., 1989).
- *nef*: the protein encoded by the *nef* gene (26 kd) is the best known negative regulation factor (Allan et al., 1985). *In vitro*, it decreases the viral replication rate. The *nef* protein is a myristylated protein (fatty acid at the amino terminus), which may suggest that it binds to the cell membrane. Recent results have demonstrated that this protein is capable of autophosphorylation, binds GTP, and exhibits GTPase activity (Guy et al., 1987). Such properties are usually described in *ras* oncogen product. All these properties are in accordance with the myristylated characteristic of the protein, and suggest that the *nef* product should interact with intracellular signal mechanisms and might be a signal transducing

protein (Haseltine, 1988). One molecular property of the *nef* product is a positive effect on the specific element belonging to Long Terminal Repeat (LTR) sequences. This Negative Regulatory Element (NRE) is located 200 nucleotides from the site of RNA initiation (Haseltine, 1988), and acts as a transcriptional repressor. This negative transcriptional activity may not be restricted to HIV genes: for example, *nef* product is able to downregulate cellular genes, like the CD4 molecule, in infected cells. *Nef* product synthesis may occur alone, in the absence of structural protein synthesis (*gag*, *pol*, and *env* products). Antibodies to *nef* protein may then be detected before full seroconversion in infected individuals. This has been demonstrated both in humans and monkeys (personal data). *Nef* gene may be responsible for the dormant stage of the proviral DNA, and for the latent clinical phase. One may notice that *nef* gene product function is being debated today: using cell line models, however, no effect of *nef* protein can be detected on transfected HIV-LTR.

- *tat*: this gene has two exons, and is a trans-activator gene: it encodes for a 86 amino acid long protein, the two exons being joined by a splicing mechanism. The *tat* protein has three main functions :
  1) it increases the efficiency of the LTR promotor (transcriptional mechanism) (Berkhoul et al., 1989; Laspia et al., 1989);
  2) it increases tremendously the traduction efficacy (Haseltine, 1988; Berkhoul et al., 1989);
  3) it acts as a transcriptional anti-termination factor (Kao et al., 1987).

  The first mechanism is a common transcriptional mechanism: *tat* protein transactivates the LTR promotor, and, consequently, proviral DNA transcription is made possible. The second mechanism requires an interaction between the *tat* gene product and a special element located at the 5' end of all the messenger RNAs and which function is to stop any traduction process. This element is the trans acting responsive region (TAR) (Rosen et al., 1985); the *tat* gene product is able to block this anti-traduction segment, and, consequently, it increases protein synthesis. Little is known today on the exact mechanism of this transactivator product on TAR (direct or not?), but one should note that, because all messenger RNAs have TAR sequences, all viral proteins synthesis, and perhaps cellular proteins synthesis, may be increased by this *tat* related mechanism. One of the consequences of this mechanism is an increase of the half life of messenger RNAs. It has also been reported that *tat* protein may increase the efficiency of ribosomal binding to mRNA. This increased stability of RNA-ribosome complexes may also contribute to the dramatic increase of viral proteins observed in the presence of the *tat* gene product. The third mechanism, which has still not been entirely elucidated, is a transcriptional anti-termination process. A sequence in the LTR region has a secondary structure which may stop the transcriptional mechanism. In the presence of *tat* gene product, the secondary structure of this sequence might be modified, allowing an efficient transcription of all the proviral DNA. Lastly, one should point out that the *tat* protein increases its own synthesis rate.

- *rev*: a mutation in *rev* is followed by a modification in the length of HIV messenger RNAs, and the synthesis of only small RNAs (1.5 to 2 kb) (Haseltine, 1988; Feinberg et al., 1986). Therefore, *rev* gene product might act as a stabilization factor of the large viral RNAs, through prevention of their splicing. *Rev* protein might also interact with specific sequences located in several genes and in the corresponding messenger RNAs, the cis acting antirepression sequences (CAR). The result of this interaction is an overriding of the anti-protein synthesis effects of other messenger RNA sequences, the cis acting repression sequences (CRS). CRS are present in *gag*, *pol*, and *rev*, but not in *tat*, *rev*, and *nef*. The major final effect of CRS is to make any protein synthesis impossible, probably by inhibiting the transportation of the RNAs through the nuclear membrane (Haseltine, 1988; Feinberg et al., 1986; Malim et al., 1989). Consequently, RNAs are left in the nucleus compartment, and are subjected to splicing and degradations. In the presence of the *rev* protein, CAR sequences permit the passage of large RNAs through the nucleus membrane, and prevent splicing events: thus, viral protein synthesis is possible. In contrast to *tat* protein, *rev* gene product regulates its own expression negatively, because *rev* prevents splicing events, and splicing is necessary to obtain *rev*, *tat*, and *nef* messenger RNAs. Nevertheless, recent results suggest that *nef* protein synthesis could be modulated by *rev* (Ahmad et al., 1989).

These regulatory genes might explain by themselves the level of viral replication. An increase of *nef* gene product synthesis may result in an inhibition of viral replication, and an accumulation of *rev-tat* proteins may induce a high rate of viral replication. A switch from latent phase to efficient viral replication may depend upon cellular condition, pointing out the extreme importance of signal-transduction and intracellular transcriptional factors. For example in viral LTR (near the TATA box), some particular sequences are able to bind transcriptional factors, like SP1, or intracellular activation-induced factors like NFκB. NFκB is present only in activated lymphocytes, and is also able to increase the level of expression of immunoglobulin genes, Interleukin 2 (IL2) gene, and IL2 receptor gene. NFκB may enter the nucleus through the nuclear membrane, in response to a membrane signal, as has been described after CD3 receptor activation. A consequence of the CD3 activation is a calcium flux, which induces a phosphokinase C activity, leading to the trans-nuclear membrane passage of NFκB, and the activation of the viral messenger RNAs. Therefore, virus-cell interactions depend upon cellular state and vice versa. Moreover, virus-cell interactions depend also on the nature of the infected cell. If it is a lymphocyte, HIV operates as an activator and may stimulate the cell growth. In macrophages, HIV could have an opposite cellular action. These two different relationships between HIV and the host cell might explain the viral reservoir characteristic of the macrophages. A switch from a macrophage predominant infection to a lymphocyte predominant infection may than explain the transition between clinical asymptomatic phase and full blown AIDS.

This regulatory network is not yet fully understood, and many questions are still unanswered. However, the differences in cell-virus relationships may explain why the biological cycle of HIV and its consequence in CD4+ lymphocytes (cytopathogenic effect) may not account by itself for the entire pathogenesis of AIDS. To understand all the mechanisms which play major roles in AIDS pathogenesis, one should recognize:

1) the importance of the macrophages as a viral reservoir;
2) the lack of efficiency of immune responses elicited by HIV infection;
3) the unfavourable characteristic of some of these immune responses, and
4) that HIV can induce immune dysregulations which may evolve by themselves quasi-independently of the viral replication level.

IMMUNOPATHOLOGY OF HIV INFECTION.

1. Immunologic and virologic parameters of HIV infection in humans

Infected patients exhibit several immunologic and virologic symptoms, which are now recognized as specific of HIV infection stages:

1) HIV infection
2) Primo infection (2 to 12 weeks):
   - Detectable antigenemia;
   - Detectable viral replication in peripheral blood lymphocytes (PBL) and/or cerebrospinal fluid (CSF) cell culture supernatants;
   - No HIV specific antibody is detectable.

3) Asymptomatic clinical phase:
   - Usually: no detectable antigenemia;
   - HIV is detectable in cell culture supernatants;
   - Full seroconversion is detectable: at least anti GP (120, 160, or 41) and anti-gag proteins antibodies must be detected by Western blot and/or radio-immunoprecipitation assay (RIPA);
   - High level of anti-P25 antibodies;
   - Neutralizing antibodies may be present (low titer);
   - Slow decrease of CD4+ cells in blood and lymph nodes;
   - Slow increase of CD8+ cells in blood and lymph nodes;
   - Increase of Antibody Dependant Cellular Cytotoxicity (ADCC) and Natural Killer (NK) activities;
   - Specific CTL are identifiable in blood and bronchoalveolar lavage fluid;
   - Polyclonal hypergammaglobulinemia;
   - Increased β_2 microglobulinemia.
4) Pre-AIDS:
   - Detectable P25 antigenemia;
   - HIV is detectable in PBL and CSF cell culture supernatants;
   - Decrease of anti-P25 antibodies titer;
   - Decrease of lymphocytes number (CD4+ and CD8+);
   - Thrombopenia, anemia and leukopenia;
   - Increased β_2 microglobulin and neopterine.

The main problem for scientists is now to understand the relationships between cellular and molecular phenomena described above and these clinical and biological symptoms. Little is known today, and most of the pathogenesis models are based on hypothesis. Nevertheless, one should take into consideration several *in vivo* and *in vitro* immunological findings, which may be incorporated in part of all pathogenesis concepts. Some of these findings are related directly to HIV or to a specific viral component and some to cellular factors secreted in direct or indirect response to HIV infection. Some are strictly related to the nature of the infected cell, others dependant on the humoral response.

2. Effects of free viral components in serum

Soluble HIV glycoprotein (GP) may be released in serum, either by incomplete viral particles, or after cell death following HIV replication. This soluble GP may bind to any CD4+ receptor. Two consequences may be observed:

i)   antibodies recognising HIV-GP can bind to this free GP attached on a non infected cell, and initiate cell killing;
ii) if the cell is a MHC class II CD4+ cell, it may process HIV-GP, and may present derived peptides in association with its class II antigens. This cell may then be a target for HIV GP specific CTL (Rosenberg and Fauci, 1989).

3. Cytokines and HIV

Cellular factors and cytokines modulate HIV replication, as demonstrated in chronically infected cell lines, and as noted above at the molecular level. Several intracellular signal factors like NFkB and SP1, and several cytokines like Interleukin 1 (IL1), epidermal cell derived thymocyte activating factor (ECTF), and tumor necrosis factor (TNF) are involved in HIV infection pathogenesis (Rosenberg and Fauci, 1989; Siekevitz et al., 1987; Matsuyama et al., 1989). It has been demonstrated that the cytokine induced up regulation of HIV replication occurs through DNA-binding proteins to the LTR of the provirus. Special attention has been devoted to TNF-α, which is oversynthetized in response to opportunistic microorganisms, and in HIV infected macrophages. This molecule, for which a role in cachexia has now been demonstrated, is toxic for neurons and glial cells subpopulations, and, because TNF belongs to the normal immune response network, it may be one of the factors of the switch from the latent phase to the active phase of HIV replication (Rosenberg and Fauci, 1989; Matsuyama et al., 1989; Okamoto et al., 1989; Griffin et al., 1989; Bielinska et al., 1989).

4. Lymphocyte related pathogenesis factors

Immunopathogenetic factors may be directly related to lymphocytes and their precursors. First to be recognized was the death of CD4+ lymphocytes induced by HIV replication after the activation of the host cell. Other lymphocyte related immune features have now been described, but their relative importances are still unclear.

- *In vitro,* antigen specific lymphocyte proliferation is inhibited by HIV-GP120.This functional impairment may partially explain the cutaneous anergy observed in infected patients. Several mechanisms may be proposed:
  1) down regulation of cellular genes (as demonstrated for CD4 and *nef* protein);
  2) abnormalities in cellular interactions via CD4 between monocyte and lymphocyte (Guy et al., 1987).
- Autoimmune processes have also been suggested, but once again, the mechanism remains unclear. For example, reaction of antibodies against HIV-GP with MHC (major histocompatibility complex) class II has been described. Autoantibodies might also be produced due to a transactivation process of cellular immunoglobulin genes, but this hypothesis needs experimental confirmation. In addition, one should note that anti-CD4 antibodies have been identified in HIV infected patients sera. These autoantibodies could play a role in the progressive disappearance of the CD4+ cells, and might be taken into consideration in the evaluation of soluble CD4 therapy results (Rosenberg and Fauci, 1989).
- Infection of bone marrow and thymic lymphocyte precursors has been also described. Progenitors (CD4+ or CD4-) may be killed through cytopathogenic effect, and, therefore, participate in the decrease of the lymphocyte populations (Alizon et al., 1984).
- As explained above, cytokines in excess might be toxic for CD4+ cells (Rosenberg and Fauci, 1989; Matsuyama et al., 1989; Okamoto et al., 1989; Griffin et al., 1989; Bielinska et al., 1989).
- HIV infection results in the production of specific CTL (cytotoxic T cell). These CTL, as demonstrated in this institute, may kill infected cells which present *env* and/or *gag* and/or *nef* proteins on their cellular membrane. Consequently, if a binding of free GP120 or free major core protein P25 or free *nef* to the membrane of a normal cell occurs, it may induce the death of these non infected cells via a CTL killing process (Hahn et al., 1989).
- A similar mechanism could be proposed in association with antibody dependant cellular cytotoxicity (ADCC); anti P25 or anti GP antibodies could bind to a normal cell coated with free P25 or free GP, and, therefore, induce an ADCC killing process. *In vivo,* it has been demonstrated that ADCC increases after HIV infection and reaches a plateau during the asymptomatic clinical phase. During ARC (AIDS related complex) and Kaposi sarcoma, ADCC levels decrease, and reach control values at the time of full blown AIDS (Ljunggren et al., 1987; Roock et al., 1987). Therefore, one might explain this evolution either by the development of an efficient immune mechanism during the asymptomatic phase, which decreases as all immune functions do in pre-AIDS and AIDS, or by an unfavourable character of ADCC in HIV infections, ADCC turning out to be a pathogenic mechanism of immune deficiency.
- Peripheral blood lymphocytes Natural Killer (NK) function is usually reduced in HIV infected individuals: CD16+ cells are reduced, Leu7+-CD8+ cells being elevated or unchanged. This decrease of the NK function is restored by Interleukin 2. On the other hand, several virus-derived synthetic peptides (inside GP41) can inhibit NK function of seronegative donors, and moreover, plasma from infected individuals are able to impair NK function. The exact nature of the NK function impairment in HIV infected patients remains unclear, and little is known on the possible

involvement of NK function in immunopathogenesis of AIDS (Brenner et al. 1989).

- Polyclonal hypergammaglobulinemia, which is often described in infected individuals, has not been elucidated. Persistent viral infections are often associated with hypergammaglobulinemia. In the case of HIV infection, the participation of HIV specific transactivation of normal cellular genes is still unclear, and the involvement of intracellular activation factors (NFκB) has yet to be tested.
- CD4+ lymphocyte number decrease is associated with CD8+ cell number increase. Mainly, suppressor clones are involved in this increase, and this fact explains the non specific immune response suppression observed in these patients.
- Idiotypic dysregulation has been also evoked, but no evidence is available at this time.
- HIV variability plays a major role in AIDS pathogenesis. Genomic variability from one isolate to an other has been estimated to 20-25 % (Haseltine, 1988; Alizon et al., 1986; Fischler et al., 1988). The sources of these variations are presently being debated. Do they represent reverse transcription errors (?), error in proviral DNA replication due to cellular replication mechanisms (?), differences in host cell type (?) etc. Furthermore, individuals are probably infected with more than one type of virus: some of these viruses could replicate at a low level and escape the immune system; others could replicate at a high level and induce a strong antibody response, which might not be efficient in protecting the individuals against the other types of viruses.

5. Macrophage related pathogenetic factors

As demonstrated before, infection of monocytesmacrophages (MM) is one of the main steps in HIV infection pathogenesis. Today, the routes of infection of this type of cell are: classical infection via CD4 molecule, phagocytosis and internalization of immune complexes "HlV-antiGP". Replication of HIV in MM is almost always intracellular: HIV does not exhibit any major cytopathic effect in these cells. Numerous biological and molecular consequences of HIV infection of MM have been described. Most of them have been previously described in this paper. To summarize:

1) synthesis of cytokines like IL1 or TNF, which are known to be toxic for several categories of cells (CNS cells, etc.),
2) probable major role in CNS infection,
3) neuroleukin is a specific cytokine which is involved in neuronal damage repair and in neuronal growth; receptors to neuroleukin can bind HIV-GP120; therefore, competition may occur, and biological effects of neuroleukin may be inhibited,
4) general inflammatory processes, and
5) defect in antigen presentation (Rosenberg and Fauci, 1989; Koenig et al.; Dormont et al., 1988).

6. Dendritic cells (DC) HIV infection consequences

Dendritic cells have been reported to be infectable with HIV. These are CD4+ cells and antigen presenting cells (APC), and are involved in both primary and secondary immune responses. When infected with HIV, the replication rate is more important than in MM (Muller et al., 1986). Several results have to be taken into consideration, and may explain part of HIV infection pathogenesis:

- DC class II level is reduced in AIDS patients (same phenomenon as described with CD4 molecule and nef in Lymphocytes?).
- DC infection suppresses their ability to present other antigens.
- DC of asymptomatic infected individuals induce a reduced allogenic stimulation.
- Because 1) DC enhance Con A response, and

2) follicular DC of the B-dependent areas may be infected and
induce a B-cell memory defect, DC may play a major role in
general immune deficit shown in AIDS.

7. Infection of Langerhans cells with HIV

Langerhans cells (LC) may be considered as a DC subpopulation. LC are
also infectable, and skin LC infection in seropositive individuals has been
reported by several laboratories. These cells are also CD4+, and they are
HLA class I and HLA class II+ (HLA: human leukocyte antigen complex). As DC
and macrophages, they are antigen presenting cells. The level of HIV
replication in infected Langerhans cells is comparable to the one observed
in DC (Tschachler et al., 1987). HIV infection induces a reduction of CD4
molecules and class II antigens on the cell membrane. *In vivo* infection of
LC may occur either in differentiated cells or in bone marrow precursors
(Rosenberg and Fauci, 1989). Due to their wide *in vivo* distribution and
their biological functions, LC may be of crucial importance in immune
function pathogenesis.

8. Infection of bone marrow progenitors with HIV

Bone marrow precursors may be infected *in vivo* and *in vitro*. Recently,
myeloid progenitors, which are CD4- cells have been shown to be infected
with HIV: no cytopathic effect could be detected in infected cells. In
common with macrophages, HIV production is mainly intracellular, and CD4-
bone marrow progenitors may act as a viral reservoir (Rosenberg and Fauci,
1989).

9. HIV and Central Nervous system

CNS infection and pathogenesis of CNS dysfunctions is still unclear;
one may suspect both a predominant role of the general immune deficiency,
as suggested by the lack of evidence of HIV presence in CNS cells in
infants with neurologic symptoms, and numerous HIV related features.
Nevertheless, MM and reticuloendothelial system (RES) cells are crucial in
the HIV infection of the CNS. This infection requires at the first stage,
abnormalities of the blood brain barrier (BBB). These abnormalities might
be induced by opportunistic agents, but, also, could be a consequence of an
immune mechanism. Endothelial cells are known to express MHC (major
histocompatibility complex) class II and class I antigens. This expression
is directly related to cytokines, including interleukin 1 and gamma
interferon, and perhaps TNF. If class I and class II antigens are expressed
on the endothelial cell membranes, these cells could become adherent to T
peripheral blood cells. This adherence may then induce specific LAK
(lymphokine-activated killer) cells, and endothelial cells might be killed
through a Natural Killer-like process. Thus, the BBB may be altered, and
HIV infected cells and/or opportunistic agents may enter the CNS
(Haseltine, 1988; Gallo, 1988; Rosenberg and Fauci, 1989; Koenig et al.,
1986; Dormont et al., 1988).

10. Neutralizing antibodies and facilitating seric factors

One of the major problems today is the efficiency of neutralizing
antibodies in HIV infected patients (Rey et al., 1987). The lack of high
neutralizing titers has to be explained before starting any vaccine trial
in seronegative individuals. Several explanations may account for this
findings, including:

a) genetic variability of HIV. The part of the external envelope protein
which induces neutralizing antibodies is one of the most variable
portions of all the HIV proteins.

b) the main functional domains of HIV-GP are related to complex sugars of very high density, building a carbohydrate coat for the viral protein. Moreover, the carbohydrate chains addition to the envelope glycoprotein is processed by cellular enzymes; this may explain the failure of antigenicity of these functional domains (Haseltine, 1988).

c) a general immune deficit related to HIV infection may also take part in this poor antibody protection.

d) lastly, facilitating serum factors of HIV infection have been reported: two different categories have been identified: thermoresistant factors, which are specific immunoglobulins, and thermosensitive factors, which are molecules of the alternative complement pathway (mainly C3) (Robinson et al., 1988; Gras et al., 1988; Montefiori et al., 1989).

11. Kaposi sarcoma pathogenesis

A special place should be made for pathogenesis of Kaposi sarcoma (KS). Gallo and colleagues reported that proviral DNA was not identifiable in KS DNA, suggesting that KS pathogenesis might not be directly related to HIV replication. They described a novel growth factor released by CD4+ lymphocytes after infection with human retroviruses. When grown in the presence of this new growth factor, KS cell populations can produce numerous cytokines (TNF, IL1, tumor growth factor TGF, fibroblast growth factor, etc.) (Gallo, 1988; Nakamura et al., 1988). When tested for ability to induce blood vessels in chorioallantoic membrane assay, these KS-derived cells induce an important increase of blood vessels growth, suggesting a high angiogenesis potentiality. Injected to the nude mouse, these cells induce tumors which resemble early KS lesions in humans. Genetic analysis of these tumors demonstrated that the tumoral tissue was of murine origin. Therefore, KS lesions in humans may be related to a cytokine-induced mechanism: one of those cytokines may promote activation and growth of endothelial cells, converting them into KS cells, which might themselves be able to produce these cytokines. Hormonal factors might explain why males develop more KS than females (Gallo, 1988).

CONCLUSION

Unfortunately, no general concept of pathogenesis may be proposed at present. Numerous hypotheses have been suggested, involving:

- either a gradual rise of viral replication along the asymptomatic phase (Haseltine et al., 1988): increase in number of latently infected CD4 cells, and, because cumulative probability of activation of CD4+ cells increases with time, large amounts of virus might be released. This process being autocatalytic, a switch occurs from a controlled replication state to a continuous high level of viral production.
- or an infection with "multiple HIV variants" (close to the hypothesis developed by L. Montagnier). Some variants are able to induce a strong immune response (variant infecting CD4+ lymphocytes?), including production of antibodies. Other variants replicate intracellularly in macrophages or other long lived-PAC, escaping from the immune system. A switch from macrophage to CD4+ lymphocytes may then occur, and could induce a progressive disease through CD4+ lymphocyte death.

REFERENCES

Ahmad, N., Maitra, R.K., and Venkatesen, S., 1989, Rev-induced modulation of nef protein underlines temporal regulation of human immunodeficiencey virus replication, Proc. Natl. Acad. Sci. USA, 86:6111.

Alizon, M., Sonigo, P., Barre-Sinoussi, F., Chermann, J.C., Tiollais, P., Montagnier, L., and Wain-Hobson, S., 1984, Molecular cloning of lymphadenopathy associated virus, Nature, 312:757.

Alizon, M., Wain-Hobson, S., Montagnier, L., and Sonigo, P., 1986, Genetic variability of the AIDS virus: nucleotide sequence analysis of two isolates from African patients, Cell, 46:63.

Allan, J.S., Coligan, J.E., Lee, T.H., McLane, M.F., Kanki, P.J., Groopman, J.E., and Essex, M., 1985, A new HTLVIII/LAV antigen detected by antibodies from AIDS patients, Science, 230:810.

Barre-Sinoussi, F., Chermann, J.C., Rey, F., Nugeyre, M.T., Chamaret, S., Gruest, J., Daguet, C., Axler-Blin, C., Vézinet-Brun, F., Rouzioux, C., Rozenbaum, W., and Montagnier, L., 1983, Isolation of a T lymphotropic retrovirus from a patient at risk for Acquired Immunodeficiency Syndrome (AIDS), Science, 220:868.

Berkhoul, B., Silverman, R.H., and Jeang, K.T., 1989, Tat trans-activates the human immunodeficiency virus through a nascent RNA target, Cell, 59:273.

Bielinska, A., Krasnow, S., and Nabel, G.J., 1989, NFkB mediated activation of the human immunodeficiency virus enhancer: site of the transcriptional initiation is independant of the TATA box, J. Viral., 63-9:4087.

Brenner, B.G., Dascal, A., Margolese, R.G., and Wainberg, M.A., 1989, Natural Killer function in patients with acquired immunodeficiency syndrome and related diseases, J. Leuk. Biol., 46:75.

Daniel, M.D., Letvin, N.L., King, N.W., Kannagi, M., Sehgal, P.K., Hunt, R.D., Kanki, P.J., Essex, M., and Desrosiers, R.C., 1985, Isolation of T-cell tropic HTLV-III-like retrovirus from macaques, Science, 228:1201.

Dormont, D., Boussin, F., and Merrouche, Y., 1988, Interactions virus VIH et système nerveux central: données récentes et hypothèses pathogéniques, Nouv. Rev. Fr. Hematol., 30:21.

Feinberg, M.B., Jarret, R.F., Aldovini, A., Gallo, R.C., and Wong-Staal, F., 1986, HTLVIII expresion and production involve complex regulation at the level of splicing and translation of viral RNA, Cell, 46:807.

Fischer, A.G., Ensoli, B., Looney, D., Rose, A., Gallo, R.C., Saag, M.S., Shaw, G.M., Hahn, B.H., and Wong-Staal, F., 1988, Biological diverse molecular variants within a single HIV 1 isolate, Nature, 334:444.

Fischer, A.G., Ensoli, B, Ivanoff, L, Chamberlain, M., Petteway, S., Ratner, L., Gallo, R.C., and Wong-Staal, F., 1987, The sor gene of HIV1 is required for efficient virus transmission in vitro, Science, 237:888.

Gallo, R.C., 1988, HlV-the cause of AIDS: an overview on its biology, mechanisms of disease induction and our attempts to control it, J. AIDS, 1-6:521.

Gartner, S., Markovits, P., Markovitz, D., Kaplan, M.H., Gallo, R.C., and Popovic, M., 1986, The role of mononuclear phagocytes in HTLV-III/LAV infection, Science, 233:215.

Gottlieb, Schroff, R., Schanker, H., Weisman, J.D., Fan, P.T., Wolf, R.A., and Saxon, A., 1981, Pneumocystis carinii pneumonia and mucosal candidiasis in previously healthy homosexual men: Evidence of a new acquired immunodeficiency, N. Eng. J. Med., 305:1425.

Griffin, G.E., Leung, K., Folks, T.M., Kunkel, S., and Nabel, G.J., 1989, Activation of HIV gene expression during monocyte differentiation by induction of NF-KB, Nature, 339:70.

Gras, G., Strub, T., and Dormont, D., 1988, Antibody dependent enhancement of HIV infection, Lancet, i:1285.

Guy, B., Kieny, M.P., Rivierre, Y., Le Peuch, C., Dott, K., Girard, M., Montagnier, L., and Lecocq, J.P., 1987, HIV F/3'ORF encodes a phosphorylated GTP-binding protein resembling an oncogene product, Nature, 330:266.

Hahn, T., Schattner, A., Handzel, Z.T., Levin, S., and Bentwich, Z., 1989, Possible role of natural cytotoxic activity in the pathogenesis of AIDS, Clin. Immunol. Immunopathol., 50:53.

Haseltine, W.A., 1988, Replication and pathogenesis of the AIDS virus, J. AIDS, 1:217.

Kao, S.Y., Calman, A.F., Luciw, P., and Peterlin, B.M., 1987, Anti-termination of transcription within the long terminal repeat of the HIV1 by tat gene product, Nature, 330:489.

Klatzmann, D., Champagne, E., Chamaret, S., Gruest, J., Guetard, D., Hercend, T., Gluckman, J.C., and Montagnier, L., 1984, T lymphocyte T4 molecule behaves as a receptor for human retrovirus LAV, Nature, 321:767.

Koenig, S., Gendelman, H.E., Orenstein, J.M., Dal Canto, M.C., Prezeshkpour, G.H., Yungbluth, M., Janotta, F., Aksamit, A., Martin, M.A., Fauci, A.S., 1986, Detection of AIDS virus in macrophages in brain tissue from AIDS patient with encephalopathy, Science, 233:1089.

Laspia, M.F., Rice, A.P., and Mathews, M.B., 1989, HIV1 tat protein, increases transcriptional initiation and stabilizes elongation, Cell, 59:283.

Ljunggren, K., Boettiger, B., Biberfeld, G., Karlson, A., Fenyoe, E.M., and Jondal, M., 1987, Antibody dependant cellular cytotoxicity inducing antibodies against human immunodeficiency virus: Preserve at different clinical stages, J. Immunol.,139:2263.

Malim, M.H., Hauber, J., Le, S.H., Maizel, J.V., and Cullen, B.R., 1989, The HIV1 rev trans activator acts through a structures target sequence to active nuclear export of unspliced viral mRNA, Nature, 338:254

Matsuda, Z., Chou, M.J., Matsuda, M., Huang, J.H., Chen, Y.M., Redfield, R., Mayer, K., Essex, M., and Lee, T.H., 1988, Human immunodeficiency virus type 1 has an additional coding sequence in the central region of the genome, Proc. Natl. Acad. Sci. USA, 85:6968.

Matsuyama, R., Yoshiyama, H., Hamamoto, Y., Yamamoto, N., Soma, G.-I., Mizuno, D., and Kobayashi, N., 1989, Enhancement of HIV replication and giant cell formation by tumor necrosis factor, AIDS Research and Human Retroviruses, 5-2:139.

Montefiori, D.C., Robinson, E., and Mitchell, W., 1989, Antibody-independent, complement-mediated enhancement of HIV1 infection by mannosidase I and II inhibitors, Antiviral Res.,11:137.

Muller, H., Falk, S., and Stutte, H.J., 1986, Accessory cells as primary target of human immunodeficiency virus HIV infection, J. Clin. Pathol., 39:1161.

Nakamura, S., Salahuddin, S.Z., Bibberfeld, P., Ensoli, B., Markham, P.D., Wong-Staal, F., and Gallo, R.C., 1988, Kaposi's sarcoma cells: Long-term culture with growthfactor from Retrovirus-infected CD4[+] T cells, Science, 242:426.

Okamoto, T., Matsuyama, T., Mori, S., Hamamoto, Y., Kobayashi, N., Yamamoto, N., Josephs, S.F., Wong-Staal, F., and Shimotohno, K., 1989, Augmentation of human immunodeficiency virus type 1 gene expression by tumor necrosis factor a, AIDS Research and Human Retroviruses, 5-2:131.

Rey, F., Barre-Sinoussi, F.C., Schmidtmayerova, H., and Chermann, J.C., 1987, Detection and titration of neutralizing antibodies to HIV using an inhibition of the cytopathic effect on MT4 cells, J. Virol. Methods., 87:1.

Robinson, W.E. Jr, Montefiori, D.C., and Mitchell, W.M., 1988, Antibody dependant enhancement of human immunodeficiency virus type I infection, Lancet, i:790.

Rook, A.H., Lane, H.C., Folks, T., McCoy, S., Alter, H., and Fauci, A.S., 1987, Sera from HTLV-III/LAV antibody positive individuals mediate antibody dependant cellular cytotoxicity against HTLVIII/LAV infected T cells, J. Immunol., 138:1064.

Rosen, C., Sodroski, J., and Haseltine, W.A., 1985, The location of cis acting regulatory sequences in the human T Iymphotropic virus type III (HTLVIII/LAV) long terminal repeat, *Cell*, 41:813.

Rosenberg, Z.F., and Fauci, A.S., 1989, Immunopathogenic mechanisms of HIV infection, *Clin. Immunol. Immunopathol.*, 50:S149.

Siekevitz, M., Joseph, S.F., Dukovitch, Peffer, N., Wong-Staal, F., and Greene, W.C., 1987, Activation of HIV-1 LTR by T cell mitogens and the transactivator protein of HTLV1, *Science*, 238:1575).

Strebel, K., Daugherty, D., Clouse, K., Cohen, D., Folks, T., and Martin, M.A., 1987, The HIV A *(sor)* gene product is essential for virus infectivity, *Nature*, 328:728.

Strebel, K., Klimkit, T., and Martin, M.A., 1988, A novel gene of HIV1 *vpu*, and its 16 kd product, *Science*, 1221.

Terwilliger, E.F., Cohen, E.A., Lu, Y., Sodroski, J.G., and Haseltine, W.A., 1989, Functional role of human immunodeficiency virus type 1 *vpu*, *Proc. Natl. Acad. Sci. USA*, 86:5163.

Tschachler, E., Groh, D.V., Popovic, M., Mann, D.L., Konrad, K., Safai, B., Eron, L., diMarzo, V., Wolff, K., and Stingl, G., 1987, Epidermal Langerhans cells - a target for HTLV III/LAV infection, *J. Invest. Dermatol.*, 88:233.

Wain-Hobson, S., Sinigo, P., Danos, O., Cole, S., and Alison, M., 1985, Nucleotide sequence of the AIDS virus, LAV. *Cell*, 40:9.

MYCOSES IN AIDS PATIENTS: AN OVERVIEW

Edouard Drouhet and Bertrand Dupont

Institut Pasteur, Unité de Mycologie, Paris, France

The acquired immunodeficiency syndrome (AIDS) induced by the retrovirus HIV identified at the Pasteur Institute by Montagnier and his group (Barré-Sinoussi et al., 1983) is characterized by a marked depression of cellular immunity due to the reduction and destruction of T4 helper lymphocytes and some macrophages by the virus. This often leads to multiple opportunistic infections including fungal infections. Since the first report (Gottlieb et al., 1981) of AIDS cases associated with pneumonia due to *Pneumocystis carinii*, which has recently been reclassified as a fungus according to molecular taxonomy (Edman et al., 1988; see Mackenzie, this book), and mucosal candidosis, the role of fungal infections has become more important. The most recent clinical definition of AIDS (revised in 1988) by CDC/WHO recognizes as important "indicator" diseases: candidosis of the esophagus, trachea, bronchi or lungs and meningeal cryptococcosis as well as other opportunistic infections induced by protozoa, bacteria and viruses (Table 1). In the presence of laboratory evidence of HIV infection, disseminated coccidioidomycosis and histoplasmosis at a site other than, or in addition, to the lungs or cervical or hilar lymph nodes, are considered indicative of AIDS. The spectrum of clinical responses to HIV as seen in adults, is schematically depicted in Figure 1 according to a WHO Immunology Workshop (Seligmann et al., 1987).

Table 1. CDC classification system for HIV (1)

Group I   : acute infection
Group II   : asymptomatic infection
Group III   : persistent generalized lymphadenopathy
Group IV   : other disease
     Subgroup A   : constitutional disease
     Subgroup B   : neurological disease
     Subgroup C   : **secondary infectious disease**
          Category C1   : major opportunistic infections*
          Category C2   : minor opportunistic infections**
     Subgroup D   : secondary cancers
     Subgroup E   : other conditions

---

(1) Revision in the case definition of AIDS by CDC/WHO (1988)
*   Includes: Candida esophagitis, cryptococcosis, histoplasmosis, coccidioidomycosis
**   Includes: thrush

Mycoses in AIDS Patients, Edited by
H. Vanden Bossche *et al.,* Plenum Press, New York, 1990

27

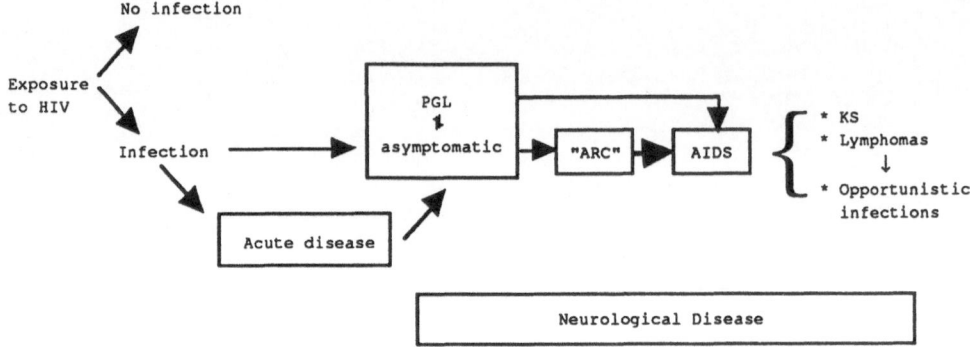

Figure 1. Course of infection with human immunodeficiency virus (HIV).
PGL=persistent generalized lymphadenopathy; "ARC"=AIDS-related
complex; AIDS=acquired immunodeficiency syndrome; KS=Kaposi
sarcoma. After Seligmann et al., 1987.

The Walter Reed Army Institute of Research (Redfield et al., 1986) has
adopted a stage classification for HIV infections which ranges from
asymptomatic (with viremia, antibody or both), through chronic generalized
lymphadenopathy to subclinical and clinical T-cell deficiency (Table 2).
The clinical presentation of patients with HIV infection depends on the
functional integrity of the CD4+T helper cells and T-cell immunity. The
appearance of thrush with less than 400 CD4+T cells/mm^3 and cutaneous
anergy to *Candida albicans* antigen indicates the WRS stage which appears
just prior to the terminal stage or classical AIDS. The Walter Reed Staging
(WRS) scheme is applicable to adults only; new-born infants and young
children may require differents base-line functional T-cell indexes.

The Bangui WHO (1985) workshop recognizes as major symptoms of AIDS in
Africa: weight loss (>10 %), chronic diarrhea (>1 month), and fever
(>1 month). Among the most important opportunistic infections are
oropharyngeal and esophageal candidosis (82% for Zaire) and cryptococcal

Table 2. Walter Reed Staging classification for HIV infection*

	HIV antibody or virus +	CLA	CD4	DHS	Thrush	Opport. infection
WR 0	−	−	> 400	N	−	−
WR 1	+	−	> 400	N	−	−
WR 2	+	+	> 400	N	−	−
WR 3	+	+	> 400	N	−	−
WR 4	+	+	< 400	P	−	−
WR 5	+	+	< 400	C or/and	+	−
WR 6	+	+	< 400	P	+	+

CLA: chronic lymphadenopathy; CD4: T helper cells/mm^3; DHS: delayed
hypersensitivity; N: normal; P: partial cutaneous anergy, defined as an
intact cutaneous response to only one of the four test antigens: *Candida*,
Trichophyton, tetanus, and mumps; C: complete cutaneous anergy to the four
antigens; *: according to Redfield et al., 1986.

meningoencephalitis (13% for Zaire, 33% for Burundi and other Central and
East African regions) (Colenbunders et al., 1987, Coulaud, 1988).
Pneumocystis carinii pneumonia, the most frequent opportunistic infection
in AIDS patients in America and Europe is more rarely observed in Africa.

Retrospective studies of AIDS all over the world show that 58-81% of
all patients contract a fungal infection at some time during the prodromal
stage or after developing AIDS and 10-20% have died as a direct consequence
of fungal infections (Holmberg and Meyer, 1986). Reports from Europe seem
to suggest that AIDS patients in the USA and some regions of Africa run a
higher risk of contracting fungal infectious than those in Europe. Several
reports review general information regarding the mycoses observed in HIV
infections during the various stages of AIDS, (Chandler 1985; Holmberg and
Meyer, 1986; Drouhet and Dupont, 1987; Dupont ,1988; Armstrong, 1988;
Cairns, 1988; Male, 1989; Spencer and Jacobson, 1989).

In addition to major fungal infections (Table 3) with their new
clinical, pathological and biological aspects, other mycoses with defects
in cell-mediated immunity appear to be associated with AIDS (Table 4).
These include penicilliosis due to *Penicillium marneffei*, a
reticuloendothelial mycosis of South East Asian origin, and sporadic deep,
visceral infections (alternariosis, mucormycosis, pseudallescheriosis,
sporotrichosis and nocardiosis) which are related to drug addiction or
other predisposing factors.

Table 3. Spectrum of mycoses related to AIDS - major opportunistic
infections

Mycoses	Agents	Main target tissues	Incidence %
Yeasts			
Candidosis	*Candida albicans*	oral mucosae esophageal mucosae bronchi, trachea lungs	70-90* 10-35 3
Cryptococcosis	*Cryptococcus neoformans*	brain meninges lungs...	3-33*
Dimorphic fungi			
Histoplasmosis	*Histoplasma capsulatum*	RES, dissemination lungs, lymph nodes	0.5-5 (2.7-29)**
Coccidioidomycosis	*Coccidioides immitis*	lungs, blood bone marrow lymph nodes, CNS	27**

*:   Cosmopolitan
**: in greographically limited endemic areas
***: RES reticuloendothelial system

Table 4. Spectrum of mycoses related to AIDS - deep rare, but severe infections

Mycoses	Agents	Main target tissues	Incidence %
	Dimorphic fungi		
Penicilliosis	Penicillium marneffei	RES, dissem. lungs, lymph nodes	sporadic*
Blastomycosis	Blastomyces dermatitidis	lungs	sporadic*
Paracoccidioido-mycosis	Paracoccidioides brasiliensis	(lungs), oral nasal + mucosae skin	sporadic*
Sporotrichosis	Sporothrix schenkii	brain skin dissem.	sporadic*
	Molds		
Aspergillosis	Aspergillus sp.	lungs (invasive) brain	0.16**
Mucormycosis	Mucor, Rhizopus	brain	sporadic**
	Yeasts		
Yeast infections	Trichosporon beigelii Rhodotorula	blood (invasive)	sporadic**
	Actinomycetes		
Nocardiosis	Nocardia asteroides	brain, lungs pericardium	0.3**

* geographically limited areas
** infections associated with IV drug injection, catheter, granulopenia, and/or corticosteroids

It is surprising that invasive aspergillosis, which is widespread among immunodepressed patients, is rare in AIDS (Asnis et al., 1988; Fournier et al., 1989; Henochowicz et al., 1989; Schaffner, 1984). This is probably due to normal or subnormal function of polymorphonuclear cells, and because cellular immunity appears to play a lesser role in *Aspergillus* than in other opportunistic fungal infections such as those due to *Candida albicans*, and *Cryptococcus neoformans*. Therefore, the CDC (1988) has removed aspergillosis from the list of typical mycoses associated with AIDS. Invasive yeast infections due to *Trichosporon* (Leaf, 1989), *Rhodotorula* (Causey, 1989), *Candida* (Alsina, 1988) or mold infections due to *Aspergillus*, or *Mucorales* (Smith et al., 1989; Mostaza et al., 1989) occur in AIDS only accidentally, and are particularly associated with i.v. drug injection, catheter, granulopenia, and/or corticosteroids. Most cases are diagnosed postmortem (Wilkes et al., 1988; Gotzsche et al., 1988).

Superficial cutaneous mycoses such as pityrosporosis (seborrheic dermatitis) and dermatophytosis (particularly nail infections) have a high incidence (46-80%) (Table 5). However their importance is controversial.

Table 5. Spectrum of mycoses related to AIDS - Cutaneous infections

Superficial mycoses	Agents	Main target tissues	Incidence %
Seborrheic dermatitis	*Malassezia furfur* *Pityrosporum ovale*	skin (face) papules	41-83 (versus) 5
Dermatophytosis	*Trichophyton rubrum* *Trichophyton interdigitale* *Epidermophyton floccosum*	nails (white) crural & feet area	17-37 (versus) 3-91
Trichosporonosis	*Trichosporon beiglii*	hair, skin genital & anal area	13 (versus) 0
Alternariosis	*Alternaria alternata*	nasal septum skin (face)	sporadic

VALUE OF NECROPSY IN AIDS

Systemic fungal infections are difficult to diagnose antemortem in AIDS patients. Necropsy has helped greatly in the recognition of AIDS as a distinct entity and in the understanding of the pathophysiology of AIDS and AIDS-related diseases (Gottlieb et al., 1981). Necropsy has revealed diagnoses other than the primary cause of death that were not suspected antemortem. Wilkes et al. (1988) in a well documented review compared retrospectively, necropsy findings with antemortem clinical diagnoses in 101 adult patients with the AIDS from two hospitals in New York. A high incidence of systemic fungal infection (20%) was observed (Table 6). Bronchoscopy and transbronchial biopsy were not useful in diagnosing fungal pneumonia in 7 out of 8 cases. Open lung biopsy or empirical amphotericin B therapy especially for patients with deteriorating pulmonary lesions is recommended by the authors. Gotzsche et al. (1988) reviewed case records from 33 deceased AIDS patients in an European hospital (Copenhagen, Denmark). In 23 autopsies, 20 cases of deep mycoses including 2 cases of disseminated and pulmonary candidosis, 8 esophageal, 7 tracheal and one lymphonodular candidosis, one case of cryptococcosis (adrenals) and one cerebral and cardiac aspergillosis were observed.

Table 6. Value of necropsy in AIDS*

Infection	Necropsy	Clinical only proven	Clinical suspected	Clinical & necropsy proven	suspected
Fungal (disseminated)					
candidosis	11	0	1	0	1
aspergillosis	7	0	0	1	0
cryptococcosis	0	0	1	5	0
histoplasmosis	2	0	0	0	0
Nocardia pneumonia	1	0	0	0	0

* according to Wilkes et al., 1988

An analysis of the spectum of mycoses in AIDS shows some obvious differences between the mycoses observed in the immunocompromized HIV host or non immunocompromized host. The altered cellular immunity due to reduced numbers of CD4+ T lymphocytes (Klatzman et al., 1984), functional changes of monocytes (Prince et al., 1985), macrophages, and Langerhans cells (Niedecken et al., 1987), polyclonal B lymphocyte activation and the relatively unaltered numbers and functions of granulocytes (at least in the first stages of AIDS), may be responsible for the selection of certain forms of opportunistic mycoses in AIDS (Table 7). This is the case for oral and esophageal candidosis, which is so frequently observed in AIDS patients even during preliminary stages. Prior to the appearance of AIDS, the incidence of esophageal candidosis in immunocompromized patients was only 0.8%. It is interesting to note that invasive candidosis, as well as invasive aspergillosis and other opportunistic mycoses are absent in AIDS patients although they are frequently observed in immunocompromized patients, particularly in granulopenic patients. In AIDS patients with chronic mucosal candidosis, the absence of invasive candidosis might be explained by the increased levels of protective antibodies against the 47 kD antigen, due to the polyclonal activation of B lymphocytes (Matthews et al., 1986).

The distribution of the various target cells for HIV virus in the different regions of the body may be responsible for the localization of some mycoses in AIDS patients. A selective pressure for a single species or group of mycoses is observed where there are usually multiple species (Table 7). For example, *C. albicans* is the only species in candidosis in AIDS patients (exceptions in special conditions not related to AIDS). *Histoplasma capsulatum* (and not *H. duboisii*) is quite exclusively observed in histoplasmosis in Africa, in spite of the coexistence of both species in locations with patients largely infected by HIV.

Table 7. Selective pressure of altered immunity in AIDS

CANDIDOSIS	clinically	oral & esophageal mucosae no systemic involvement except in presence of other favorable factors (IV drug, catheter)
	mycologically	only *C. albicans* species serotype B is more frequent role of mannan as immunosuppressor high level of Abs (anti 47 KD)
CRYPTOCOCCOSIS	clinically	CNS – extra CNS – dissemination
	mycologically	no B-C (except 2 cases)
HISTOPLASMOSIS		only *H. capsulatum* no *H. duboisii* (only one exception)
PITYROSPOROSIS		only seborrheic dermatitis no pityriasis versicolor *Pityrosporum ovale / Malassezia furfur*
DERMATOPHYTOSIS		tinea pedis & unguis no tinea capitis
TRICHOSPORONOSIS		carriage of *Trichosporon beigelii* on scrotal & anal skin few white piedra, non systemic infection

A selection of serotypes of the same species, is also observed. The most interesting example is given by the serotypes of *C. albicans*. Previous studies in France, and in other parts of Europe (Drouhet et al., 1975; Drouhet, 1981) showed that serotype A predominates (92.1%), while the B serotype varied between 4.9% (hemoculture's strains), 4.2% (oral and digestive tract), and 13.3% (genitourinary tract). This is in contrast with American statistics which indicate that the incidence of serotype A and B ranges between 50.7 to 67.4% and 49.9 to 32.6%, respectively. The distribution of the *C.albicans* serotypes in 135 HIV+ patients was (Table 8): 71 (52.59%) strains of serotype A and 58 strains (42.96%) of serotype B

Table 8. Serotypes of *C. albicans* in AIDS and non AIDS patients

Patients		No.	Serotype A	Serotype B
AIDS	patients	135	71 (52.59%)	58 (42.96%)
(HIV+)	strains isolated*	214	106 (49.53%)	108 (50.47%)
non AIDS	patients	32	25 (78.12%)	7 (21.87%)
(HIV-)	strains isolated	41	28 (68.29%)	13 (31.70%)

* Total number of strains from the 135 patients. In 6 patients the serotype changed during the evolution of the disease: 5 strains A were replaced by B strains, 1 strain B was replaced by A serotype. Final totals: 76 strains A (52.59%), 50 serotypes B (43.70%).

(Drouhet et al., 1989). In 6 patients the serotype changed during the evolution of the disease: 5 strains of serotype A were replaced by B and one serotype B was replaced by A (Table 8). The end result was 76 strains of serotype A (56.29%) and 59 of serotype B (43.70%). In a group of 32 non-AIDS patients, 25 patients had serotype A (78.12%) and in 7 patients, serotype B (21.87%). Of 112 strains from 83 HIV+ patients, 66 were serotype B: 46 strains were resistant to 5-FC (69.60%) and 20 strains (30.30%) were sensitive (Table 9). The major phenotypes in serotype A and B strains was, respectively, 5-FC^S 5-FU^R 5-FUri^S (77%) and (5-FC^R 5-FU^R 5-FUri^S (53%) (Table 10). In the USA, the studies of Brawner and Cutler (1989) (summarized in Table 11) and of Shadomy et al. (1989) confirmed our data. However the difference between the incidence of B serotypes in AIDS and non-AIDS is less marked than in our study.

Table 9. Phenotypes of 5-FC resistance

	Aids patients (112)					
	Serotype A (46 strains)			Serotype B (66 strains)		
MIC µg/ml	No. strains	5-FC*	MIC µg/ml	No. strains	5-FC
---	---	---	---	---	---
0.09	11		0.09	7	
0.18	17		0.18	4	20 S
0.39	7	39 S	0.39	5	(30.30%)
0.78	2	(84.78%)	0.78	2	
1.56	1		1.56	2	
3.12			3.12		
6.25	1		6.25		
12.50	3		12.50		
25	3	7 R	25	2	46 R
50	3	(15.2%)	50	27	(69.60%)
100	1		100	17	

	Non Aids patients (28)					
1	28	28 S (100%)	1	13	13 (100%)

*S= sensitivity; R= resistance; MIC 12.5 µg/ml at 48 h

Table 10. Resistant phenotypes to 5-fluoropyrimidines

Phenotype			AIDS patients (112)*		non AIDS patients (345)**	
			Serotypes		Serotypes	
5-FC	5-FU	5-FURI	A	B	A	B
			46 (41.07%)	66 (58.92%)	319 (90.43%)	26 (7.53%)
S	R	S	36 (78.26%)	20 (30.30%)	312 (97.80%)	5 (19.23%)
S	R	R	3 (6.52%)	–	2 (0.62%)	–
R	R	S	6 (13.04%)	42 (63.63%)	3 (0.94%)	21 (80.77%)
R	R	R	1 (2.17%)	4 (6.06%)	2 (0.62%)	–

MIC read at 48 h; * strains tested in 1989; ** strains tested in 1974

Table 11. Serotypes of C. albicans in U.S.A.*

	No. strains	A		B	
HIV+	20	7	(35%)	13	(65%)
HIV- non hospitalized	49	23	(47%)	26	(53%)
HIV- hospitalized**	25	12	(48%)	13	(52%)
HIV- hospitalized immunosuppressed	19	6	(31.6%)	13	(68%)

*   according to D.L. Brawner & J.E. Cutler (1989)
** non immunosuppressed patients

MYCOSES CAN PLAY A ROLE AS A COFACTOR IN AIDS
FUNGAL ANTIGENS AS IMMUNOSUPPRESSORS

Mycoses can play a role as a cofactor in the development of AIDS. The activation of T lymphocytes by antigenic stimulation can favour infection of these cells by HIV (Margolick et al., 1987). The fungal infection itself (cryptococcosis, candidosis, and histoplasmosis) is often a source of immunosuppressive effects by polysaccharidic circulating antigens. These antigens play an important role in the immune response (Drouhet, 1988). Although small amounts stimulate the immune response, an excess of antigen (Ag) may block cellular immunity. T-lymphocyte stimulation requires Ag presentation on the surface of the macrophage. Specifically sensitized T-cells respond to Ag by proliferation and release of soluble lymphokines which have multiple effects. These include: localization and activation of macrophages, lymphocytes, blastogenesis or mediate killing or inactivation of target cells. Delayed-type hypersensitivity and chronic granuloma are expressions of T-cell mediated immunity. Fungal antigens may be in excess in acute fungal infections such as in invasive forms of candidosis, cryptococcosis, and the dimorphic systemic mycoses (histoplasmosis) or in chronic infections caused by some dermatophytes. Among these circulating antigens, the better known are mannan in candidosis, and glucuronoxylomannan in cryptococcosis which are the most frequently observed fungal diseases in AIDS patients.

The candidal antigen, mannan, exerts an immunosuppressive effect in most patients with chronic mucocutaneous candidosis (CMCC) (Fischer et al., 1978; Durandy et al., 1986, 1987, 1988; Drouhet and Dupont, 1980, 1983, 1985). A selective cellular immunodeficiency in T cell proliferation towards *Candida* antigens in CMCC patients is due to mannan, which has been identified as a serum inhibitor in these patients. The immunosuppressive effect was only observed during the acute phase of the disease and disappeared after treatment. Evidence of T suppressor (TS) activity in patients with CMCC was shown during the acute phase (Durandy et al., 1987, 1988). In order to better understand the generation of such TS lymphocytes in CMCC patients, T lymphocytes from normal subjects were preincubated with excess mannan which resulted in a strong suppressive effect of both T and B cell responses to mannan. These results demonstrate that mannan can suppress normal subjects' T lymphocytes. The characteristics of the *in vitro* mannan induced TS cells were shown to be comparable to the *in vivo* induced TS cells found in CMCC patients.

In vivo, the Ag specific T cell mediated suppression correlates well with disease since it disappeared in patients treated with an appropriate effective therapy of ketoconazole (Drouhet and Dupont 1980, 1983). Suppression of mannan production restored the depressed cellular immunity. An analogy between CMCC and oral and esophageal candidosis in AIDS has been

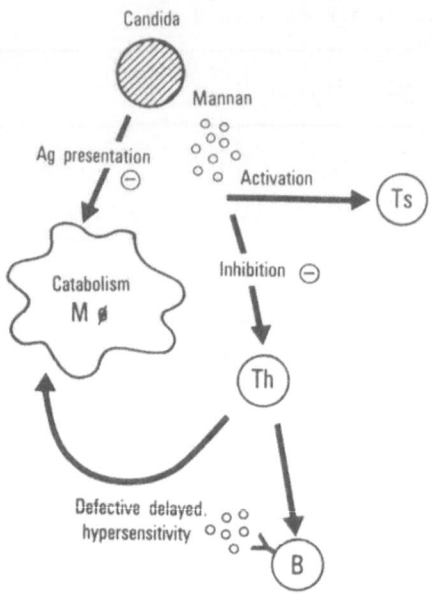

Figure 2. Model for immune abnormalities in patients with CMCC (according
to Durandy et al., 1988). The same scheme can be proposed in
mucous candidosis of AIDS patients.

reported (Matthews et al., 1989). Candidal antigens might be important
cofactors in AIDS. For this reason it is necessary to treat rapidly, with
energy and efficiency, the oral and esophageal conditions of AIDS patients.
As in CMCC, this focalized form does not lead to invasive conditions and
therefore, the same scheme of the specific immune deficiency observed in
CMCC can be proposed (Fig. 2).

The polyclonal activition of B lymphocytes by HIV may be reinforced by
C. albicans antigens. This polyclonal B cell activation may be responsible
for the frequency with which antibody against the 47 kD antigen of
C. albicans is produced in AIDS patients with oral and esophageal
candidosis and in patients with CMCC (Matthews et al., 1988). This might
explain the absence or the rarity of disseminated candidal infections which
occur only in AIDS patients with special risk factors. Nonetheless, the B
cell hyperactivation is not usually beneficial.

Recent data (Domer et al., 1989) confirmed that in an animal model
(mice), mannan introduced via the lymphoid tissue suppresses the
development of cell-mediated immune responses.

Cryptococcal antigens. The capsular polysaccharide
(glucuronoxylomannan) of *Cryptococcus neoformans* is recognized as a
virulence factor (Drouhet et al., 1950), an inhibitor of leukocyte
migration (Drouhet and Segretain, 1951) and of phagocytosis (Kozel, 1977),
and inducer of immunologic paralysis or tolerance. When the optimal
immunogenic dose is exceeded, splenic antibody producing cells and
circulating antibodies are suppressed (Kozel et al., 1977). Decreased
lymphocyte transformation in response to *C. neoformans* is observed in
cryptococcosis in man (Diamond and Bennett, 1973). The capsular

polysaccharide is a T-independent elictor of IgM, (Reiss, 1986), but the antibody level is regulated by T-suppressor cells (Murphy and Moorhead, 1982). A suppressive cellular immune response to cryptococcal antigen was shown by immunocyte migration inhibition in mice infected by *C. neoformans* or submitted to high doses of cryptococcal antigen (Gauthier-Rahman et al., 1988).

Other fungal Ags. The immunosuppressive effect of fungal antigens by induction of T-suppressor activity in chronic dermatophytic infections, histoplasmosis, coccidioidomycosis, and paracoccidioidomycosis have been reported in recent reviews (Drouhet, 1988; Reiss 1986).

OTHER GENERAL ASPECTS OF MYCOSES IN AIDS

The opportunistic fungal superinfections sometimes reveal a HIV infection by their sudden appearance and by their resistance to recurrent treatments. Prognosis is poor for AIDS patients with extremely severe mycoses such as cryptococcosis. Sixty to 80% of all AIDS patients have at least one fungal infection during the course of their illness, and 10-20% die as a direct result of fungal infection. The diagnosis of opportunistic infections in AIDS may be difficult. The flora varies between different countries and patients may be infected without obvious symptoms. The clinical course of fungal infection in AIDS patients is often unusual, and diagnosis may require a high degree of perspicacity on the part of the clinician. When compared to other immunocompromised hosts, AIDS patients with fungal infections may have nonspecific symptoms over a long period of time. Diagnosis may depend on histologic identification of organisms obtained through invasive procedures, or upon isolation of the organism in culture, usually a time-consuming process (Cairns, 1988). For that reason numerous systemic fungal infection in AIDS are diagnosed only post mortem.

Treatment of fungal infections in this patient population may also be frustrating. Some AIDS patients respond quickly to appropriate therapy, but these individuals are rarely cured of their infections. Since their underlying immunologic abnormality is not correctable at this time, AIDS patients often require life-long antifungal therapy to prevent recurrence of their infections. Other AIDS patients may have widely disseminated fungal infections that respond poorly even to appropriate therapy. Improved care of AIDS patients, therefore, rests on early diagnosis of fungal infections, early institution of appropriate antifungal therapy, and attempts at immunologic restoration through antiviral chemotherapy. Immunomodulating prophylactic and maintenance therapy for AIDS patients with opportunist life-threatening mycoses is absolutely necessary.

PARTICULAR ASPECTS OF PRINCIPAL MYCOSES IN AIDS PATIENTS

Candidosis. The first clinical descriptions of AIDS, mentioned not only pneumonia due to *Pneumocystis carinii* but also oral candidosis (Gottlieb et al., 1981). Klein et al. (1984) showed that oral candidosis is an initial manifestation of AIDS in high-risk patients. Thrush, which is not common in healthy adults who have not received broad-spectrum antibiotics or corticosteroids is a sensitive indicator of immunodeficiency due to a HIV infection and is used as a marker of disease severity in the classification of HIV infection. In patients at high risk for AIDS, the presence of unexplained oral candidosis predicts more than 50 per cent of the time the development of serious opportunistic infections (Klein et al., 1984). The combination of oral candidosis and a CD4 + count below $400/mm^3$ is a hallmark of advanced HIV infection, and the risk of progression to AIDS in three years is 90% (Moss, 1988). The importance of thrush in Walter Reed stage classification for HIV is emphasized due to its correlation with low numbers of T helper cells (less than $400/mm^3$) and the severe anergy to *C. albicans* antigen. In CDC/WHO classification, thrush is placed at the stage ARC or IVC2. The clinical definition of thrush as oral candidosis ("creamy-white, curd-like patches on the tongue or other oral mucosal surfaces which are removed by scraping") has to be confirmed by positive KOH preparation or Gram stain which shows masses of hyphae, pseudohyphae and yeast forms with positive culture. Thrush has to be distinguished from oral hairy/viral leukoplakia which is very common in young homo or bisexual males in groups at high risk for AIDS. An intimate association between oral candidosis and hairy oral viral leukoplakia is well documented (Greenspan, 1985). A severe chronic oral candidosis with redued CD4 + T cells was observed in a young woman in the absence of HIV infection (Pankhurst and Peakman, 1989). Therefore we have to be careful to avoid a misdiagnosis of a HIV infection even when the clinical picture suggests an acquired immunodeficiency.

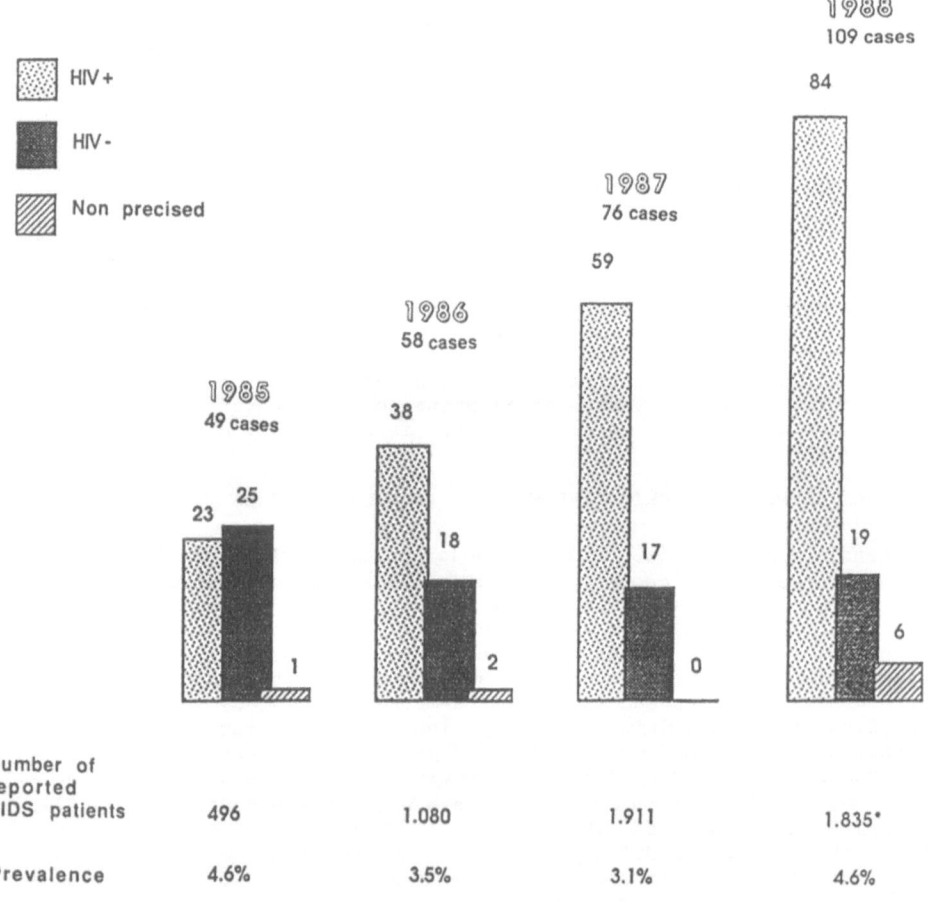

Figure 3. Prevalence of cryptococcosis in AIDS patients in France between 1985–1988

Cryptococcosis, the "sleeping giant" (Ajello, 1970), became the "awakening giant" (Kaufman, 1978) just prior to AIDS . This was probably due to the increase of iatrogenic immunodepressive conditions (corticotherapy, organ transplants, etc.). However, since the introduction of AIDS, the incidence of cryptococcosis has increased considerably and can now be considered a mycosis of the future (Drouhet et al., 1986). The prevalence of cryptococcosis in AIDS patients varies within different countries, regions, and risk groups. The reported frequency in USA ranges from 1.9 to 9% with an average of 6-7%. However, the prevalence can rise up to 33% in Central Africa. This high prevalence may be due to the high exposure to *C. neoformans* which have been isolated inside or in the environment of houses of AIDS-associated cryptococcosis in Burundi (Swinne et al., 1989). Before AIDS made its appearance, only 5-10 cases were reported annually in France. Since 1985 (Fig. 3), this number has increased 5 to 10 times, with a prevalence in AIDS patients between 3.1 to 4.6 %, approximating the incidence in the U.K (3.2%). Table 12 details the varied factors in cryptococcosis seen in 283 cases studied in the last 4 years in France. Since AIDS represents 72.08% of these cases, we must remark that in 34.17% of the HIV negative cases, the predisposing factor is unknown. No underlying diseases are mentioned in 15 to 30% of European or American cases of cryptococcosis and these figures reach 70 % in African patients. In our series, 100% of African and Haitian patients with cryptococcosis were HIV +. The majority of patients were homosexual in 1985 (21/23), but progressively, cases have appeared in heroin addicts, in patients who have undergone transfusions and in the heterosexual population. Table 13 shows the cryptococcal localizations in the AIDS period, compared to the period before AIDS. The CNS alone or the CNS with other manifestations is a principal target of *C. neoformans* in HIV + patients, meningitis being the most common manifestation in any stage of AIDS. Cryptococcosis represents the third most common CNS infection of AIDS, behind HIV and *Toxoplasma gondii*.

Table 12. Cryptococcosis in France (1985-1988)

Predisposing factors	
AIDS	204 cases
HIV-negative	79 cases
lymphoma, Hodgkin	8
malignant hemopathia	8
kidney transplant	15
sarcoidosis	4
cirrhosis	6
various (*)	10
not precised	1
unknown factors	27
Total	283 cases

* Cases of lupus, diabetes mellitus, cancer, hyper IgE syndrome, liver
  transplant, renal failure, periphigus, trauma

Table 13. Clinical spectrum of cryptococcosis before (1970-80) and after
        AIDS introduction (1985-88)

Organs involved	1970 - 1980 39		1985 - 1988 (*)			
			CNS only 127	CNS + other 120	Extra CNS 37	Total (%) 284**
CSF	25	(64.10%)	127	120		247 (86.97)
Blood				80	15	95 (33.45)
Urine				56	7	63 (22.18)
Lungs	8	(20.51%)		50	11	61 (21.47)
Skin	10	(25.64%)		9	6	15 (5.28)
Liver				7		7 (2.46)
Lymph nodes				6		6 (2.11)
Bone marrow	2	(5.12%)		4		4 (1.40)
Bones	7	(17.94%)				7 (17.94)
Other				8	1	9 (3.16)

* Data from Centre National de Références des mycoses et des antifongiques
   (Unité de mycologie, Institut Pasteur) reported by B. Dupont (1989)
** Including 204 cases of AIDS patients

The CSF is often normal or subnormal despite the presence of a large
number of cryptococci. Extra meningeal, visceral dissemination occurs
frequently with positive hemocultures in 50-60% of cases. Serum and CSF
antigen titers are often very high (up to 1:1,000,000) if cryptococcosis is
systematically looked for in AIDS patients presented with fever, headaches
or an unexplained pneumopathy. The vital diagnosis is very bleak and the
average survival time is about 6-9 months (Dupont et al., 1989). Relapses
occur in more than 50 % of cases and require prophylactic treatment
(Dismukes, 1988).

   Histoplasmosis. Histoplasmosis due to *Histoplasma capsulatum*, a true
pathogenic fungus, is not usually considered an opportunistic infection.
However, cases of histoplasmosis among immunocompromised patients due to
either systemic malignancy or steroid therapy have been reported recently
(Wheat et al., 1982). Cell-mediated immunity is thought to be responsible
for limiting proliferation of *H. capsulatum* in tissue. Attention has been
focused on the numerous cases of histoplasmosis occurring in AIDS patients
in USA, which is reported as high as 6% and is present not only in endemic
areas. This suggests that in many cases a previous quiescent histoplasmosis
which had been caught elsewhere was reactivated (Johnson et al., 1988;
Graybill et al., 1988). In Europe (Table 14) 13 cases have been reported:
in France (9 cases) (Brivet et al., 1986; Dupont et al., 1985; Drouhet and
Dupont 1987; Dupouy-Camet et al., 1987; Deluol et al., 1987, Datry et al.,
1989; Dallot et al., 1988; Dupont, unpublished observation; Schuh, 1989,
personal communication), Belgium (Depré et al., 1987; Arendt et al., 1989),
Switzerland (Dietrich et al., 1986), and in Sweden (Petrini et al., 1989).
In most of these cases, Europeans travelled in endemic areas and developed
an infection shortly after their return. However some, such as the
Cambodian patient reported by Dallot et al., (1988) had lived in France for
7 years. The clinical manifestations, as seen in American cases, were
characterized by a progressive disseminated histoplasmosis. Fever,
septicemia, bone marrow and peripheral blood smears with characteristic
yeast forms of *H.capsulatum* inside of polymorphonuclear cells (usually very
rare) were observed in 3 cases. Generalized skin lesions, erythematous
maculae, papules, nodules, even granulomatous lesions on face, trunk, and
extremities were observed in 4 cases. An abdominal form with intestinal

ulcerations was observed (Brivet et al., 1986). In all cases, only the species *H. capsulatum* was reported, with the exception of a 38 year old, heterosexual Belgian. He had lived in Zaire (1972-78) and later in Belgium in 1986 developed African histoplasmosis due to *H. capsulatum var. duboisii* (Depré et al., 1987) as a first manifestation of AIDS. In spite of a relatively high incidence of histoplasmosis duboisii in Central Africa, no cases of such form were reported in regions where prevalence of AIDS is very high.

Coccidioidomycosis. Because T-lymphocyte-dependent immunity is implicated in infections caused by *Coccidioides immitis*, it would be expected that persons with AIDS in endemic areas would be seriously infected with this fungus. Recent case reports are consistent with this hypothesis (Bronnimann et al., 1987; Graham et al., 1988). Incidence is as high as 27% in areas such as Tucson, even though the rates of annual infections are only 4% or less. As in histoplasmosis, the primary pulmonary infection usually occurs in immunocompetent subjects, and disseminated coccidioidomycosis is reported in AIDS as a principal characteristic. Not surprisingly, AIDS patients with disseminated coccidioidomycosis had a significantly higher incidence of other opportunistic infections which were documented postmortem, than patients without AIDS. As with candidosis, cryptococcosis and histoplasmosis in AIDS, poorly developed granuloma formation is observed in disseminated coccidioidomycosis. The intimate interaction between T-lymphocytes and macrophages is responsible for the histologic pattern of granuloma formation and numbers of organisms found in tissue; the alterations in lymphocyte number and function in AIDS, explains the poorly formed granulomas and overload of *C. immitis* spherules observed in these patients (Graham et al., 1988).

OTHER RARE SYSTEMIC MYCOSES DUE TO DIMORPHIC FUNGI

Penicilliosis marneffei. This mycotic infection is increasingly observed in AIDS. *Penicillium marneffei* is the only known dimorphic species of *Penicillium*, first described as the agent of a reticuloendothelial disease of bamboo rats in Vietnam (Segretain, 1959) and as a potential human pathogen (inoculation accident). Not until, 1973 was a natural human infections reported in two immunocompromised American patients travelling

Table 14. Histoplasmosis in AIDS patients in Europe (13 cases)

Country	France 9, Belgium 2, Switzerland 1, Sweden 1
Race	white 7, black 5, Asian 1
Age, Sex	24-66 years (mean: 37.25)
Risk factors	homosexuality (4), bisex (1) I.M. drug injection, others: not determined
Endemic area of living or travelling	Franch Guyana (3), Martinique (3), Haiti (2), Mexico, Guatemala (1), Cambodgia (1), Zaire (3)
Clinical manifestations	disseminated histoplasmosis (septicemia, meningitis, pneumonia, pericarditis, cutaneous lesions) (6), abdominal form, intestinal ulcerations (2), pneumopathia, lymphadenopathy (5)

in Far East of Asia (DiSalvo et al., 1973; Paultler et al., 1984).
Additional cases include 21 autochthonous patients from the South East of
Asia: Thailand (5 cases), S-E of China (14), and Hong-Kong (2) (Deng et
al., 1988; Chan et al., 1988). The first cases reported in AIDS were
observed among Europeans in France (Ancelle et al., 1988; Romana et al.,
1989), Great Britain (Peto et al., 1988), and Italy (Coen et al., 1989). At
the same time, one American case with AIDS was reported by Piehl (1988) and
one Thailandese women (Tanphaichitra et al., 1988). In all cases, the
disease was rapidly acquired after a short stay in endemic regions. The
risk factors in these cases were homosexuality (4), transfusion (1) and
i.v. drug addiction (1). A clinically disseminated form predominates with
involvment of lungs (5), blood (5) bone marrow (3), and skin (5). The
histological diagnosis is based on the typical oval, round or elongated
septate intracellular parasitic forms which have to be differentiated from
the intracellular budding yeasts of *H. capsulatum*. The culture can be
identified by the production of red pigment and by the characteristic
intracellular septated forms obtained when the fungus is inoculated into
golden hamsters (Drouhet et al., 1988). The predominance of the fungus in
the soil is most likely the source for both human and animal infections.
Due to the high incidence of *P. marneffei* among the wild bamboo rats (Deng
et al., 1988) and the human cases recently reported,further cases of
penicilliosis are to be expected among AIDS patients (see also Viviani and
Tortorano, this book).

Paracoccidioidomycosis. In spite of the large number of AIDS cases in
risk regions of Brazil (Sao Paulo), only rare cases of disseminated
paracoccidioidomycosis are reported (Bakos, 1989; Goldani et al., 1989;
Pedro et al., 1989). This may be explained (Lacaz et al., 1989) by the fact
that AIDS is essentially a rural phenomenon. In only one case were specific
HIV antibodies detected out of the sera of 50 patients with
paracoccidioidomycosis from São Paulo (Lacaz et al., 1989). The absence of
blastomycosis due to *Blastomyces dermatitidis* among the North American
patients with AIDS is also surprising.

Sporotrichosis. *Sporothrix schenckii* is cosmopolitan, but predominant
in some sub- and tropical regions. It is the agent of a lymphocutaneous and
subcutaneous mycosis which is rarely disseminated. Less than one hundred
cases of disseminated sporotrichosis have been reported in the literature.
In AIDS patients, several cases of disseminated sporotrichosis with
multiple cutaneous and subcutaneous abscesses and osteoarticular lesions
have been reported (Bibler et al., 1986; Fitzpatrick et al., 1988; Kurosawa
et al, 1985; Lipstein-Kresh et al., 1985). In spite of the relation of
this fungus to cell-mediated immunity, the place of sporotrichosis in AIDS
patients is limited to anecdotal cases due to the rural limited
distribution of this mycosis. In contrast to typical lymphocutaneous
sporotrichosis in non AIDS patients, a history of exposure to soil or
plants is not common in AIDS. The portal of entry is presumed to be the
lungs, with secondary hematogenous dissemination.

Actinomycotic infections. Aerobic actinomycetales infections in AIDS are
due particularly to *Nocardia asteroides* with an incidence of 0.3%,
principally among parenteral drug abusers (Holtz et al., 1985). Pericardial
involvment, lymphadenopathy, pulmonary and brain abcesses (Adair et al.,
1987) are among the principal localizations.

PRINCIPAL CUTANEOUS MYCOTIC INFECTIONS

Seborrheic dermatitis. Pityrosporosis. Seborrheic dermatitis, which
most authors consider to be closely related to a *Pityrosporum* infection
(Belew, 1981; Drouhet, 1988; Groshans, 1988) is the most common skin
manifestation reported in HIV-infected individuals. Mathes and Douglass
(1985) found a high prevalence (83%) of seborrheic dermatitis in patients
with AIDS compared to 1% to 3% in control subjects. Patients with AIDS

related complex (ARC) also have an increased incidence of seborrheic dermatitis (42%). In this population, infections are more often explosive, inflammatory, and severe than usually seen in otherwise healthy patients. Soeprono et al. (1986) compared 25 skin biopsies from 21 patients with AIDS or its prodrome with "seborrheic dermatitis" to biopsies from patients with seborrheic dermatitis but who had neither AIDS or its risk factors. They found many features in common, but also some distinctive differences. Among them, spotty keratinocytic necrosis, leukoexocytosis and a superficial perivascular infiltrate of plasma cells and frequently, neutrophils, which indicate occasional leukocytoclasis. The increased numbers of plasma cells correlate with hypergammaglobulinemia. The decreased number of Langerhans cells in the skin of patients with AIDS (Belsito et al., 1984), and the necrotic keratinocytes surrounded by lymphocytes are suggestive of graft-versus-host reactions, and analogous with the histological alterations that we obtained in experimental pityrosporosis (Drouhet et al., 1980; Drouhet et al., 1988). In only a minority of the biopsies of Soeprono et al. (1986) were an excessive number of *Pityrosporum* organisms observed. The only study on the lipophilic yeast *Pityrosporum ovale* in HIV-seropositive and HIV-seronegative men was reported by Hakansson et al. (1988). The mean number of cultured *P.ovale* yeast cells/ml was 131 ± 332 in the HIV + group, and 22 ± 31 in HIV-, but this difference is not statistically significant. Also, in each group of 12 men studied, seborrheic dermatitits was present only in the scalp, face and/or the external ears.

The role of *Pityrosporum ovale* in the pathogenesis of dandruff (pityriasis capitis) and seborrheic dermatitis, emphasized by Sabouraud as early as 1904, has been controversial for a long time (Belew, 1981) However, numerous ecological and mycological studies on lipophilic cutaneous flora experimental model studies of seborrheic dermatitidis due to *Pityrosporum* (Drouhet et al., 1980; Faergeman, 1981; Drouhet et al., 1988) and immunological studies demonstrating significant antibody levels (Midgley and Hay, 1988) and CMI (Civilla et al., 1980) have convinced most dermatologists that *P.ovale* is the causative agent (Groshans, 1988). Additionally, the spectacular therapeutic results with oral ketoconazole in seborrheic dermatitis of HIV negative or HIV positive patients (Ford et al., 1984, Satellite Symposium to the 2d International Skin Therapy Symposium, Antwerp, Belgium, May 1988) are in favor of this hypothesis. It is noteworthy that pityriasis versicolor has never been reported in AIDS patients. *Malassezia (Pityrosporum) furfur* is undoubtedly the agent of pityriasis versicolor. *Pityrosporum ovale* is considered now as a synonym of *M. furfur* due to their identical genomic and ultrastructural characteristics (Guého and Meyer, 1988). However, some physiopathological differences, which are not yet well defined, suggests this latter lipophilic yeast is a variety of *Malassezia furfur*.

Dermatophytosis. Several authors such as Badillet and Traore (1987) found a high incidence of cutaneous dermatophytosis due to *Trichophyton rubrum, T. interdigitale* and *Epidermophyton floccosum* (17%, 25 dermatophytes carriers) among 142 patients with positive HIV serology. The incidence in the normal population is about 2.8%. Additionally in AIDS, the incidence for nail mycoses is quite high. Clinical nail changes are observed in 50% of cases, in contrast with normal controls presenting nail alteration in only 5% of the patients (Carvalho et al., 1988). The colour of nails varies from white to yellow (Chernosky et al., 1985; Deluol et al., 1988). Tinea pedis, tinea cruris and tinea corporis are common in AIDS, but no tinea capitis is observed.

Trichosporonosis. A high carriage of *Trichosporon beigelii* (13%) has been observed in the anal region of 343 Danish homosexual men compared to an absence among normal individuals. Pubic white piedra was observed in 2 Danish patients with AIDS and similar results were reported by Torsander et al. (1985) in Sweden, and by Fischman et al (1988) in Brazil regarding occurrence of genital white piedra. Nevertheless, white piedra with *T. beigelii* is not exclusive to HIV + homosexuals. Preliminary reports of Guého (1989) based on genome comparison and other characteristics of

T. beigelii showed that this species, which is distinct from *T. cutaneum*, was isolated from superficial surfaces (skin, nail, hair) as well as from deeper localisations. Several species of *Trichosporon* can develop white piedra nodules. However, there is a relationship between each species and the hair type. The pubic white piedra is caused by *T. inkin*, a species closely related to *T. beigelii*, but specifically located in the genital area. An invasive trichosporonosis was reported in a patient with AIDS, but the risk factor in this case was neutropenia (100/mm^3) induced by an immunosuppressive treatment (Leaf and Simberkoff, 1989).

Alternariosis. *Alternaria* spp., particularly *A. alternata*, responsible for recent cases of necrotic cutaneous, mucous, ocular, osseous in non AIDS immunocompromized patients (Viviani et al., 1986) is rarely reported in AIDS patients. A French case (Levy-Klotz et al., 1985) was reported in a 7 year old girl with acute lymphoblastic leukemia who developed a necrotic lesion of the leg due to a *A. tenuissima* after a HIV contamination by blood transfusion. Additionally, a 31 year old homosexual American man with AIDS, developed a necrotic lesion of his nasal septum due to *A. alternata* (Wiest et al., 1987).

CURATIVE THERAPY AND MAINTENANCE (SUPPRESSIVE) THERAPY

In order to ascertain the impact of antifungal agents on the immune system and to determine the evolution of the disease: cure, protection or failure, it is necessary to understand the relationship between the causative fungi, the host, and the antifungal agent (Fig. 4).

In AIDS patients, the profound and irreversible defect of cell-mediated immunity makes them susceptible to a wide range of opportunist fungi which are normally contained by the immune system. Additionally, as observed in other immunodeficiency states, the level of susceptibility and hence the range of potential pathogens increases with time (Pinching, 1988). In spite of powerful antifungal treatment, even after a successfully

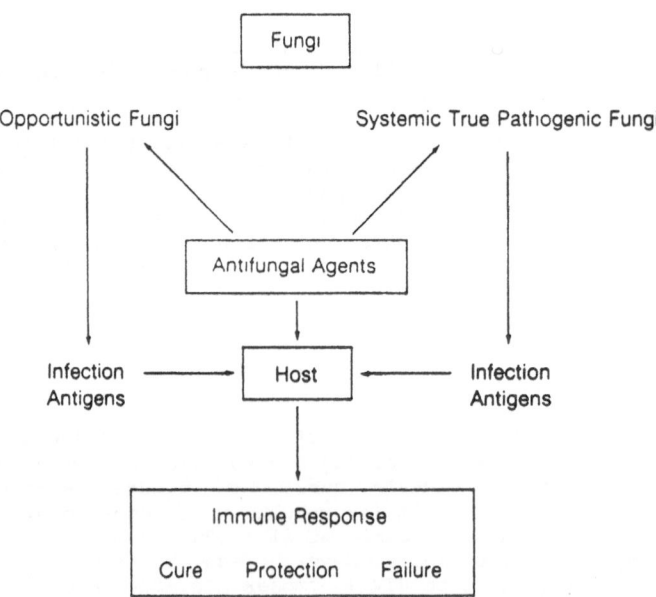

Figure 4. Interaction fungi-host antifungal agents and immune response

treated episode of a fungal infection, invasive fungi can persist and re-emerge to cause a further episode. Patients with AIDS are increasingly refractory to classical treatments in systemic life-threatening opportunistic mycoses such as cryptococcosis, histoplasmosis, or coccidioidomycosis. Although amphotericin B is regarded as the drug of choice for systemic fungal disease in most immunocompromised hosts, the efficacy of this antibiotic has not been impressive in AIDS patients. Most patients with AIDS cannot tolerate flucytosine associated with amphotericin B.

The high rate of relapse between 50 to 65% of systemic mycoses such as cryptococcosis, histoplasmosis and coccidioidomycosis in AIDS patients requires new strategies in their long-term management. Prolonged maintenance or suppressive therapy after initial intensive treatment is necessary.

The important problems of prophylaxis and maintenance therapy for opportunistic infections in AIDS have been discussed by Pinching (1988). Recent developments of new antifungal molecules have been the subject of important symposia (Fromtling, 1987; Georgiev,1988). The new investigative triazole derivatives such as itraconazole and fluconazole have been shown to be effective in the ambulatory treatment of some of these systemic mycoses in AIDS. Fluconazole is a water-soluble, bifluorinated triazole, with a low binding affinity for plasma proteins. It distributes extensively throughout the body, and readily diffuses into CSF and saliva. A dramatic success in the treatment of an AIDS patient with cryptococcal meningitis who had relapsed after amphotericin B and flucytosine treatment (Dupont and Drouhet 1986) has been observed. It proved to be a very promising drug for cryptococcosis according to the results of an open French multicentre study where it was initial therapy in 52 AIDS patients (Dupont et al., 1989). Itraconazole, a highly lipophilic triazole, with an extensive tissue distribution with concentrations many times higher in plasma, in spite of an absence of diffusion in CSF, proved to be effective as curative treatment or as suppressive treatment after amphotericin-flucytosine therapy in AIDS patients (Viviani et al., 1989; Denning et al., this book).

Table 15. Future directions of research in mycoses in AIDS

- New antifungal agents
      including   . triazoles   - fluconazole
                                    - itraconazole
              . liposomal amphotericin B
              . cilofungin
              . other new antifungal agents

- Development of novel prostaglandins
      may preserve integrity of mycosae

- Immunotherapy
      . interferrons, lymphokines, thymic hormones
      . colony - stimulating factors
      . activated monocyte transfusions
            - adjuncts to fungal therapy

- Antiretroviral agents
      may indirectly augment cellular immunity in AIDS

- More studies on relationship   - fungi
                                        - host organism
                                        - antifungal agents

Encouraging results were obtained with itraconazole in histoplasmosis of AIDS patients (Graybill, 1988; Graybill et al., this book). Ongoing trials include analysis of the new triazole molecules as both curative or maintenance therapy for the life-threatening systemic mycoses in AIDS.

Even for non life-threatening mycoses such as oral and esophageal candidosis, oral systemic treatment with the new azoles such as fluconazole and itraconazole avoids the difficulties of compliancy associated with topical agents (Dupont and Drouhet, 1988).

In spite of remarkable progress, there is a need for the development of new antifungal agents and novel approaches for the treatment of mycoses in AIDS patients (Table 15). The increased knowledge of host defences due to immunomodulating agents such as interferon, interleukins and colony stimulating factors may become important adjuncts to specific antifungal therapy.

REFERENCES

Adair, J.C., Beck, A.C., Apfelbaum, R.I., and Baringer, R., 1987, Nocardial cerebral abscess in the acquired immunodeficiency syndrome, Arch. Neurol., 44:548.

Ajello, L., 1970, The medical mycological iceberg, in "Proceedings International Symposium on Mycoses", Pan American Health Organization, Washington, DC, USA, Sci. Publ.

Alsina, A., Mason, M., Uphoff, R.A., Riggsky, W.S., Becker, J.M., and Murphy, D., 1988, Catheter-associated Candida utilis fungemia in a patient with acquired immunodeficiency syndrome: species verification with a molecular probe, J. Clin. Microbiol., 26:621.

Ancelle, T., Dupouy-Camet, J., Pujol, F., Nassif, X., Ferradini, L., Choudat, L., Bièvre, C. de, Dupont, B., Drouhet, E., and Lapierre, J., 1988, Un cas de pénicilliose disséminée à Penicillium marneffei chez un malade atteint d'un syndrome immunodéficitaire acquis, Presse Méd., 17:1005.

Armstrong, D., 1988, Life-threatening opportunistic fungal infection in patients with acquired immunodeficiency syndrome, Ann. N.Y. Acad. Sci., 544.

Asnis, D.S., Chitkara, R.K., Jacobson, M., and Goldstein, J.A., 1988, Invasive aspergillosis: an unusual manifestation of AIDS, in: "N.Y. St. J. Med.".

Badillet, G., and Traore, F., 1987, SIDA et dermatophytes, Bull. Soc. Fr. Mycol. Med., 16:95.

Bakos, L., Kronteld, M., Hampe, S., Castro, I., and Zampese, M., 1989, Disseminated paracoccidioidomycosis with skin lesions in a patient with acquired immunodeficiency syndrome, J. Am. Acad. Dermatol., 20:854.

Barré-Sinoussi, F., Chermann, J.C., Rey, F., Nugeyre, M.T., Chamarat, S., Gruest, J., Dauguet, C., Asler-Blin, C., Brun-Vezinet, F., Rouzioux C., Rosenbaum, W., and Montagnier, C., 1983, Isolation of a T-lymphotrophic retrovirus from a patient at risk of acquired immunodeficiency, Science, 220:868.

Belew, P.W., 1981, Sabouraud and Rivolta were right. Seborrhoeic dermatitis is microbial, Cosmetics & Toiletries, 96:28.

Bibler, M.R., Luber, H.J., Glueck, H.I., and Estes, S.A., 1986, Disseminated sporotrichosis in a patient with HIV infection after treatment for acquired factor VIII inhibitor, JAMA, 256:3125.

Bonner, J.R., Alexander, W.J., Dismukes, E., App, W., Griffin, F.M., Little, R., and Shin, M.S., 1984, Disseminated histoplasmosis in patients with AIDS, Arch. Intern. Med, 144:2178.

Bottone, E.J., Toma, M., Johansson, B.E., Szporn, A., Poon, M., and Wormser, G.P., 1985, Capsule-deficient Cryptococcus neoformans in AIDS patients, Lancet, 1:400.

Brawner, D.L., and Cutler, J.E., 1989, Oral *Candida albicans* isolates form non hospitalized normal carriers, immunocompetent hospitalized patients and immunocompromised patients with or without acquired immunodeficiency syndrome, J. Clin. Microbiol., 27:1335.

Brivet, F., Roulot, D., Naveau, S., Delfraissy, J.F., Goujard, C., Tertian, G., Cartier, I., Drouhet, E., and Dupont, B., 1986, The acquired immunodeficiency syndrome : histoplasmosis, Ann. Intern. Med, 104:447.

Bronnimann, D.A., Adam, R.D., Galgiani, J.N., Habib, M.P., Peterson, E.A., Poster, B., and Bloom, J.W., 1987, Coccidioidomycosis in the acquired immunodeficiency syndrome, Ann. Intern. Med., 106:372.

Byrne, W.R., and Dietrich, R.A., 1989, Disseminated coccidioidomycosis with peritonitis in a patient with acquired immunodeficiency syndrome, Arch. Intern. Med., 149:947.

Cairns, M.R., 1988, Fungal infections in the acquired immunodeficiency syndrome, J. Electron. Microsc. Tech., 8:115.

Carme, B., Mbitsi, A., and Nigolet, A., 1989, Cryptococcose, histoplasmose africaine et SIDA au Congo. Abstracts Vth Intern.Conf. AIDS, Montréal, poster MBP91.

Carvalho, M.T.F., Fischman, O., Acceturi, C.A., Salomao, R., Silva, T.M.J., Turcato, G., and Castelo, A., 1988, Nail mycoses in human immunodeficiency virus infection. Abstract Book. Intern. Congress Infect. Dis., Rio de Janeiro (Brazil), Poster N°662.

Causey, D.M., Leedom, J.M., and Olsen, C., 1989, *Rhodotorula rubra* septicemia in an AIDS patient with an in-dwelling Hickman catheter, Abstracts V Intern. Conf. AIDS, Montréal.

CDC, 1988, AIDS. Revision of CDC/WHO case definition, Weekly Epidemiol. Record., 63:1.

Chan, J.K.C., and Tsan, D.N.C., 1988, *Penicillium marneffei* in bronchoalveolar lavage fluid, Acta Cytol. (in press).

Chandler, F.W., 1985, Pathology of the mycoses in patients with the acquired immunodeficiency syndrome, in: "Current topics in medical mycology", vol. 1, M.R. McGinnis, ed., New York, Berlin.

Chernosky, M.E., and Finley, V.K., 1985, Yellow nail syndrome in patients with acquired immunodeficiency disease, J. Am. Acad. Dermatol, 13:731.

Civila, E.S., Vignale R.A., and Conti-Diaz, I.A., 1980, *Malassezia ovalis* : mycologic and immuno-antigenic aspects and probable pathogenic role, in: "Proceeding Fifth International Conference on the Mycoses", Pan American Health Organization, Washington, D.C., Scient. Publ., N°396,

Coen, M., Viviani, M.A., Rizzardini, G., Tortorano, A.M., Bonaccorso, C., and Quirino, T., 1989, Disseminated infection due to *Penicillium marneffei* in a HIV positive patient, Abstracts Vth Intern. Conf. AIDS, Montréal, poster MBP 94.

Colenbunders, R., Mann, J.M., Francis, H., and Bila, K., 1987, Evaluation of clinical case-definition of acquired immunodeficiency syndrome in Africa, Lancet, i:492.

Concus, A.P., Hefland, R.F., Imber, J., Lerner, E.A., and Sharpe, R.J., 1988, Cutaneous cryptococcosis mimicking molluscum contagiosum in a patient with AIDS, J. Infect. Dis., 158:897.

Coulaud, J.P., 1988, Manifestations cliniques de l'infection par le virus de l'immuno-déficience humaine HIV chez l'africain, Med. Trop, 48:327.

Dallot, A., Monsuez, J.J., Chanu, B., Vittecocq, D., Verola, O., Badillet, G., Rouffy, J., Morel, P., and Puissant, A., 1988, Localisations cutanées d'une histoplasmose disséminée à *Histoplasma capsulatum* au cours d'un cas d'immunodéficience acquise, Ann. Dermatol. Venereol., 115:441.

Deluol, A.M., Marche, C., Bizet, C., and Katlama, C, 1987, Histoplasmose disséminée au cours du syndrome d'immunodépression acquise, Bull. Soc. Fr. Mycol. Méd., 16:363.

Deluol, A.M., Dompmartin, D., Kerestedjian, S., Groshans, E., Basset, A., Serinot, A.G., and Coulaud, J.P., 1988, Onychomycoses et SIDA, Bull. Soc. Fr. Mycol. Med., 17:127.

Deng, Z., Ribas, J.L., Gibson, D.W., and Connor, D.H., 1988, Infections caused by *Penicillium marneffei* in China and Southeast Asia : review of eighteen published cases and report of four more Chinese cases, Rev. Infect. Dis., 10:640.

Depré, G., Coremans-Pelseneer, J., Peeters, P., Rickaert, F., Struelens, M., and Serruys, E., 1987, Histoplasmose africaine disséminée associée à un syndrome d'immunodéficience acquise, Bull. Soc. Fr. Mycol. Méd., 16:75.

Diamond, I.D., and Bennett, J.E., 1973, Disseminated cryptococcosis in man: decreased lymphocyte transformation in response to *Cryptococcus neoformans*, J. Infect. Dis., 127:694.

Dietrich, P.Y., Pugin, P., Regamey, C., and Bille , 1986, Disseminated histoplasmosis in AIDS in Switzerland, Lancet, 7-52.

DiSalvo, A.F., Ficking, A.M., and Ajello, L., 1973, Infection caused by *Penicillium marneffei* : description of first natural infection in man, Am. J. Clin. Pathol., 60:259.

Domer, J.E., Garner, R.E., and Befidi-Mengue, R.N., 1989, Mannan as an antigen in cell-mediated immunity (CMI) assays and as a modulator of mannan-specific CMI, Infect. Immun., 57:693.

Drouhet, E., Segretain, G., and Aubert, J.P., 1950, Polyoside capsulaire d'un champignon pathogène, *Torulopsis neoformans*. Relation avec la virulence, Ann. Inst. Pasteur, 79:891.

Drouhet, E., and Segretain, G., 1951, Inhibition de la migration leucocytaire *in vitro* par un polyoside capsulaire du *Torulopsis neoformans*, Ann. Inst. Pasteur, 81:674.

Drouhet, E., Mercier-Soucy, L., and Montplaisir, S., 1975, Sensibilité et résistance des levures pathogènes aux 5-fluoropyrimidines I. Relation entre les phénotypes de résistance à la 5-fluorocytosine, le sérotype de *Candida albicans* et l'écologie de différentes espèces de *Candida* d'origine humaine, Ann. Inst. Pasteur (Microbiol.), 126B:25.

Drouhet, E., 1981, Ecology of the serotypes of *Candida albicans* and phenotypes of resistance to 5-fluorocytosine, in: "Sexuality and Pathogenicity of Fungi", R. Vanbreuseghem and C. de Vroey, eds., Masson, Paris.

Drouhet, E., Dompmartin, D., Papachristou-Moraiti, A., and Ravisse, P., 1980, Dermatite expérimentale à *Pityrosporum ovale* et (ou) *Pityrosporum orbiculare* chez lecobaye et la souris, Sabouraudia, 18:149.

Drouhet, E., and Dupont, B., 1983, Laboratory and clinical assessment of ketoconzole in deep seated mycoses, Ann. J. Med., 74:30.

Drouhet, E., Dupont, B., and Dikeacou, T., 1985, Antifungal agents and immunity, Zbl. Bakt. Suppl., 13, Gustav Fischer Verlag, Stuttgart, New York.

Drouhet, E., Dupont, B., and Reyes, G., 1986, Antigens de *Cryptococcus neoformans* et réponse immunologique dans la cryptococcose, mycose d'actualité et mycose de l'avenir, Bull. Soc. Fr. Mycol. Med., 15:21.

Drouhet, E., and Dupont, B., 1987, Mycotic infections complicating heroin addicts, AIDS and other immunocompromised host conditions, Ann. Ist. Super. Sanità, 23:735.

Drouhet, E., 1987, Les histoplasmoses, mycoses d'importation en 1986. Rôle du SIDA, Bull. Soc. Fr. Mycol. Méd., 16:29.

Drouhet, E., Ravisse, P., Avé, P., de Bièvre, C., Dupont, B., and Pietfroid, A. 1988, Etude mycologique, ultrastructurale et expérimentale sur *Penicilium marneffei* isolé d'une pénicilliose disséminée chez un SIDA, Bull. Soc. Fr. Mycol. Méd., 17:77.

Drouhet, E., 1988, Overview on fungal antigens, in: "Fungal antigens", E. Drouhet, G.T. Cole, L. de Repentigny, J.P. Latgé, B. Dupont ,eds., Plenum Press, New York.

Drouhet, E., Dupont, B., and Ronin, O., 1989, Serotypes and resistant phenotypes to 5-fluorocytosine in *Candida albicans* strains from patients infected with human immunodeficiency virus (HIV). Abstracts of 3rd Symposium topics in Medical Mycology. Mycoses in AIDS patients, Pasteur Institute, Paris, P40.

Dupont, B., Sansonetti, P., Fleury, J., Ravisse, P., and Drouhet, E., 1987, Histoplasmose septicémique au cours du SIDA. Présence inhabituelle d'*Histoplasma capsulatum* dans les polynucléaires, Med. Mal. Infect., 14:621.

Dupont, B., and Drouhet, E., 1987, Cryptococcal meningitis and fluconazole, Ann. Intern. Med., 106:778.

Dupont, B., and Drouhet, E., 1988, Fluconazole in the management of oropharyngeal candidosis in a predominantly HIV antibody-positive group of patients, J. Med. Vet. Mycol., 26:67.

Dupont, B., 1990, Cryptococcal meningitis in AIDS patients - A pilot study of fluconazole therapy in 52 patients, in: "Mycoses in AIDS patients", H. Vanden Bossche, D.W.R. Mackenzie, G. Cauwenbergh, J. Van Cutsem, E. Drouhet, B. Dupont, eds., Plenum Press, New York, London.

Dupouy-Camet, J., Vilette, B., Noizat, F., Moyal, M., Vedel, G., and Lapierre, J., 1987, Histoplasmose disséminée chez un malade grand voyageur, atteint de SIDA, Bull. Soc. Fr. Mycol. Med., 16:81.

Durandy, A., Fischer, A., Le Deist, F., Drouhet, E., and Griscelli, C., 1987, Mannan - specific and mannan - induced T-cell suppressive activity in patients with chronic mucocutaneous candidiasis, J. Clin. Immunol., 7:400.

Durandy, A., Fischer, A., Drouhet, E., and Griscelli, C., 1988, Mannan antigen of *Candida albicans* and cellular immune response *in vitro* and *in vivo*, in: "Fungal antigens", E. Drouhet, G.T. Cole, L. de Repentigny, J.P. Latgé, B. Dupont, eds., Plenum Press, New York.

Edman, J.C., Kovacs, J.A., Masur, H. Santi, D.V., and Elwood, H.J., 1988, Ribosomal RNA sequence shows *Pneumocystis carinii* to be a member of the fungi, Nature, 334:519.

Fischer, A., Ballet, J.J., and Griscelli, C., 1978, Specific inhibition of *in vitro Candida* induced lymphocyte proliferation by polysaccharide antigens present in the serum of patients with chronic mucocutaneous candidiasis, J. Clin. Invest., 62:1005.

Ford, G.R., Farr, P.M., and Shuster, S., 1984, The response of seborrheic dermatitis to ketoconazole, Br. J. Dermatol., 111:603.

Fournier, J.P., Bernardin, G., Hoffman, P., Marty, P., Gari-Toussaint, M., Chavaillon, J.M., Le Fichoux, Y., and Mattei, M., 1989, Cardiopulmonary aspergillosis in an AIDS patients, in: "3rd Symposium Topics in Mycology. Mycoses in AIDS Patients", Pasteur Institute, Paris, Abstract P32.

Fromtling, R.A. (ed.), 1987, "International Symposium on Recent Trends in the Discovery, Development and Evaluation of Antifungal Agent", J.R. Prous Sci., Publ., Barcelona.

Gauthier-Rahman, S., Wahab, S., and Drouhet, E., 1988, Kinetic study of humoral and cellular immune responses to *Cryptococcus neoformans* measured by ELISA and immunocyte migration inhibition, in: "Fungal antigens", E. Drouhet, G. Cole, L. de Repentigny, J.P. Latgé, B. Dupont, eds., Plenum Press, New York.

Georgier, V. St. (ed.), 1988, "Antifungal Drugs", Ann. N.Y. Acad. Sc., 544.

Goldani, L.Z., Martinez, R., Landell, G.A.M., Machardo, A.A., and Continho, V., 1989, Paracoccidioidomycosis in a patient with acquired immunodeficiency syndrome, Mycopathologia, 105:71.

Gottlieb, M.S., Schrodd, R., Schanker, H.M., Weisman, J.D., Fan, P.T., Wolf, R.A., and Sawon, A., 1981, *Pneumocystis carinii* pneumonia and mucosal candidiasis in previously healthy homosexual men. Evidence of a new acquired cellular immunodeficiency, N. Engl. J. Med. A, 305:1425.

Götzsche, P.C., Bygbjerberg, I.B.C., Olesen, B., Möller, L.H., Salim, Y.S., and Faber, V., 1988, Yield of diagnostic tests for opportunistic infections in AIDS : A survey of 33 patients, Scand. J. Infect. Dis., 20:395.

Graham, A.R., Sobonya, R.E., Bronnimann, and Galgiani, J.N., 1988, Quantitative pathology of coccidioidomycosis in acquired immunodeficiency syndrome, Human Pathol., 19:800.

Greenspan, D, 1985, Oral viral leukoplakia ("hairy" leukoplakia) : a new oral lesion in association with AIDS, Contin. Educ., 6:204.

Graybill, J.R., 1988, Histoplasmosis and AIDS, J. Infect. Dis., 158:623.

Grosshans, E., and Bressieux, A., 1988, L'eczéma séborrhéique (la pityrosporose), Ann. Dermatol. Vénérol., 115:79.

Guého, E., and Delga, J.M., 1989, Trichosporonosis and AIDS, in: "3rd Symposium Topics in Mycology, Mycoses on AIDS patients", Pasteur Institute, Paris, Abstract, P.55.

Hakansson, C., Faergemann, J., and Löwhagen, G.B., 1988, Studies on the lipophilic yeast Pityrosporum ovale in HIV-seropositive and seronegative homosexual men, Acta Dermatol. Venereol., (Stockh), 68:422.

Henochowicz, S., Mustafa, M., and Lawrinson, W.E., 1985, Cardiac aspergillosis in acquired immune deficiency syndrome, Am. J. Cardiol., 155:1239.

Henochowicz, S., Sahovic, E., Pistole, M., Rodrigues, M., and Mashe, A., 1985, Histoplasmosis diagnosed on peripheral blood smear from a patient with AIDS, JAMA, 253:3148.

Holmberg, K., and Meyer, R.D., 1986, Fungal infections in patients with AIDS and AIDS related complex, Scand. J. Infect. Dis., 18:179.

Holtz, H., Lavery, D.P., and Kapila, R., 1985, Actinomycetales infection in the acquired immunodeficiency syndrome, Ann. Intern. Med., 102:203.

Jimenez-Finkel, B.E., and Murphy, J.W., 1988, Characterization of efferent T-suppressor cell induced by Paracoccidioides brasiliensis antigen, Infect. Immun., 56:737.

Johnson, P.C., Khardori, N., Najjar, A.F., Butt, F., Mansell, P.W.A., and Sarosi, G.A., 1988, Progressive disseminated histoplasmosis in patients with acquired immunodeficiency syndrome, Am. J. Med., 85:152.

Kapend'a, K., Komichelo, K., Swinne, D., and Vandepitte, J., 1987, Meningitis due to Cryptococcus neoformans biovar gatti in a Zairean patient, Eur. J. Clin. Microbiol. Infect. Dis., 7:587.

Kaufman, L., and Blumer, S., 1978, Cryptococcosis : the awakening giant, in: "Proceedings of the 4th Intern. Conference on the mycoses. The black and white yeasts" Brasilia, Pan Americ. Health Org., Washington, D.C.

Klatzmann, D., Champagne, E., and Chamaret, S., 1984, The T4 molecule behaves as the receptor for human retrovirus LAV, Nature, 312:767.

Klein, R.S., Harris, C.A., Small, C.B., Moll, B., Lesser, M., and Friedland, G.H., 1984, Oral candidiasis in high-risk patients as the initial manifestation of acquired immunodeficiency syndrome, N. Engl. J. Med., 9:354.

Kozel, T.R., 1977, Nonencapsulated variant of Cryptococcus neoformans. II Surface receptors for cryptococcal polysaccharide and their role in inhibition of phagocytosis by polysaccharide, Infect. Immun., 16:99.

Kozel, T.R., Gulley, W.F., and Cazin, Jr J., 1977, Immune response to Cryptococcus neoformans soluble polysaccharide : Immunological unresponsiveness, Infect. Immun., 18:701.

Kurosawa, A., Pollock, S.C., Collins, M.P., Kraff, C.R., and Tso, M.O.M., 1988, Sporothrix schenckii endophthalmitis in a patient with human immunodeficiency virus infection, Arch. Ophthalmol., 106:376.

Lane, H.C., Masur, H., Edgar, L.C., Whalen, G., Rook, A.H., and Fauci, A.S., Abnormalities of B cell activation and immunregulation in patients with acquired immunodeficiency syndrome, N. Engl. J. Med., 309:453.

Larsen, R., Bozzette, S., McCutchan J.A., Chiu, J., Leal, M., and Richman, D., 1989, Persistent *Cryptococcus neoformans* prostatic infection after successful treatment of meningitis, Ann. Intern. Med., 111:125.

Leaf, H., and Simberkoft, M.S., 1989, Invasive trichosporonosis in a patient with the acquired immunodeficiency syndrome, J. Infect. Dis., 160:356.

Levy-Klotz, B., Badillet, G., Cavalier-Balloy, B., Chemaly, P., Leverger, G., and Civatte, J., 1985, Alternariose cutanée au cours d'un SIDA, Ann. Dermatol. Vénérol., 112:739.

Lipstein-Kresch, E., Isenberg, H.D., and Singer, C., 1985, Disseminated *Sporothrix schenckii* infection with arthritis in a patient with acquired immunodeficiency syndrome, J. Rheumatol., 12:805.

Male, O., 1989, Synopsis of mycotic infections in AIDS, in: "Immunodeficiency and skin. Current problems in dermatology", P. Fritsch, G. Schuler, H. Hinter, eds., S. Karger Ag. Basel.

Margolick, J.B., Volkman, D.J., Folks, T.M., and Fauci A.S., 1987, Amplification of HTLV infection by antigen-induced activation of T cells and direct suppression by virus of lymphocyte blastogenic responses, J. Immunol., 138:1719.

Mathes, B.M., and Douglass, M.C., 1985, Seborrheic dermatitis in patients with acquired immunodeficiency syndrome, J. Am. Acad. Dermatol., 13:947.

Matthews, R., Burnie, J., Smith, D., Clark, J., Conolly, M., and Gazzard, B., 1988, Candida and AIDS : evidence for protective antibody, Lancet, i:263.

Midgley, G., and Hay, R.J., 1988, Serological responses to *Pityrosporum (Malassezia)*, in seborrheic dermatitis demonstrated by ELISA and western blotting, Bull. Soc. Fr. Mycol. Med., 17:267.

Mostaza, J.M., Barbado, J., Fernando-Martin, J., Pena-Yanez, J., and Vasquez-Rodriguez, J.J., 1989, Cutaneoarticular mucormycosis due to *Cunninghamella bertholletiae* in a patient with AIDS, Rev. Infect. Dis., 11:316.

Niedecken, H., Lutz, G., Bauer, R., and Kreysel, H.W., 1987, Langerhans cells as primary target and vehicle for transmission of HIV, Lancet, ii:519.

Odds, F.C., 1988, "*Candida* and candidosis. A review and bibliography", Second Edition, Baillière Tindall, London.

Pankhurst, C., and Peakman, M., 1989, Reduced CD4 + T cells and severe oral candidiasis in absence of HIV infection, Lancet, i:672.

Pedro, R.J., Aoki, F.H., Branchini, M.L., Lima, J.L., Abreu, W.B., and Dias, M.B.S., 1989, Paracoccidioidomycosis and HIV infection, Abst. Vth International Conference on AIDS MBP 93.

Pautler, K.B., Padhye, A.A., and Ajello, 1984, Imported penicilliosis in the United States, report of a second human infection, Sabouraudia, J. Med. Vet. Mycol., 22:433.

Peto, T.E.A., Bull, R., Millard, P.R., Mackenzie, D.W.R., Campbell, C.K., Haines, M.E., and Mitchell, R.G., 1988, Systemic mycosis due to *Penicillium marneffei* in a patient with antibody to human immunodeficiency virus, J. Infect., 16:285.

Petrini, B., Gyllensten, K., Jorup-Rönström, C., and Rosen, M.L. von, 1989, Första fallet i sverige av generaliserad histoplasmos hos AIDS patient, Läkartidningen, 86:4101.

Pielh, M.R., Kaplan, R.L., and Haber, M.H., 1988, Disseminated penicilliosis in a patient with acuired immunodeficiency syndrome. Arch. Pathol. Lab. Med., 112:1262.

Pinching, A., 1988, Prophylactic and maintenance therapy for opportunist infections in AIDS, AIDS, 2:335.

Prince, H.E., Moody, D.J., Shubin, B.I., and Fahey, I.L., 1985, Defective monocyte function in acquired immune deficiency syndrome (AIDS) : evidence from a monocyte-dependent T-cell proliferative system, J. Clin. Immunol., 5:21.

Redfield, R.R., Wright, D.C., and Tramont, E.C., 1986, The Walter Reed staging classification for HTLV III/LAV infection, N. Engl. J. Med., 314:131.

Reiss, E., 1986, Molecular immunology of mycotic and actinomycotic infections, Elsevier Science Publ., New York;

Salzman, S.H., Smith, R.L., and Aranda, C.P., 1988, Histoplasmosis in patients at risk for the acquired immunodeficiency syndrome in a nonendemic setting, Chest, 93:916.

Schaffner, A., 1984, Acquired immune deficiency syndrome : is disseminated aspergillosis predictive of underlying cellular immune deficiency ? J. Infect. Dis., 149:828.

Schimpff, S.C., and Bennett, J.E., 1975, Abnormalities in cell-mediated immunity in patients with Cryptococcus neoformans infection, J. Clin. Immunol., 55:430.

Segretain, G., 1959, Penicillium marneffei n.sp., agent d'une mycose du système réticuloendothélial, Mycopathol. Mycol., 11:327.

Seligmann, M., Pinching, A.J., Rosen, F.S., Fahey, J.L., Khaitov, R.M., Klatzmann, D., Koenig, Luo, N., Ngu, J., Riethmüller, G., and Spira, T.J., 1987, Immunology of human imunodeficiency virus infections and the acquired immunodeficiency syndrome, Ann. Intern. Med., 107:234.

Shadomy, H.J., De Prada, P., Davis, B.A., Friedman, R.B., Svirsky, J.A., and Shadomy, S., 1989, Distribution of different biotypes and serotypes of Candida albicans in HIV-positive patients, Abstracts 29th ICAAC, Houston, PO373.

Sindrup, J.H., Weismann, K., Petersen, C.S., Rindum, J., Pedersen, C., Mathiesen, L., Worm, H.M., Kron, S., Söndergaard, J., and Wantzin, G.L., 1989, Skin and oral mucosal changes in patients infected with human immunodeficiency virus, Acta Dermatol. Venereol. (Stock), 68:440.

Smith, A.G., Bustamante, C.I., and Gilmor, G.D., 1989, Zygomycosis (absidiomycosis) in AIDS patients, Mycopathologia, 105:7.

Soeprono, F.F., Schinella, R.A., Cockerell, C.I., Comite, S.L., and Linda, L., 1986, Seborrheic-like dermatitis of acquired immunodeficiency syndrome, J. Am. Acad. Dermatol., 14:242.

Spencer, P.M., and Jackson, G.G., 1989, Fungal and mycobacterial infections in patients infected with the human immunodeficiency virus, J. Antimicrob. Chemother., 23, suppl.A/107.

Stenderup, A., Schönheyder, H., Ebbesen, P., and Melbye, M., 1986, White piedra and Trichosporon beigelii carriage in homosexual men, J. Med. Vet. Mycol., 24:401.

St-Germain, G., Noël, G., and Kwon Chung, K.J., 1988, Disseminated cryptococcosis due to Cryptococcus neoformans variety gatii in a Canadian patient with AIDS, Eur. J. Clin. Microbiol. Infect. Dis., 7:587

Stenderup, A., and Schönheyder, H., 1984, Mycoses complicating AIDS, Microbiol. Sci., 1:219.

Tanphaichitra, D., and Srimnang, S., 1984, Cellular immunity in tuberculosis melioidosis, pasteurellosis and penicilliosis : role of levamisole and isoprinosine, in: "Symposium on monoclonal antibody", Pasteur Institute, Dev. Biol. Stand., 57:117.

Tavitian, A., Raufman, J.P., and Rosenthal, E., 1986, Oral candidiasis as a marker for esophageal candidiasis in the acquired immunodeficiency syndrome, Ann. Intern. Med., 104:54.

Torsander, J., Carlsson, B., and von Krogh, G., 1985, Trichosporon beigelii: increased occurrence in homosexual men, Mykosen, 28:355.

Van Cutsem, J., Van Gerven, F., Van Peer, A., Woestenborghs, R., Fransen, J., and Janssen, P.A.J., 1988, Kétoconazole : activité in vitro sur Pityrosporum. Efficacité dans la pityrosporose expérimentale du cobaye et dans le pityriasis capitis humain, Bull. Soc. Fr. Mycol. Méd., 17:283.

Viviani, M.A., and Tortorano, A.M., 1986, Two new cases of cutaneous alternariosis with a review of the literature, Mycopathologia, 96:3.

Viviani, M.A., Tortorano, A.M., and Giani, P.C., 1987, Itraconazole for
    cryptococcal infection in the acquired immunodeficiency syndrome.
    Ann. Intern. Med., 106:166.

Walsh, T.J., Hamilton, S.R., and Belitros, N., 1988, Esophageal
    candidiasis, Postgrad. Med., 84:193.

Weisman, K., Knudsen, E.A., and Pedersen, C., 1988, White nails in AIDS/ARC
    due to *Trichophyton rubrum* infection, Clin. Exper. Dermatol., 13:24.

Wheat, L.J., Slama, T.G., and Zeckel, M., 1985, Histoplasmosis in the
    acquired immune deficiency syndrome, Am. J. Med., 78:203.

Wiest, P.M., Wiese, K., Jacobs, M.R., Morrisssey, A.B., Abelson, T.I.,
    Witt, N., and Lederman, M.M., 1987, *Alternaria* infection in a
    patient with acquired immunodeficiency syndrome : case reported
    review of invasive *Alternaria* infections, Rev. Infect. Dis., 9:799.

Wilkes, M.S., Fortin, A.H., Felix, J.C., Godwin, T.A., and Thompson, W.G.,
    1988, Value of necropsy in acquired immunodeficiency syndrome,
    Lancet, 85.

PNEUMOCYSTIS CARINII: A NOMADIC TAXON

Donald W.R. Mackenzie

Mycological Reference Laboratory, Central Public Health
Laboratory, 61 Colindale Avenue, LONDON NW9 5HT, England

Pneumocystis carinii was first described 80 years ago in Brazil by Chagas (1909), who considered it to be a sporogonous form of trypanosome. The agent was named by the Delanoes in France in 1914.

The associated disease, pneumocystosis was initially rare. It first became prominent as a disease of malnourished infants in central and eastern Europe during the 1940s. *P. carinii* was implicated as the aetiological agent by Vanek and Jirovec in 1952. The condition was refractory to current antimicrobial therapy and had a high mortality. Case fatality rates declined markedly following the introduction of pentamidine, and later, trimethoprim-sulphamethoxazole.

PCP (P. carinii pneumonia) became an increasingly important cause of pneumonia amongst immunosuppressed patients in the 1960s and 1970s, but with the advent of AIDS an explosive increase was seen in its prevalence. It is the commonest opportunistic pathogen in AIDS patients, affecting up to 85% of cases, and also the most immediate threat to life. Although mortality rates have been reduced, they are still, at about 20-30%, unacceptably high.

P. carinii is primarily a parasite of the alveolar lumen. Its occurrence in extrapulmonary sites is uncommon, but widespread dissemination has been documented (Matsuda et al., 1988). A proposed life cycle is shown in Fig.1.

The agent may exist in the lung in two forms. Cysts (Figs. 2, 3) are about 4-6 μm diameter. Their organization is eukaryotic, with one or more mitochondria containing lamellate cristae, a single nucleus and nucleolus, a bilayered nuclear membrane with pores, and 13 chromosomes. As they mature, the wall becomes thicker (up to 160 nm) by interpolation of an extra electronlucent layer. The single nucleus divides to give up to 8 daughter nuclei. Uni-nucleate intracystic bodies (sporozoites) 1-2 μm diameter are then formed, each surrounded by a double plasma membrane. These may be released into the alveoli, where they develop into trophozoites, or they may undergo morphological changes *in situ*.

Mycoses in AIDS Patients, Edited by
H. Vanden Bossche *et al.,* Plenum Press, New York, 1990

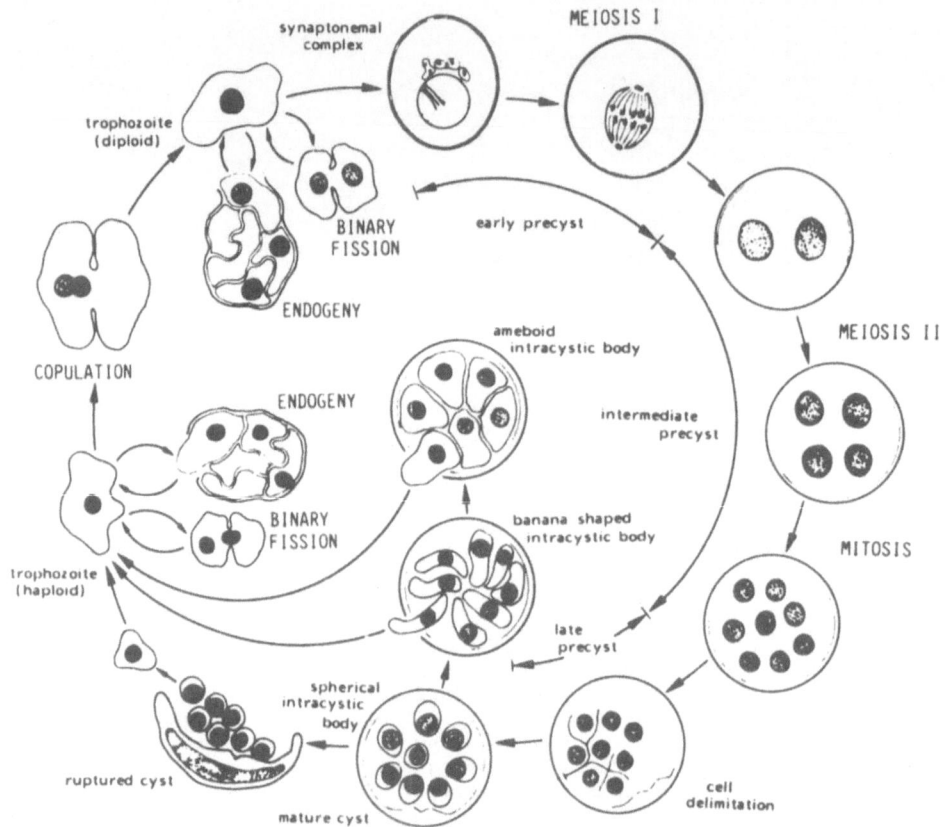

Figure 1. Proposed life cycle of *Pneumocystis carinii* - (Yoshida, 1989)

Trophozoites (Fig. 2) are polymorphic mostly 2-8 µm long, and lack a cell wall, being invested with a narrow pellicle about 25 nm thick. Transmission EM studies suggest that trophozoites may reproduce by binary fission. It is suspected that this may be associated with sexual reproduction, but evidence to date is inconclusive. Of particular interest is the nature of the sporogeny occurring within the cyst. Its similarity to spore formation in ascosporogenous yeasts was noted almost 20 years ago (Vavra and Kucera, 1970). Synaptonemal complexes, characteristic of prophase, have been reported by Yoshida (1989) in the early precyst. This suggests that meiosis takes place within the cyst. Although nuclear division within the cyst is considered to be synchronous (and therefore fungus-like) by some investigators (Yoshida, 1989), a contradictory view is taken by others (ul Haque et al.,1987).

A distinctive ultrastructural feature of *P. carinii* trophozoites is the presence of tubular extensions, slender filamentous outgrowths from the cell margins, which become increasingly abundant as the trophozoite ages (Fig. 3). Such structures have not, to my knowledge, been described amongst the fungi.

In clinical specimens such as bronchial lavage or induced sputum samples, *P. carinii* is diagnosed by its appearance when stained with Giemsa, immunofluorescence with labelled monoclonal antibody, or Grocott methenamine silver stain. Cysts (not trophozoites) of *P. carinii* stain well with Grocott, and can often resemble non-budding yeast cells in size and appearance.

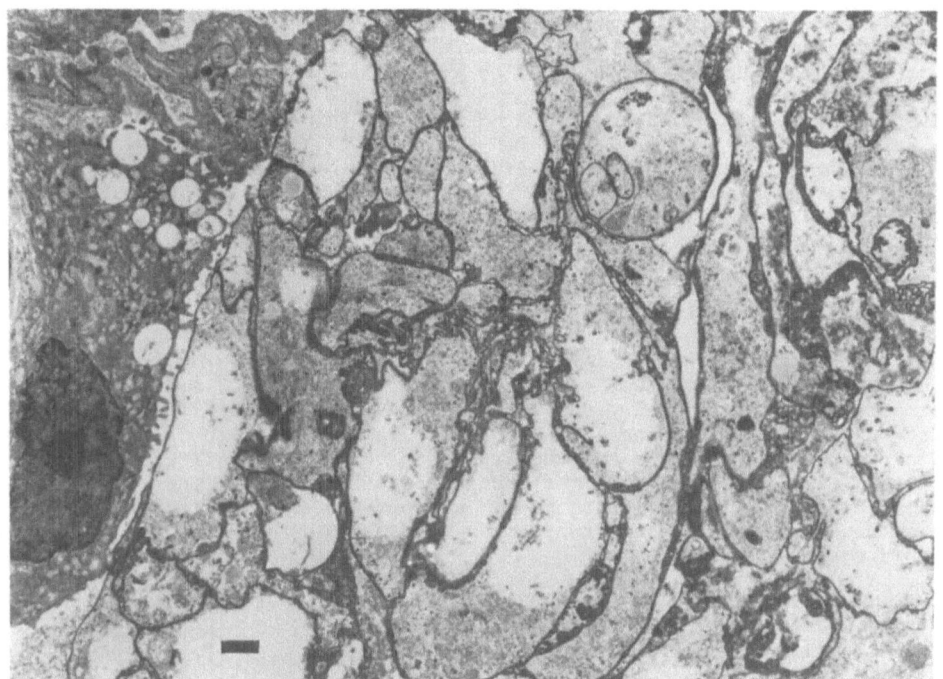

Figure 2. Human lung infected with *P. carinii*. Alveolar wall to the left.
Alveolus packed with thinwalled, polymorphic trophozoites. A
single rounded cyst to present containing two intracystic bodies.
(Courtesy Dr. A. Curry and Dr. P.S. Hasleton). Bar = 1 μm.

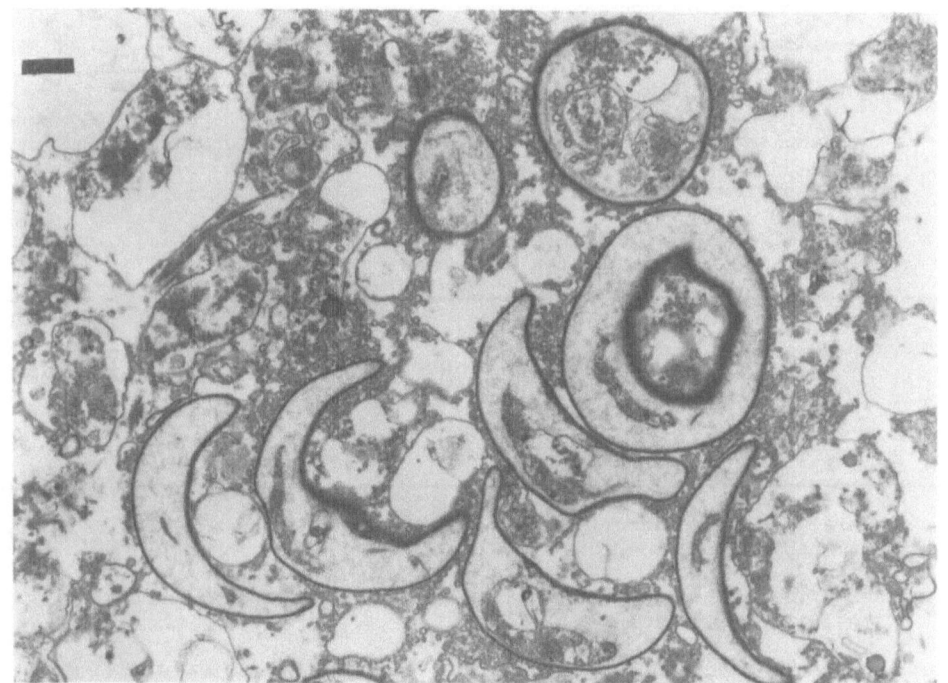

Figure 3. Collapsed intra-alveolar cysts of P.*carinii in lung* of AIDS
patient. A few degenerated trophozoites are also present. Note
abundant tubular extensions. (Courtesy Dr. A. Curry) Bar = 1 μm.

Results of tests for antibodies to *P. carinii* suggest that it is very widespread and is acquired early in life. Whilst there is general agreement that the agent becomes established by the aerial route, the nature of the infective particles is unknown. Nor is it known if the agent has a free-living state or an asyet undiscovered stage in its life cycle.

P. carinii has never been cultivated in cell-free media, and only limited success has been achieved in tissue culture systems (Cushion, 1989). The source of *Pneumocystis* for most published studies on its biology are the lungs of rats whose latent infections have been activated by administration of steroids such as a dexamethasone, usually in association with tetracycline and a low protein diet. In these circumstances, active pneumocystosis invariably results.

Although published comments on the nature of *P. carinii* before 1988 were infrequent, there has been general agreement, based largely on its ultrastructural morphology, that the agent was a eukaryote, and that it must be a protozoan, probably belonging to the Sporozoa, or a fungus of uncertain affinity. Before 1988, arguments for its classification in the Sporozoa were vague and at times contradictory (Table 1). They included inability to grow in cell-free media, susceptibility to sulphonamides and other drugs effective against protozoa, ultrastructural organization, and simply "general features".

Supporters of a fungal identity advanced arguments which were largely negative (Table 2), concentrating on features which made a Sporozoan identity unlikely or unacceptable.

Table 1. *P. carinii* : resemblances to protozoa (pre-1988)

---

1. General features including protozoa-like life cycle.

2. Membrane complexes and "pseudopodia" of trophozoites.

3. Microtubules below sporozooite cell membrane-associated with motility?

4. Unable to grow in cell-free media.

5. Susceptibility to sulphonamides and other chemotherapeutic agents effective against protozoa.

Table 2. *Pneumocystis carinii* : resemblances to fungi (pre-1988)

---

1. General features including absence of protozoal characteristics (polar filaments, micronemes, rhoptries, sub-pellicular tubules).

2. Lack of organelles of motility (pseudopodia, filopodia, cilia, flagella).

3. Absence of phagocytic system or golgi apparatus, and reliance on low molecular weight nutrients.

4. Cell wall inner thin plasma membrane adherent to outer thick wall.

Since 1988, however, there has been a dramatic increase in the number of reports which shed new light on the problems of its basic nature and relationships. These include the milestone article by Edman et al. in Nature (1988), and the proceedings of an international conference devoted to the basic science of *P. carinii* (Journal of Protozoology, **36**:(1), 1989).

Data based on rRNA sequencing reported by two groups of investigators in the United States have provided evidence for a fungal identity of *P. carinii* which is both impressive and convincing. Edman et al. (1988) isolated genomic clones encoding rRNA from *P. carinii*, subcloned the 16s-like rRNA coding sequence, and compared this with sequences from a range of other life forms. Comparisons of nucleotide residues in what appears to be the complete rRNA sequence shows (Fig. 4) that *P. carinii* clusters closely with the fungi rather than with the protozoa.

A similar approach by Stringer et al. (1989) yielded comparable data and results. This group analyzed nearly 300 nucleotides of *P. carinii* obtained by reverse transcription from oligonucleotide primers and matched these with sequences from 8 other taxa, including *Cryptococcus diffluens* and *Candida albicans*. Greatest homology (about 92%) was obtained between *P. carinii* and *Saccharomyces cerevisiae* compared with 81% for sequences from protozoa.

Watanabe et al. (1989) in Japan studied 5s rRNA, which is not only smaller, but even more highly conserved. Their conclusion was that at what would have been an earlier stage of evolution, *P. carinii* was phylogenetically linked to amoebae, slime moulds and zygomycetes.

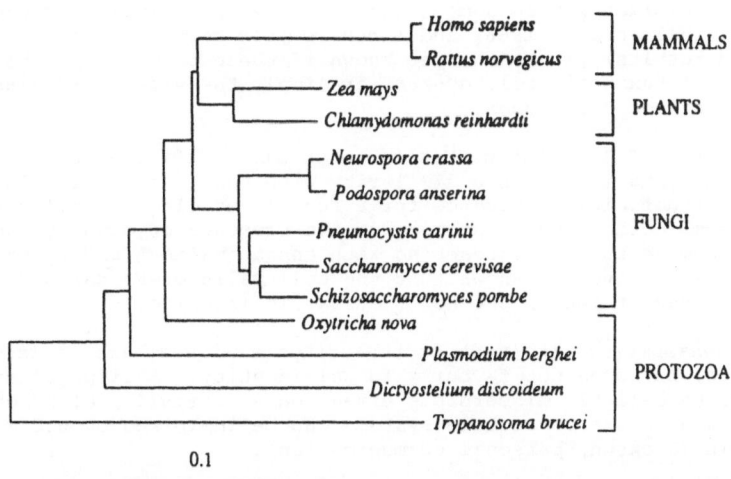

Figure 4. Phylogenetic tree based on similarities of rRNA sequences. (Edman, J.C. et al., 1989).

Table 3. Affinities of *Pneumocystis carinii*

Types of Studies	Fungal	Protozoal
Ultrastructural studies	(−)	(−)
Peptide Elongation Factor-3	(−)	
DNA content/cell		(+)
Antigen analysis	Inconclusive	
Cell wall chemistry	(+)	
Drug sensitivity		+
Thymidilate synthase		−
RNA sequencing	+	

+ = indicating affinity, − = lack of affinity, ( ) = inconclusive

Transcription of tRNA by eukaryote ribosomes requires the participation of soluble proteins called Elongation Factors (Uritani and Miyazaki, 1988). All systems investigated to date involve two distinct factors, EF-1 and EF-2. Fungi differ in requiring a third factor EF-3. At a Workshop held in London last month, Dr. Helen Jackson reported that this fungal elongation factor could not be detected in *P. carinii* using monoclonal antibody directed against EF-3 as a probe. It is not yet known if this could be explained by the absence of protein synthesis by the *P. carinii in vitro*, or if the absence of EF-3 is innate and not artefactual.

Data from other recent studies (Table 3) also have some bearing on the fungal versus non-fungal identity of *Pneumocystis.*

DNA content of cells has been proposed as a means of characterising major taxons, and has been determined for several protozoan pathogens. It has also been determined for *Saccharomyces* and more recently for *P. carinii* (Gradus et al., 1988). *P. carinii* was found to contain 0.22-0.34 pg DNA per cell, levels comparable to those found for protozoa. The difference between cell DNA content of protozoa and *Saccharomyces* (0.02 pg) was 2-30 fold. The authors suggest that these findings support the classification of *P. carinii* as a protozoan.

Analysis of *P. carinii* antigens suggests the existence of common antigenic determinants with some fungal pathogens, including *Aspergillus fumigatus* (Trull et al., 1986) and *Saccharomyces cerevisiae* (Camus et al., unpublished results). It is not yet known if these findings can be viewed as evidence of taxonomic relatedness, or simply the widespread distribution of certain antigenic epitopes.

Presence of ß-1,3 glucan (Matsumoto et al., 1989) in the wall of *Pneumocystis* cysts indicates a similarity with the yeast cell wall, but is not by itself definitive evidence for a fungal aetiology. Similarly, evidence from lectin studies suggesting the presence of mannosyl and N-acetylglucosamine residues (reacting with concanavalin A and wheat germ agglutinin respectively), shows that the cell walls of *P. carinii* cysts are similar to those of fungi, but not necessarily identical.

Drug sensitivity like most of the criteria listed here is largely inconclusive, but does not favour a fungal identity. Antifungal drugs in current use have little discernible effect on *P. carinii* , although its limited growth in tissue culture systems can be inhibited by 0.5 mg/l of amphotericin (Jackson, personal communication).

Perhaps the most convincing additional evidence for an identity other than a protozoal one comes from a report by Edman, et al. (1989). This group described the isolation of the thymidilate synthase gene from *P. carinii* DNA libraries and its expression in *E. coli.* In all protozoa studied to date, this enzyme co-exists as a bifunctional protein on the same polypeptide chain as dihydrofolate reductase. For *P. carinii*, however, these two enzymes were found to be located on different chromosomes. This is considered by the authors to be evidence against *P. carinii* being a protozoan.

This is an intriguing finding, but it has to be placed in the context of the extreme variability known to exist within the protozoa. It has been shown, for example, that intrageneric variation in rRNA sequences of the amoeba *Naegleria* is substantial, being as great as those which exist between mammal and frog (Johnson and Baverstock, 1989).

If *P. carinii* is a fungus, what are its closest relatives? Vavra and Kucera (1970) thought it was a yeast with an ascospore stage (Subclass Hemiascomycetes; order Endomycetales). Although interesting to note that one of the original candidate identities was *Schizosaccharomyces octosporus,* and that rRNA sequencing shows *P. carinii* to be quite close to this genus, it is difficult to accept P. *carinii* as a yeast on the basis of our current understanding of this diverse group of fungi. It is even more difficult to suggest a convincing alternative when so many details of its life cycle remain obscure.

Protagonists of a fungal aetiology for *P. carinii* base their case almost exclusively on data from RNA sequencing. But how valid are these data? Do alternative explanations exist for the apparent relatedness of *P. carinii* and the fungi with which it has been compared?

Is it possible that changes in a relatively small number of regulatory genes, in circumstances of high selection pressure, could lead to the emergence of a new taxon, markedly different in morphology and life style from descendants of its nearest relatives, but retaining a high level of overall resemblance in its nucleic acid sequences? If so, then the apparent affinity between *Pneumocystis* and the fungi may also be a real one.

However, according to Johnson and Baverstock (1989), rRNA sequencing, whilst providing new insights into phylogenetic relationships of the protista, should not be seen as a panacea for definitive construction of evolutionary trees, but as a means for testing traditional hypotheses, confirming some, but challenging others.

Taxonomy of microorganisms still depends on detailed knowledge of the full range of their characteristics, including morphology, genetics, biochemistry, physiology, ecology, and other criteria. Data from analysis of nucleic acids may provide vital clues to taxonomy, but classification should not be based on a single feature.

It has recently been pointed out by Hughes (1989) that definitions of fungi and protozoa are so imprecise that *P. carinii* could be accommodated in both groups. Not all known features of *P. carinii* can be readily accommodated in the protozoa. The same is true if placement in the fungi is attempted.

The best verdict based on evidence available to date that *P. carinii* is a fungus is perhaps NOT PROVEN. Its exclusion from the present symposium may be justified on this count. Search for its true taxonomic affinities should continue. If it eventually comes to rest within the fungi, the definition of the fungal kingdom might have to be amended. This could be done, but the case for doing so is not yet a compelling one.

I cannot foretell the future, but it is greatly to be hoped that this dramatic claim of a fungal identity for *P. carinii* will provide the stimulus for further studies. Many key questions remain unanswered

Table 4. *Pneumocystis carinii* the key questions.

---

1. How can the *in vivo* phases be reproduced *in vitro*?

2. Does meiosis take place within the cyst?

3. Do the trophozoites function sexually?

4. Does *P. carinii* proliferate outside the body?

5. What is the nature of the infectious propagule?

6. What are the true taxonomic affinities of *P. carinii?*

(Table 4). Mycologists have an important role in addressing themselves to these questions, and in reappraising the taxonomic status of this perplexing and fascinating microorganism.

It remains to be seen if additional evidence will confirm a fungal aetiology for *P. carinii,* and if the guardians of the mycological kingdom will be required in time to raise their barriers, and admit this taxonomic vagrant into full citizenship.

ACKNOWLEDGEMENTS

Figures 1 and 4 are published by permission of the Journal of Protozoology. I acknowledge with thanks photomicrographs supplied by Dr. Alan Curry, Public Health Laboratory, Manchester, UK. and Dr. Philip S. Hasleton, Dept. of Histopathology, Wythenshawe Hospital, Manchester, England (Figure 2), and Dr. A. Curry, (Figure 3).

REFERENCES

Chagas, C, 1909, Nova tripanozoma humana. Estudos sobre a morfologia e o ciclo evolutivo do Schizotrypanum cruzi n.gen.,n.sp., agente etiologio de nova entidade morbida de homen, Mem. Inst. Oswaldo. Cruz, 1:159.
Cushion, M., 1989, *In vitro* studies of *Pneumocystis carinii, J. Protozool.,* 36:45.
Delanoe, P., and Delanoe, Mme., 1914, De la rarete de *Pneumocystis carinii* chez des cobayes de la region de Paris; absence de kystes chez d'autres animaux lapin, grenouille, 3 anguilles, Bull. Soc. Path. exot., 7:271.
Edman, J.C., Kovacs, J.A., and Masur, H., 1988, Ribosomal RNA sequencing shows *Pneumocystis carinii* to be a member of the fungi, Nature (Lond), 334:519.
Edman, U., Edman, J.C., Jundgren, B., and Santi, D.V., 1989, Isolation and expression of the *Pneumocystis carinii* thymidilate synthase gene, Proc. Nat. Acad. Sci. USA., 86:6503.
Gradus, M.S., Gilmore, M., and Lerner, M., 1988, An isolation method of DNA from *Pneumocystis carinii:* a quantitative comparison to known protozoan DNA, Comp. Biochem. Physiol., 89B:75.
ul Haque, A., Plattner, S.B., Cook, R.T., and Hart, M.N., 1987, *Pneumocystis carinii.* Taxonomy as viewed by electron microscopy, Am. J. Clin. Path., 87:504.
Hughes, W.T., 1989, *Pneumocystis carinii:* taxing taxonomy, Eur. J. Epidem., 5:265.

Johnson, A.M., and Baverstock, P.R., 1989. Rapid ribosomal RNA sequencing and the phylogenetic analysis of protists, Parasitol. Today, 5:102.

Linke, M.J., Cushion, M.T., and Walzer, P.D., 1989. Properties of the major antigens of rat and human *Pneumocystis carinii*, Infect. Imm., 57:1547.

Matsuda, S., Shiota, T., Yoshikawa, H., Tegoshi, T., Yamada, Y., Okada, S., Nakamura, H., Kitaoka, T., Urata, Y., and Ashihara, T., 1988, Dissemination of *Pneumocystis carinii* in a case of AIDS, Jap. J. Trop. Med. Hyg., 16:159.

Matsumoto, Y., Matsuda, S., and Tegashi, T., 1989. Yeast glucan in the cyst wall of *Pneumocystis carinii*, J. Protozool., 36:21S.

Stringer, S.L., Stringer, J.R.., Blase, M.A., Ewalzer, P.D., and Cushion, M.T., 1989, *Pneumocystis carinii*: sequence from ribosomal RNA implies a close relationship with fungi, Exp. Parasitol., 68:450.

Trull, A.K., Warren, R.E., and Thiru, S., 1986, Novel immunofluorescence test for *Pneumocystis carinii*, Lancet, i:271.

Uritani, M., and Miyazaki, M., 1988, Role of yeast peptide elongation factor 3 (EF-3) at the AA-tRNA binding step, J. Biochem., 118:126.

Vanek, J., and Jirovec, O., 1952, "Interstitielle" Plazmazellenpneumonie der Fruehgeborenen, verursacht durch Pneumocystis carinii, Z. Bakt., 1.Abt.Orig., 158:120.

Vavra, J., and Kucera, K., 1970, *Pneumocystis carinii* Delanoe, its ultrastructure and ultrastructural affinities, J. Protozool., 17:463.

Watanabe, J-L., Hori, H., Tanabe, K., and Nakamura, Y., 1989, 5S ribosomal RNA sequence of *Pneumocystis carinii* and its phylogenetic association with Rhizopoda/Myxomycota/ Zygomycota group, J. Protozool., 36:16S.

Yoshida, Y., 1989, Ultrastructural studies of *Pneumocystis carinii*, J. Protozool., 36:53.

Candida and Candidosis

EPIDEMIOLOGY OF *CANDIDA* INFECTIONS IN AIDS

Frank C. Odds[a], Jan Schmid[b] and David R. Soll[b]

[a]Department of Bacteriology and Mycology
 Janssen Pharmaceutica, B-2340 Beerse, Belgium
[b]Department of Biology, University of Iowa
 Iowa City, Iowa 52242, U.S.A.

ABSTRACT

   Oral, vaginal and oesophageal candidosis occur commonly in patients
with HIV infection. The incidence of oesophageal candidosis varies between
subgroups of the HIV-infected population, indicating that factors more
complex than the presence or absence of HIV infection determine the
pathogenic status of *Candida*. Almost all *Candida* infections in AIDS are
caused by the species *C. albicans*. Some surveys, but not all, show an
increased prevalence of *C. albicans* serotype B strains in AIDS patients.
The possibility that the immunosuppression and protracted antifungal
chemotherapy associated with HIV infection may create a selective pressure
for evolution of unusual *C. albicans* strains is supported by the finding of
unusually closely related DNA types among four of nine AIDS patients in one
survey.

INTRODUCTION

   Candida infections of the mouth, oesophagus and vagina occur very
commonly in patients with AIDS, while candidosis of skin and nails and
involvement of deeper tissues such as lung, meninges, peritoneum and
multiple viscera have been noted only rarely. It is likely that the
particular distribution of candidosis lesions associated with HIV infection
is primarily a reflection of the specific alteration of host immune
responses engendered by the virus; however, both the immunosuppression and
the long-term treatment with antifungal agents associated with HIV
infection create selective pressures that may influence the nature of the
Candida strains found in AIDS patients. In this article the prevalence and
the types of *Candida* found in HIV-infected individuals is examined.

PREVALENCE OF *CANDIDA* INFECTIONS AMONG DIFFERENT TYPES OF AIDS PATIENTS

   Among the several published surveys that indicate the prevalence of
Candida infections in AIDS, the one published by Selik et al. in 1987 is
probably the most informative so far. It is based on the Centers for
Disease Control database of 30632 AIDS patients. Selik et al. (1987) warn
of problems that must be considered with any prevalence statistics. They
note, for example, that up to 1983, when physicians were required to write
in longhand the occurrence of oral candidosis in their patients, the

Mycoses in AIDS Patients, Edited by
H. Vanden Bossche *et al.*, Plenum Press, New York, 1990

prevalence of this condition in AIDS appeared to be only 7.0%. However, when reporting of oral candidosis subsequently became a matter of merely checking a box on a printed form, the infection was reported in 44.8% of the next 6545 patients.

Oesophageal candidosis

Table 1 shows the incidence of oesophageal candidosis among various AIDS subgroups, according to Selik et al. (1987). It is striking that the condition occurred significantly less often than average among male homosexuals who were not intravenous drug abusers and significantly more often than average among drug abusers of both sexes and blood recipients infected with HIV.

Table 1. Incidence of *Candida* oesophagitis in AIDS patients[a]

Patient category	Number	Percent with oesophagitis
All patients	30632	10.6
Male homosexuals, non-IVDA[b]	19795	8.5
Male homosexuals, IVDA	2332	10.0
Heterosexual IVDA, male & female	5128	15.4
Blood recipients	622	19.0

[a] data from Selik et al. (1987)
[b] IVDA: intravenous drug abusers

Even within these subgroups there were apparent differences in incidence of oesophageal candidosis: it was twice as common among black, male homosexuals as in white, male homosexuals, and 50% more common among female intravenous drug abusers than among male drug abusers (Selik et al., 1987). These data seem to offer support for a hypothesis that the capability of *Candida* to invade an HIV-infected host is regulated by more complex factors than the immunosuppressive effects of the virus *per se*.

Sources of *Candida* in AIDS and selective pressures on fungal strains

It is normally assumed that most *Candida* infections arise from a patient's endogenous commensal flora (Odds, 1988) and there is no reason to suppose endogenous infection is not similarly common among patients infected with HIV. *Candida* carried in the digestive tract can be transferred to the skin (from the mouth and the anus) and to the bloodstream (by passage through the gut wall). However, in the context of HIV infection there is always a possibility of exogenous transfer of *Candida*, notably by sexual transmission or by contaminated intravenous needles. No experimental data are available to indicate the relative occurrence of endogenous and exogenous *Candida* infection in AIDS, but it is possible that the latter route might become important if strains particularly well adapted to survival within an HIV-infected host became widespread among the AIDS population.

The two major selective pressures that could influence the types of *Candida* strains in AIDS patients are the highly specific immunodeficiency resulting from HIV-1 infection and the prophylactic use of antifungals over long periods of time. In addition, the apparently differential vulnerability of different categories of AIDS victims to oesophageal candidosis (Table 1) suggests that properties such as race, sex and other socio-environmental factors may place further selective pressures on *Candida* strains associated with AIDS.

Candida species and strains isolated from HIV-infected patients

Evidence is emerging that HIV infection is indeed leading to the selection of unusual *Candida* types in patients with AIDS. It is clear that *C. albicans* is the species most commonly found in candidosis lesions in AIDS: as in superficial forms of candidosis among all types of patients, *C. albicans* accounts for more than 95% of reported infections (Odds, 1988). Since this species makes up only 70% of oral yeast isolates and 50% of anal yeast isolates from healthy individuals (Odds, 1988) it must possess a greater potential than other yeast species to cause clinically apparent lesions.

However, within the species *C. albicans* there may be certain strain types that have a particular affinity for the HIV-infected patient. The most striking example of this is the appearance of reports of a high prevalence of *C. albicans* serotype B strains among AIDS patients. Table 2 summarizes the statistics available so far on the occurrence of *C. albicans* serotype B in AIDS. Many of the surveys listed have shown a comparatively high prevalence of serotype B strains among AIDS patients. However, only one of the studies in Table 2 was directly controlled by means of comparative patient groups: Brawner & Cutler (1989) found a prevalence of type B strains of 53.1% among non-immunocompromised patients in their study. They also found an elevated proportion of type B strains among therapeutically immunocompromised patients, as well as patients with AIDS.

Table 2. Prevalence of *C. albicans* serotype B in AIDS patients

AIDS patients in...	No.	% Serotype B	Reference
Paris, France	74	43.2	Drouhet et al. (1989)
U.S.A. (various)	20	65.0	Brawner & Cutler (1989)
San Francisco, U.S.A.	32	31.2	Odds (unpublished)[b]
Barcelona, Spain	10	40.0	Saballs et al. (1989)[a]
Barcelona, Spain	42	7.1	Torres-Rodriguez et al (1989)[a]
Leicester, U.K.	10	50.0	Odds (unpublished)[c]
London, U.K.	14	0.0	Midgley, J. (personal communication)

[a] data presented as posters at the 3rd symposium "Topics in Mycology: Mycosis in AIDS patients"
[b] oral isolates kindly provided by Dr. C. Halde
[c] oral isolates kindly provided by Drs. K. Nicholson and M. Wiselka

To assess the significance of the data in Table 2 for other, uncontrolled reports it is necessary to compare the listed prevalences of *C. albicans* serotype B in AIDS patients with historical data on the prevalence of this serotype in the countries involved. In the U.S.A., a survey done in 1982 showed a mean prevalence of 48.9% for type B isolates obtained from five different regions within the country (Stiller et al., 1982). The finding of 31.2% serotype B isolates among AIDS patients in San Francisco (Table 2) therefore indicates no exceptional prevalence of this type, while that of Brawner & Cutler shows an increase. For European isolates, historical data on the prevalence of serotype B are less recent than for the U.S.A., but they indicate a background prevalence of from 6 to 20% (Odds, 1988). Thus, the absence of type B among the 14 *C. albicans* isolates tested in London is consistent with the historical prevalence for the U.K., while the 50% prevalence of type B strains in Leicester is considerably higher. Similarly, the two surveys in a single city in Spain show a large increase in type B in one case, no change in the other. The prevalence of serotype B in AIDS in the study from France certainly suggests an increase from previous data (Drouhet et al., 1975). Moreover, the higher frequency of *C. albicans* strains resistant to flucytosine among AIDS patients reported by Korting et al. (1988) is compatible with an increased prevalence of serotype B strains, since the property of flucytosine resistance is substantially more common among type B than type A isolates (Odds, 1988).

The significance of the data in Table 2 is difficult to evaluate with confidence. It would seem that in some centres there is an unequivocally high preponderance of *C. albicans* type B among the AIDS population, while no change is found in others. A serotype B prevalence of 75.9% was recently reported among *C. albicans* isolates from the Ivory Coast in Africa (Penali et al., 1989), which suggests that there is enormous geographic and epidemiologic diversity in the occurence of this strain type. Such considerable differences in prevalence of a readily definable *C. albicans* phenotype is surely worthy of further study.

If there are discernible changes in phenotype of *C. albicans* strains from AIDS patients, it might be predicted that an associated change in genotype should occur. The only published survey of DNA types among *C. albicans* strains from AIDS patients so far found that the distribution of types was essentially similar to that found in other patient groups (Matthews et al., 1988). However, this study was based only on DNA typing by gross restriction fragment length polymorphisms, which are inherently less discriminatory than typing based on the Southern blot hybridization patterns of endonuclease-digested whole cell DNA probed with cloned moderately repetitive genomic sequences (Scherer & Stevens, 1988; Soll et al., 1987, 1988, 1989; Fox et al., 1989; Soll, 1990).

A preliminary study of *C. albicans* DNA types among isolates from nine AIDS patients in Leicester (U.K.) has been carried out in our laboratories. The results strongly suggest that particular strain types may be selected for among the AIDS population. The nine isolates were typed with the moderately repetitive sequence probe Ca3 according to the methods described in detail elsewhere (Schmid et al., 1990; Soll, 1990). The Ca3 hybridization patterns obtained (Figure 1) were analysed by the Dendron computer program, which allows computation of the similarity values (S_{AB}'s) for all pairs of strains based on the intensity and position of the bands that hybridize with Ca3 and generates histograms and dendrograms based on S_{AB} values (Schmid et al., 1990; Soll, 1990). The S_{AB} value is calculated by the formula:

$$S_{AB} = \frac{\sum\limits_{i=1}^{K}(a_i + b_i - |a_i - b_i|)}{\sum\limits_{i=1}^{K}(a_i + b_i)}$$

where a_i and b_i are the class strengths of band i for strain a and b (no band is counted as class 0), and K is the number of bands. If the S_{AB} value for two strains is zero, no bands in the Southern blot hybridization patterns of two strains correlate for molecular weight, and if the S_{AB} for two strains is 1.0, every band correlates in molecular weight and intensity (Schmid et al., 1990; Soll, 1990).

The results show several striking features. First, 5 of the 9 strains (89/011, 89/016, 89/010, 89/012 and 89/007, each from a different patient) gave S_{AB} values greater than 0.88, and 3 (89/011, 89/016 and 89/010) gave S_{AB} values greater than 0.94 (Figure 2). The mean S_{AB} value for 1035 possible comparisons between 46 presumably unrelated test strains has been shown to be 0.69± s.d. 0.11, and the mean S_{AB} value for identical strains analysed in repeat gels has been shown to be 0.96±0.02 (Schmid et al., 1990). Therefore, the 5 strains with S_{AB} values above 0.88 represent a set of highly related strains which cluster in the dendrogram in Figure 2.

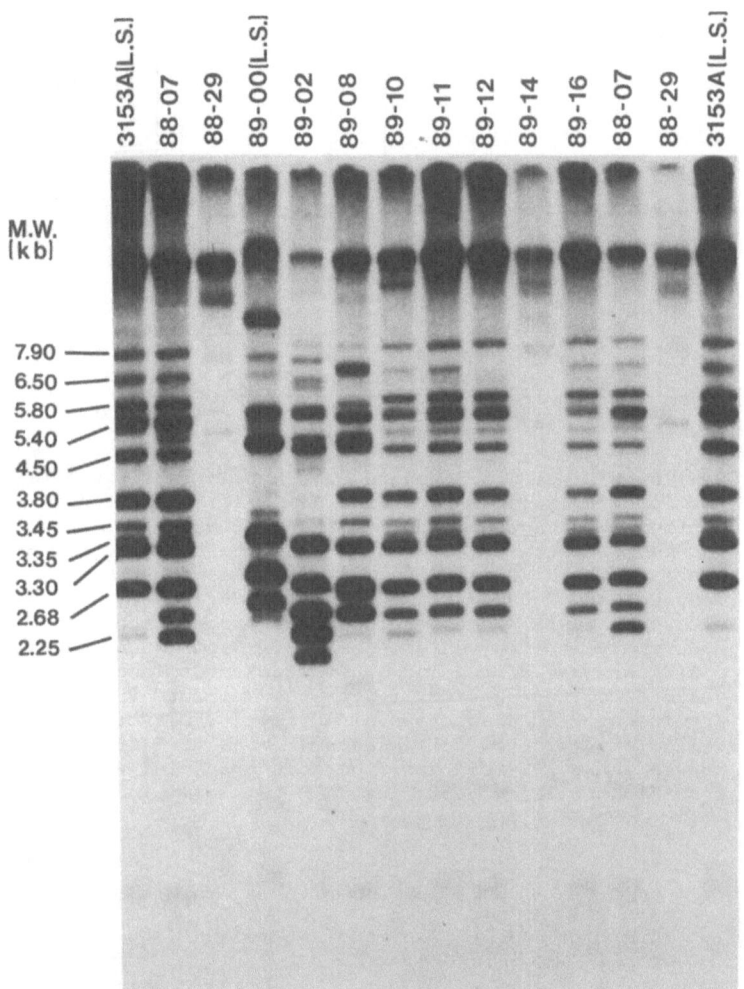

Fig. 1. Southern blot hybridization patterns of EcoRI digested
cellular DNAs of *Candida* strains from AIDS patients
probed with the cloned, moderately repetitive sequence
Ca3. The molecular weights of the major bands of
laboratory strain 3153A are noted to the left of the gel.
Strains are labelled at the top of each lane. Laboratory
strains include 3153A and 89-00 (L.S.). AIDS isolates
include 88-07, 88-29, 89-02, 89-08, 89-10, 89-11, 89-12,
89-14, 89-16, 89-07. Note that 3153A, 88-07 and 88-29
were run twice.

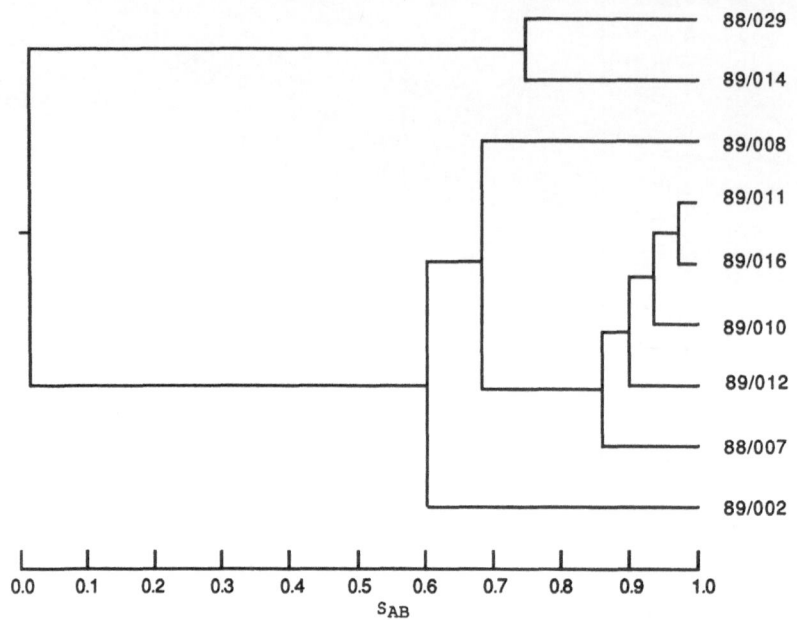

Fig. 2. Similarities of Ca3 Southern blot hybridization patterns
for *Candida* isolates from nine AIDS patients. The
dendrogram was generated from the similarity values
between the strains through the Dendron software package.
DNA from the nine isolates was digested with the
endonuclease EcoRI, fragments were separated by
electrophoresis, and the blots were probed with the
moderately repetitive sequence Ca3. Band intensities and
positions were scored and SAB values determined. An SAB
value of 0.96 or greater represents absolute identity, and
a value of 0.00 absolute nonidentity. The average
similarity value of *C. albicans* strains is 0.69. Strains
88/029 and 89/014 did not type as the species *C. albicans*
according to Ca3 specificity.

Furthermore, the 4 most highly related strains in this cluster
(89/010, 89/011, 89/012 and 89/016) were *C. albicans* serotype B, precisely
the type whose prevalence may be greater than normal among AIDS patients.
There is no known contact between the four patients of a nature likely to
lead to dissemination of a common strain type. In addition, because of the
relatively slow divergence of the Ca3 pattern within (Schmid et al., 1990)
and the very clear, although minor, differences in banding patterns between
these strains, it is likely that these strains are highly related, but not
identical.

Another striking feature in Fig. 1 is the occurrence of strains from
two AIDS patients — 88/029 and 89/014 — that reacted minimally with the Ca3
probe, which appears to be specific for the yeast species *C. albicans*
(Soll, 1990). These two yeast isolates displayed unusual phenotypic
properties. Both formed no germ tubes in serum, both gave sugar
assimilation patterns in API 20C strips that did not conform precisely with
the patterns of any known species, but both formed copious hyphae and

chlamydospores on rice-Tween 80 agar, a property consistent with their tentative identification as *C. albicans*.

The true significance of the findings illustrated in Figs. 1 and 2 will become apparent only after the strains have been further studied and the survey extended to a greater number of patients. However, when they are considered together with the reports of increases of prevalence of serotype B strains among AIDS patients, there does seem to be a basis for the general conclusion that selective pressures may operate on *C. albicans* within the AIDS population that lead to the evolution of unusual types of the fungus. This conclusion may have important consequences for future clinical management of candidosis in AIDS.

ACKNOWLEDGEMENTS

Grateful thanks are due to Drs. Carlyn Halde, Karl Nicholson and Martin Wiselka for providing isolates from AIDS patients. Fingerprinting experiments were supported by grant AI23922 from the National Institutes of Health to DRS.

REFERENCES

Brawner, D.T.,and Cutler, J.E., 1989, Oral *Candida albicans* isolates from nonhospitalized normal carriers, immunocompetent hospitalized patients, and immunocompromised patients with or without acquired immunodeficiency syndrome, J. Clin. Microbiol., 27:1335.

Drouhet, E., Mercier-Soucy, L., and Montplaisir, S., 1975, Sensibilité et résistance des levures pathogènes avec 5-fluoropyrimidines, Ann. Microbiol., 126B:25.

Drouhet, E., Dupont, B., Improvisi, L., Lesourd, M., and Provost, M.C., 1989, La cilofungine (LY121019), nouvel antifongique lipopolypeptidique agissant sur la paroi fongique. Activité *in vitro* sur les levures, Bull. Soc. Fr. Mycol. Med., 18:31.

Korting, H.C., Ollert, M., Georgii, A., and Froeschl, M., 1988, *In-vitro* susceptibilities and biotypes of *Candida albicans* isolates from the oral cavities of patients infected with human immunodeficiency virus, J. Clin. Microbiol., 26:2626.

Matthews, R., Burnie, J., Smith, D., Clark, I., Midgley, J., Connolly, M., and Gazzard, B., 1988, *Candida* and AIDS: evidence for protective antibody, Lancet, 2:263.

Odds, F.C., 1988, "*Candida* and candidosis", 2nd edition, Bailliere Tindall, London.

Penali, L.K., Kone, M., Codo-Agbo, Faye-Kette, H., Kouassi, M.T., Boue Tan, A., and Kouakou, B., 1989, Sensibilité aux antifongiques des sérotypes de *Candida albicans* isolaés à Abidjan (Côte d'Ivoire), Bull. Soc. Fr. Mycol. Med., 18:371.

Scherer, S, and Stevens, D.A., 1988, A *Candida albicans* dispersed, repeated gene family and its epidemiologic application, Proc. Nat. Acad. Sci. USA, 85: 1452.

Schmid, J., Voss, E., and Soll, D.R., 1990, Computer-assisted methods for assessing strain relatedness in *C. albicans* by fingerprinting with the moderately repetitive sequence Ca3, J. Clin. Microbiol., (submitted).

Selik, R.M., Starcher, E.T., and Curran, J.W., 1987, Opportunistic diseases reported in AIDS patients: frequencies, associations, and trends, AIDS, 1:175.

Soll, D.R., 1990. Current status of the molecular basis of *Candida* pathogenicity, in "The Fungal Spore and Disease Inititiation in Plants and Animals", G.T. Cole and H.G. Hoch, eds., Plenum Publishing Corp., New York (in press).

Soll, D.R., Langtimm, C.A., McDowell, J., Hicks, J., and Galask, R., 1987, High frequency switching in *Candida* strains isolated from vaginitis patients, *J. Clin. Microbiol.*, 25:1611.

Soll, D.R., Staebell, M., Langtimm, C., Pfaller, M., Hicks, J., and Rao, T.V.G., 1988, Multiple *Candida* strains in the course of a single systemic infection, *J. Clin. Microbiol.*, 26:1448.

Soll, D.R., Galask, R., Isley, S., Rao, T.V.G., Stone, D., Hicks, J., Schmid, J., Mac, K., and Hanna, C., 1989, "Switching" of *Candida albicans* during successive episodes of recurrent vaginitis, *J. Clin. Microbiol.*, 27: 681.

CANDIDEMIA IN PATIENTS WITH ACQUIRED IMMUNODEFICIENCY SYNDROME

Freda E. Chu, Melanie Carrow, Ann Blevins, Donald Armstrong*

*Infectious Disease Service, Memorial Sloan-Kettering Cancer
Center, 1275 York Avenue, New York, New York 10021, USA

ABSTRACT

Positive blood cultures for *Candida* species from January 1981 to
November 1989 were reviewed at Memorial Sloan-Kettering Cancer Center.
Thirteen patients with AIDS were found to have candidemia. The most common
species were *Candida albicans* isolated in 8 patients, *Candida parapsilosis*
in 4 patients and *Candida krusei* in 1 patient. Intravascular catheters
were present in 12 of the 13 patients, 9 of which were indwelling or
central catheters. Neutropenia was present in 7 patients. Treatment was
instituted in 10 patients of whom 6 responded. Candidemia is a rare
occurrence in AIDS patients who are very frequently colonized with *Candida*.
The presence of factors such as intravascular catheters and neutropenia
appear to predispose AIDS patients to candidemia.

INTRODUCTION

Infection with the human immunodeficiency virus (HIV) has been
associated with persistent mucosal candidiasis since the first reports of
the acquired immunodeficiency syndrome (AIDS) in 1981 (Gottlieb et al.,
1981; Masur et al., 1981; Siegal et al., 1981).
Since then pharyngeal candidiasis (thrush) and esophageal candidiasis
have been reported regularly and esophageal candidiasis has become an AIDS-
defining opportunistic infection (OI) (CDC, 1987).
Disseminated candidiasis has not been reported with any regularity and
in our experience has been associated with other risk factors such as
intravenous catheters or neutropenia. We evaluated this clinical
observation by a retrospective study of HIV-infected patients who were
fungemic with a *Candida* species.

PATIENTS AND METHODS

We reviewed all positive blood cultures for *Candida* species performed
at Memorial Sloan-Kettering Cancer Center from January 1, 1981 to November
15, 1989. Blood cultures were performed using a standard two-bottle broth
system prior to September 1981. After that time a lysis-centrifugation
system was also used in special circumstances but not routinely until April
1984.

Medical records of HIV and AIDS patients with positive blood cultures for *Candida* species from January 1, 1981 to November 15, 1989 were reviewed for the following information: HIV risk factor; AIDS-related diagnoses; Oral thrush by clinical exam and/or culture; *Candida* esophagitis by esophagoscopy or based on clinical impression; Documentation of other *Candida* infection by culture or biopsy; anti-candidal therapy at time of candidemia; presence (and type) of any intravascular catheter; evidence of neutropenia as defined by an absolute neutrophil count of 1000 or less at the time blood cultures were drawn and the reason for neutropenia; use of broad-spectrum antibiotics when candidemia was documented; corticosteroid therapy and indication for its use; sites of positive cultures for *Candida*; candidemia defined as any positive blood culture for *Candida* species and quantitative colony counts if available; record of clinical diagnosis of candidiasis prior to death; treatment of candidiasis and outcome. Response to treatment was defined as clearing of candidemia and resolution of symptoms and signs of infection and/or no evidence of *Candida* by culture or histopathology on autopsy. Failure of treatment was defined as persistent candidemia, focal or disseminated organ involvement by *Candida* on autopsy, or death during therapy.

RESULTS

There were 1667 positive blood cultures for *Candida* species in 638 patients at Memorial Sloan-Kettering Cancer Center from January 1, 1981 until November 15, 1989. Thirty of the positive cultures were found in 13 patients with CDC-defined AIDS (Table 1.). These patients were male. Risk factor for AIDS was homosexuality in 10 patients, IVDU in 1 patient, homosexuality and IVDU in 1 patient, and unidentified in 1 patient. All patients had history or evidence at autopsy of opportunistic infections. The most common were cytomegalovirus (CMV) in 8 patients, *Pneumocystis carinii* pneumonia (PCP) in 7, *Mycobacterium avium* (MAI) in 4, tuberculosis (TB) in 2, toxoplasmosis (Toxo) of the central nervous system in 2 and one patient each with disseminated herpes simplex and disseminated histoplasmosis (Histo). Kaposi's sarcoma (KS) was diagnosed in 7 patients. Other malignancies included CNS lymphoma in 2 patients and T-cell lymphoblastic lymphoma/leukemia in 1 patient. Eight of these patients were examined post-mortem.

Colonization with *Candida* (*Candida albicans* in all cases) was seen in 11 patients with oral thrush. Only one patient had both oral thrush and *Candida* esophagitis. Other culture sites which grew *Candida* species were expectorated sputum or bronchial washings obtained during bronchoscopy, throat or pharynx, urine and stool. Of note were 3 catheter tip cultures positive for the same species of *Candida* obtained by blood cultures.

Intravascular catheters were used in 12 patients: 4 patients with Broviacs, 2 with Mediports, 3 with central lines for venous access and/or parenteral nutrition, and the remaining 3 patients had peripheral catheters only. Only one patient did not have any catheters at the time of positive blood culture. His isolate was a *Candida parapsilosis*. He was treated with only 11 milligrams of amphotericin B and showed no further evidence of an invasive *Candida* infection.

Neutropenia occurred in 7 patients. It was due to ganciclovir therapy for CMV retinitis in 2 patients and treatment with interferon, antibiotics, zidovudine, foscarnet or cytotoxic, anti-cancer chemotherapy was given in the others. Antibiotics were being given to 11 of the 13 patients at the time of candidemia. Three patients received adrenocorticosteroids, respectively for respiratory failure associated with *Pneumocystis carinii* pneumonia, possible adrenal insufficiency, and as part of an anti-lymphoma chemotherapeutic regimen. Three patients were being treated with ketoconazole for oral thrush at the time of fungemia. Serum levels were not obtained.

Candida albicans was isolated in 8 patients, *Candida parapsilosis* in 4 patients and *Candida krusei* in 1. There were 5 patients with positive Broviac blood cultures, 2 with positive Mediport blood cultures,

Table 1. AIDS patients with candidemia

| Patient No. | Year | Age | HIV Risk Factor | Oral Thrush | Candida esophagitis | Candida Risk Factors | | | | Use of Ketoconazole | No. (+) Blood cultures |
						IV catheter	Neutropenia	Antibiotics	Steroids		
1	1982	37	Homosexual	+	-	T	+I	+	-	+	3
2	1984	36	Homosexual	+	-	B	-	+	-	-	8
3	1986	23	Unknown	-	-	+	+K	+	+	-	10
4	1986	31	Homosexual, IVDU	+	-	B	-	+	+	-	4
5	1987	45	Homosexual	-	-	B	+G	+	-	-	1
6	1987	35	Homosexual	+	-	+	-	+	-	-	1
7	1989	44	Homosexual	+	-	S	+A	+	+	+	2
8	1989	54	Homosexual	+	+	C	+Z	+	-	-	6
9	1989	52	Homosexual	+	-	+	-	+	-	-	1
10	1989	32	IVDU	+	-	B	-	+	-	-	1
11	1989	37	Homosexual	+	-	-	-	-	-	-	1
12	1989	35	Homosexual	+	-	M	+G	-	-	+	2
13	1989	34	Homosexual	+	-	M	+F	+	-	-	3

Key : * see text
T=TPN; B=Broviac; M=Mediport; S=Intensive care; C=Central line; I=Interferon; K=Chemo; G=Ganciclovir; A=Antibiotics; Z=Zidovudine; F=Foscarnet; Am=Ampho B; R=Removal of catheter; 5=Flucytosine; CMV=cytomegalovirus; KS=Kaposi's sarcoma; PCP=*Pneumocystis carinii* pneumonia; MAI=*Mycobacterium avium*; Toxo=toxoplasmosis; TB=tuberculosis; Histo=histoplasmosis

Table 1. AIDS patients with candidemia (continued)

Type (CFU/ml)	Candida species	Other (+) cultures	Ampho/ Removal Treatment	Sites of Candidiasis on Autopsy	Other Diagnoses/ Autopsy Findings
Peripheral (Unquan)	C.albicans	Catheter tip, Urine	+Am,R*	Disseminated by culture and histopathology	CMV, KS, PCP, Toxo
Broviac and Peripheral (Unquan)	C.albicans	Nasopharynx, Stool	+Am	None	Cryptosporidia, CMV, Histo, Isospora, Toxo
Peripheral (0.1-0.9)	C.krusei	Arterial catheter tip, Urine	+Am,5	Disseminated by culture and histopathology	Lymphoma, MAI
3 Broviac (13,0.9,0.4), 1 Peripheral (Unquan)	C.albicans	Bronch wash, Throat, Sputum, Stool	-	Kidneys	CMV, KS, CNS Lymphoma, MAI
Broviac (81)	C.albicans	None	+R	None	CMV, KS
Broviac (0.2)	C.parapsilosis	Palate, Sputum (C.albicans)	-	None	MAI, PCP
Peripheral (4,4)	C.albicans	Femoral catheter tip, Sputum, Urine	+Am,R	Lung	CMV, KS, PCP
4 Peripheral (0.1,0.6,1,2), 2 CVP (0.2,0.4)	C.albicans	Stool, Urine	-	Disseminated by culture and histopathology, Blood	CMV, MAI, TB, PCP
Peripheral (0.3)	C.albicans	Bronch wash	+Am	No autopsy	CMV, HSV, PCP
Broviac (0.1)	C.parapsilosis	Throat	+Am,R	No autopsy	Cryptococcal meningitis
Peripheral (2)	C.parapsilosis	None	+Am*	Alive	KS, PCP
Mediport (56,3)	C.albicans	None	+Am	No autopsy	CMV, KS, ?CNS lymphoma, TB, PCP
2 Mediport (1,133), 1 Peripheral (>1000)	C.parapsilosis	None	+Am,R	Alive	Cryptosporidia, CMV, KS

and 1 patient with positive central line culture for a total of 8 patients with positive cultures from indwelling intravenous catheters. Three other patients had positive cultures of catheter tips although no positive cultures were obtained through these catheters. Only 2 patients had a single positive peripheral blood culture for *C. albicans* and *C. parapsilosis* respectively.

Treatment for candidemia was instituted in 10 patients (Table 2). Seven patients responded to treatment with clearing of subsequent blood cultures. Of these, two had no evidence of candidiasis at autopsy and a third patient remains alive without evidence of candidiasis. Four of the responders received amphotericin B (Patient # 11 received only 11 milligrams of amphotericin B because his isolate was considered a contaminant). Both amphotericin B and catheter removal were instituted in 2 patients who responded to treatment. One treated patient had removal of the catheter as the only therapeutic intervention.

Three patients were deemed treatment failures. Two died of disseminated candidiasis and the third patient had pulmonary candidiasis at autopsy (Patient #1 received <50 mg of amphotericin B which was stopped at the request of the patient's family).

No treatment was given to 2 patients with candidemia, one with renal microabscesses discovered on autopsy and another (whose isolate was considered a contaminant) with no evidence of candidiasis at autopsy.

Candida appeared to cause the death of 3 patients with candidemia. It contributed to the deaths of 2 candidemic patients.

Seven of the 13 candidemias occurred during 1989, two each in 1987 and 1986, and one each in 1984 and 1982 (Table 1).

Table 2. Outcome of AIDS patients with candidemia

	Total	Response/Failure
Treated: (N=10)		
Amphotericin B only	4	4 / -
Ampho B and Flucytosine	1	- / 1
Removal of catheter only	1	1 / -
Both Amphotericin B		
and removal of catheter	4	2 / 2
Untreated	3	- / -
Candidemia as cause of death	3 of 13	
Candidemia as contributing to death	2 of 13	

DISCUSSION

We have followed 903 AIDS patients of whom 13 patients were documented to have fungemia with *Candida* species. Intravenous catheters were present in 12 of 13 patients and use of broad-spectrum antibiotic therapy was present in 11 of 13. One patient (#11) did not have a catheter. An admission blood culture drawn because of fever was positive for *Candida parapsilosis*. He was treated with only 11 mg of amphotericin B and had no subsequent positive cultures. This was assumed to be a contaminant. Seven of the patients were also neutropenic prior to or during the episode of candidemia. This data is consistent with previous reports of candidemia occurring in 1% of patients with AIDS (Whimbey et al., 1986).

Mucocutaneous candidiasis due to impaired cell-mediated immunity is a frequent complication of HIV infection. The rare instances of disseminated candidasis in AIDS patients suggest that mechanisms other than HIV infection (i.e. intravenous catheters, neutropenia, parenteral nutrition) must be present for fungemia to occur (Whimbey et al., 1987; Karabinis et al., 1988)

As expected, *Candida albicans* was the most frequently isolated species in blood as well as from sites of colonization such as the oropharynx, sputum and stool. *Candida parapsilosis* was the second most common blood culture isolate. Previous studies associate this species with indwelling catheters, particularly those used for total parenteral nutrition (Horn et al., 1985). *Candida krusei* was isolated in multiple cultures of the patient receiving chemotherapy for lymphoma, a common setting for fungemia with this species (Horn et al., 1985).

Simultaneous blood cultures from catheter and peripheral blood were helpful in determining probable sources of candidemia. In patients #5, #6, #10, and #12 positive cultures from the catheter only are suggestive that the catheter was the source of candidemia. In patient #13, a peripheral blood culture with >1000 CFU/ml was drawn simultaneously with a Mediport blood culture which was sterile. We suspect that mislabelling of the specimens at the time they were drawn led to this unexpected result. The patient's only other positive cultures were both from the Mediport and suggest a primary catheter infection. Patients #4 and #8 had varying colony counts from Broviac catheters and peripheral blood which did not elucidate the source of fungemia. The patients (#1, #3, #7, #9, #11) with only peripheral blood cultures positive for *Candida* could not definitively be identified as catheter-related although patients #1, #3, and #7 had growth of *Candida* from catheter tips. These catheters could have been colonized during transient fungemia.

Ketoconazole therapy prior to the development of candidemia did not appear to alter the course of illness in 3 patients. Blood levels were not measured to assure absorption of the drug. Treatment with antibiotics alone or catheter removal alone or both appeared to be effective in 7 of 10 patients. Two patients died despite institution of appropriate therapy. Patient #1 was treated with removal of the catheter but, at his family's request, amphotericin B therapy was discontinued. The mortality of 5 out of 13 patients (3 of 10 treated patients, 2 of 2 untreated patients) is not significantly different from that of fungemia in patients with neoplastic disease (Whimbey et al., 1987; Karabinis, 1988; Horn et al., 1985)

A review of the literature reveals 47 reported cases of disseminated candidiasis in AIDS (Whimbey et al., 1986; *). This includes only 5 cases with documented candidemia in AIDS patients, one of which is catheter-related (Whimbey et al., 1986; Alsina et al., 1988). Detailed information regarding the risk factors for fungemia was not available in the other 4 cases.

Of note is the cluster of 7 cases in 1989 compared to 6 cases in the years 1981-1988. It is likely to be related to the increased use of indwelling catheters for therapy of CMV retinitis or cryptococcal meningitis in the past two years and the concomitant neutropenia secondary to treatments (ganciclovir, foscarnet) for CMV.

Candidemia occurs rarely in AIDS patients despite frequent colonization. It appears to be more common in patients with indwelling catheters and neutropenia. *Candida albicans* and *Candida parapsilosis* account for the majority of isolates. Treatment with any modality (amphotericin B alone or in combination with catheter removal) was successful in 7 of 10 patients with candidemia. Candidemia may be occurring more frequently with the increasing use of indwelling catheters and myelosuppressive therapies for AIDS and opportunistic infections.

---

*Heinemann et al., 1987; Horn et al., 1985; Hui et al., 1984; Karabinis et al., 1988; Klatt, 1988; Levy et al., 1985; Loureiro et al., 1988; Marchevsky et al., 1985; Masur et al., 1981; Matthews et al., 1988; Mobley et al., 1985; Moskowitz et al., 1985; Niedt and Schinella, 1985; Oleske et al., 1983; Reichert et al., 1983; Siegal et al., 1981; Welch et al., 1984; Whimbey et al., 1986; Whimbey et al., 1987; Wilkes et al., 1988; Zimmerli et al., 1988.

REFERENCES

Alsina, A., Mason, M., Uphoff, R.A., Riggsby, W.S., Becker, J.M., and Murphy, D., 1988, Catheter-associated *Candida utilis* fungemia in a patient with acquired immunodeficiency syndrome: Species verification with a molecular probe, J. Clin. Microbiol., 26:621.

Ambros, R.A., Lee, E.Y., Sharer, L.R., Khan, M.Y., and Robboy, S.J., 1987, The acquired immunodeficiency syndrome in intravenous drug abusers and patients with a sexual risk: Clinical and postmortem comparisons, Hum. Pathol., 18:1109.

Armstrong, D., 1989, Problems in management of opportunistic fungal diseases, Rev. Infect. Dis., 11:S1591.

Belman, A.L., Ultmann, M.H., Horoupian, D., Novick, B., Spiro, A.J., Rubinstein, A., Kurtzberg, D., and Cone-Wesson, B., 1985, Neurological complications in infants and children with acquired immune deficiency syndrome, Ann. Neurol., 18:560.

Centers for Disease Control, 1987, Revision of the CDC Surveillance Case Definition for Acquired Immunodeficiency Syndrome, M.M.W.R., 36:3S.

Ehni, W.F., Ellison, R.T., 1987, Spontaneous *Candida albicans* meningitis in a patient with the acquired immune deficiency syndrome, Am. J. Med., 83:806.

Eng, R.H.K., Bichburg, E., Smith, S.M., Geller, H., and Kapila, R., 1986, Bacteremia and fungemia in patients with acquired immune deficiency syndrome, Am. J. Clin. Pathol., 86:105.

Goetz, D.W., Hall, S.E., Harbison, R.W., and Reid, M.J., 1988, Pediatric acquired immunodeficiency syndrome with negative human immunodeficiency virus antibody response by enzyme-linked immunosorbent assay and Western blot, Pediatrics, 81:356.

Gottlieb, M.S., Schroff, R., Schanker, H.M., Weisman, J.D., Fan, P.T., Wolf, R.A., and Saxon, A., 1981, *Pneumocystis carinii* pneumonia and mucosal candidiasis in previously healthy homosexual men, New Engl. J. Med., 305:1425.

Heinemann, M.H., Bloom, A.F., Horowitz, J., 1987, *Candida albicans* endophthalmitis in a patient with AIDS, Arch. Ophthal., 105:1172.

Horn, R., Wong, B., Kiehn, T.E., Armstrong, D., 1985, Fungemia in a cancer hospital: Changing frequency, earlier onset and results of therapy, Rev. Infect. Dis., 7:646.

Hui, A.N., Koss, M.N., Meyer, P.R., 1984, Necropsy findings in acquired immuno-deficiency syndrome: A comparison of premortem diagnoses with postmortem findings, Hum. Pathol., 15:670.

Karabinis, A., Hill, C., Leclercq, B., Tancrede, C., Baume, D., and Andremont, A., 1988, Risk factors for candidemia in cancer patients: A case-control study, J. Clin. Microbiol., 26:429.

Klatt, E.C., 1988, Diagnostic findings in patients with acquired immune deficiency syndrome (AIDS), 1988, J. AIDS, 1:459.

Levy, R.M., Bredesen, D.E., Rosenblum, M.L., 1985, Neurological manifestations of the acquired immunodeficiency syndrome (AIDS): Experience at UCSF and review of the literature, J. Neurosurg., 62:475.

Loureiro, C., Gill, P.S., Meyer, P.R., Rhodes, R., Rarick, M.U., and Levine, A.M., 1988, Autopsy findings in AIDS-related lymphoma, Cancer, 62:735.

Marchevsky, A., Rosen, M.J., Chrystal, G., and Kleinerman, J., 1985, Pulmonary complications of the acquired immunodeficiency syndrome: A clinicopathologic study of 70 cases, Hum. Pathol., 16:659.

Masur, H., Michelis, M.A., Greene, J.B., Onorato, I., Stouwe, R.A., Holzman, R.S., Wormser, G., Brettman, L., Lange, M., Murray, H.W., and Cunningham-Rundles, S., 1981, An outbreak of community-acquired *Pneumocystis carinii* pneumonia: Initial manifestation of cellular immune dysfunction, New Engl. J. Med., 305:1431.

Matthews, R., Burnie, J., Smith, D., Clark, I., and Midgley, J., Conolly, M., and Gazzard, B., 1988, *Candida* and AIDS: Evidence for a protective antibody, Lancet, 2:263.

Mobley, K., Rotterdam, H.Z., Lerner, C.W., and Tapper, M.L., 1985 Autopsy findings in the acquired immune deficiency syndrome, Pathol. Ann., 20(Pt 1):45.

Moskowitz, L., Hensley, G.T., Chan, J.C., and Adams, K., 1985, Immediate causes of death in acquired immunodeficiency syndrome, Arch. Pathol. Lab. Med., 109:735.

Niedt, G.W., Schinella, R.A., 1985, Acquired immunodeficiency syndrome: Clinicopathologic study of 56 autopsies, Arch. Pathol. Lab. Med., 109:727.

Oleske, J., Minnefor, A., Cooper, R., Thomas, K., dela Cruz, A., and Ahdieh, H., 1983, Immune deficiency syndrome in children, J.A.M.A., 249:2345.

Reichert, C.M., O'Leary, T.J., Levens, D.L., Simrell, C.R., and Macher, A.M., 1983, Autopsy pathology in the acquired immune deficiency syndrome, Am. J. Pathol., 112:357.

Siegal, F.P., Lopez, C., Hammer, G.S., Brown, A.E., Kornfeld, S.J., Gold, J., Hassett, J., Hirschman, S.Z., Cunningham-Rundless, C., Adelsberg, B.R., Parham, D.M., Siegal, M., Cunningham-Rundles, S., and Armstrong, D., 1981, Severe acquired immune deficiency in male homosexuals, manifested by chronic perianal ulcerative *Herpes simplex* lesions, New Engl. J. Med., 305:1439.

Welch, K., Finkbeiner, W., Alpers, C.E., Blumenfeld, W., Davis, R.L., Smuckler, E.A., and Beckstead, J.H., 1984, Autopsy findings in the acquired immune deficiency syndrome, J.A.M.A., 252:1152.

Whimbey, E., Gold, J.W.M., Polsky, B., Dryjanski, J., Hawkins, C., and Blevins, A., 1986, Bacteremia and fungemia in patients with the acquired immunodeficiency sydrome, Ann. Intern. Med., 104:511.

Whimbey, E., Kiehn, T.E., Brannon, P., Blevins, A., and Armstrong, D., 1987, Bacteremia amd fungemia in patients with neoplastic disease, Am. J. Med., 82:723.

Wilkes, M.S., Fortin, A.H., Felix, J.C., Godwin, T.A., and Thompson, W.G., 1988, Value of necropsy in acquired immunodeficiency syndrome, Lancet, 2:85.

Zimmerli, W., Bianchi, L., Gudat, F., Spichtin, H., Erb, P., von Planta, M, and Heitz, P.U., 1988, Disseminate *Herpes simplex* type 2 and systemic *Candida* infection in a patient with previous asymptomatic human immunodeficiency virus infection, J. Infect. Dis., 157:597.

IMNUNOLOGICAL ASPECTS OF CANDIDOSIS IN AIDS PATIENTS

David W. Warnock

Department of Microbiology Bristol Royal Infirmary Bristol, U.K.

INTRODUCTION

Candida albicans, the most common cause of human candidosis, is found as a commensal in the mouth and gastrointestinal tract of a substantial proportion of the normal population. This opportunist pathogen can cause deep-seated infection, which can be localized or disseminated, in immunosuppressed or debilitated individuals, but is more often seen causing superficial cutaneous or mucosal infection.

Mucosal forms of candidosis are among the most common and persistent infections encountered in HIV-infected men and women. Oral candidosis occurs in 40-90% of HIV-infected individuals at some time during the course of their illness, becoming more prevalent as patients progress towards AIDS (Torssander et al., 1987; Korting et al., 1988). The development of oral infection with *C. albicans* is often the initial clinical manifestation in asymptomatic HIV-infected patients and is one of several clinical signs that have been associated with an increased likelihood of progression to AIDS (Klein et al., 1984). Vaginal candidosis is another condition which appears to become more common as HIV-infected women progress towards AIDS. One recent report stated that 70% of a group of 24 women with AIDS suffered from this form of *C. albicans* infection (Carpenter et al., 1989). According to an earlier report, almost all HIV-infected women with vaginal candidosis will progress to AIDS (Rhoads et al., 1987).

Infection with *C. albicans* is the most common cause of oesophagitis in patients with AIDS. Oral infection often spreads to the oesophagus in HIV-infected individuals; oesophageal candidosis can occur without oral involvement, but this is unusual (Tavitian et al., 1986; Levine et al., 1987). In one recent report, oesophageal infection with *C. albicans* was the most common AIDS-defining condition among a group of HIV-infected women (Carpenter et al., 1989). Other localized forms of deep candidosis have been described in occasional AIDS patients, but disseminated deep infection has seldom been encountered. In the few AIDS patients who have developed disseminated candidosis, prolonged vascular catheterization has been a major predisposing factor.

The mechanisms that defend the normal host against *C. albicans* infection are complicated, diversified, and often interdependent. Efficient protection against this ubiquitous resident of the gastrointestinal tract is believed to involve both cell-mediated and humoral immunological mechanisms. Nonspecific mechanisms are also important, but it is well recognized that the contribution of particular elements to protection against mucosal and deep forms of candidosis is different.

Mycoses in AIDS Patients, Edited by
H. Vanden Bossche *et al.,* Plenum Press, New York, 1990

IMMUNOLOGICAL ASPECTS OF MUCOSAL CANDIDOSIS IN AIDS PATIENTS

It has long been recognized that normal T-cell function is required for efficient host protection against superficial forms of *C. albicans* infection, because individuals with congenital disorders in which T-cell proliferation or function are impaired often develop persistent mucosal, cutaneous or ungual infection with this fungus (Kirkpatrick, 1984). Like patients with HIV infection, individuals with congenital T-cell defects often develop persistent oral and oesophageal candidosis, but seldom develop deep candidosis. Unlike HIV infected individuals, however, most patients with chronic mucocutaneous candidosis have normal numbers of circulating T-cells (Kirkpatrick, 1984). The most common defects in these patients involve subnormal T-cell activation or subnormal production of T-cell factors needed for macrophage activation (Kirkpatrick, 1984). These defects are often limited to antigens of *C. albicans*, but some patients have more profound defects that involve the T-cell-mediated response to other antigens as well.

The pseudomembranous form of candidosis is the most common orofacial manifestation of *C. albicans* infection in HIV-infected individuals (Schulten et al., 1989). The lesions are persistent and often spread to affect all parts of the mouth. Other less prevalent forms of oral infection with *C. albicans* in AIDS patients have included chronic erythematous (atrophic) candidosis and chronic hyperplastic candidosis; angular cheilitis has also been seen (Schulten et al., 1989). In HIV-infected individuals with the pseudomembranous form of oral candidosis, as in other patients with this infection, penetration of *C. albicans* is limited to the stratum corneum of the host epithelium. This suggests that the mechanisms that prevent the fungus from penetrating beyond the host epithelium are intact. However, the mechanisms that protect the mucosal epithelium itself are impaired.

In the normal host, numerous non-specific mechanisms, such as epithelial cell desquamation, help to defend the mouth against infection. Like the mucosal secretions of the intestinal tract and lower genital tract, the fluid bathing the mouth contains high concentrations of IgA. This immunoglobulin forms an important barrier against infection, regulating the resident microbial population and helping to prevent potential pathogens from becoming established on the mucosal epithelium (McNabb and Tomasi, 1981).

It has long been recognized that an important function of the IgA found in mucosal secretions is the prevention of bacterial attachment to mucosal epithelial cells (Williams and Gibbons, 1972). It has been observed that *C. albicans* cells recovered from the mouth or lower genital tract of colonized or infected individuals are often coated with IgA (Epstein et al., 1982; Gough et al., 1984). Moreover, fungal cells coated with anti-*C. albicans* IgA have been found to attach in much lower numbers to human oral epithelial cells (Epstein et al., 1982; Vudhichamnong et al., 1982). These observations suggest that IgA secretion might be an important element in host protection against mucosal *C. albicans* infection. Additional support for this suggestion is derived from reports that parotid secretions obtained from patients with oral candidosis often contain higher concentrations of anti-*C. albicans* IgA than similar specimens from colonized or non-colonized individuals (Epstein et al., 1982). In contrast, reduced concentrations of specific IgA have been detected in parotid secretions from patients with chronic mucocutaneous candidosis (Lehner et al., 1972).

Numerous different cell lines are involved in the differentiation and maturation of mucosal IgA-secreting cells and in the subsequent regulation of immunoglobulin production (Bienenstock and Befus, 1980; Lamm et al., 1982). Helper T-cells are central to the immunological mechanisms that defend the host against mucosal infection, recognizing antigen on antigen-presenting cells; and on stimulation secreting factors that regulate B-cell proliferation and maturation into immunoglobulin-secreting cells. Following antigen penetration of the mucosal epithelium and the action of antigen-presenting cells, a distinct subset of helper T-cells induces the differentiation of IgA-committed B-cells. It is believed that these stimulated B-cells then migrate from their mucosal site of origin, entering the circulation and spreading throughout the host before returning to their site of origin or settling in other mucosal sites. Terminal differentiation to IgA-secreting cells then occurs. In addition to helper T-cells, mucosal tissue also contains IgA-specific suppressor T-cells that increase in number as IgA levels rise and which help to limit the production of this class of immunoglobulin following mucosal stimulation with antigen.

HIV infection results in numerous immunological defects. In particular, it leads to profound depletion of the helper T-cell population and this results in a marked reduction in the ratio of T helper to suppressor cells in the circulation (Lane and Fauci, 1985). Oral colonization with *C. albicans* is more common in individuals with lower than normal helper T-cell numbers or low helper/suppressor T-cell ratios (Schonheyder et al., 1984; Melbye et al., 1985). Oral infection is more prevalent in individuals with low helper/suppressor T-cell ratios in the circulation: one recent report noted that 21 of 22 HIV-infected patients with oral candidosis had a ratio of less than 1; 15 of these patients had a ratio of less than 0.5 (Korting et al., 1988). These observations suggest that T-cell-regulated immunological mechanisms are an important element in host protection against mucosal colonization and infection with *C. albicans*.

The oral mucosal epithelium and underlying corium of HIV-infected patients contain higher than normal numbers of helper and suppressor T-cells (Becker et al., 1988). However, the oral mucosal helper/suppressor T-cell ratio is much lower than normal in HIV-infected patients. The intestinal mucosal helper/suppressor T-cell ratio is also lower than normal in patients with HIV infection (Rodgers et al., 1986; Budhraja et al., 1987).

These alterations might well account for the reduced number of IgA-secreting cells found in mucosal specimens from the intestinal tract of HIV-infected individuals (Kotler et al., 1987). This group studied 39 men and women with AIDS or ARC, all of whom were being treated for oral candidosis, and found that the number of intestinal IgA-secreting cells was more than 50% lower than normal in two thirds, and more than 90% lower in one third. In addition, Kotler et al. (1987) investigated parotid IgA secretion in a further group of 10 AIDS patients with oral candidosis. It was found that although the total output of IgA was greater than normal, the proportion secreted in an unbound functional form was much reduced.

This disruption of mucosal IgA production appears to result in impairment of the barrier function of the mucosal secretions. This could well lead to increased attachment of *C. albicans* cells to the host epithelium and to increased antigen absorption which might further compound the underlying immunological impairment (see below). Taken together, these alterations in mucosal T- and B-cell numbers and function might well explain the development of persistent mucosal infection with *C. albicans* in HIV-infected individuals. Added to this immunological impairment, about 10-15% of AIDS patients suffer from reduced salivation (Silverman et al., 1986; Schiodt and Pindborg, 1987). It is well established that dryness of the mouth can lead to increased levels of oral colonization and infection with *C. albicans* (MacFarlane and Mason, 1974).

MN cells of the monocyte-macrophage lineage have numerous important functions in host protection against infection, including the processing and presentation of antigen to T-cells, and the secretion of factors that induce T-cell activation. In addition to their non-specific phagocytic function, MN cells function as microbicidal effector cells in T-cell-mediated immunological reactions. HIV infection is associated with a significant reduction in the number and function of antigen-presenting MN cells in the circulation (Shannon et al., 1985; Tsang et al., 1987). It is also associated with a marked reduction in the number of antigen-presenting dendritic cells in lymphoid organs and in the number of Langerhans cells in the epidermis (Belsito et al., 1984).

HIV infection is associated with other alterations in MN cell function. Tests on peripheral blood MN cells from HIV infected individals have indicated that their chemotactic, phagocytic and microbicidal function is often impaired (Smith et al., 1984; Poli et al., 1985; Estevez et al., 1986; Bender et al., 1988). These alterations could be a direct result of the virus infection (like helper T-cells, MN cells often bear the CD4 antigen which acts as the receptor for HIV attachment), a direct effect of another pathogen through either infection or antigen stimulation, or an indirect effect owing to the loss of the normal T-cell regulation.

It is well established that the development of certain infections in patients with AIDS is a direct result of abnormal T-cell-macrophage interactions and function which permit the multiplication of the pathogen within MN phagocytic cells that have not been activated. The extent to which defects in these host cell interactions are implicated in the development of candidosis in HIV-infected individuals is less well understood, but it is clear from other groups of patients that abnormal interactions could be involved in the development of persistent mucosal infection with *C. albicans*. Witkin et al. (1986) found that although mitogen-induced T-cell proliferation was normal in women with recurrent vaginal candidosis, specific *C. albicans*-induced proliferation was at least 70% below normal in most patients. This effect appeared to be due to abnormal macrophage function which resulted in the secretion of prostaglandin E2 in sufficient concentrations to inhibit T-cell proliferation.

IMMUNOLOGICAL ASPECTS OF DEEP CANDIDOSIS IN AIDS PATIENTS

It has long been recognized that individuals with congenital or acquired defects of T-cell function seldom develop lethal deep forms of *C. albicans* infection. These infections occur in two distinct groups of patients. The first consists of individuals rendered neutropenic as the result of an underlying malignant condition or its treatment: the gastrointestinal tract is the principal source of *C. albicans* infection in this group and the liver, spleen and lungs are often involved. Most of the second group are surgical patients: these individuals are debilitated, but not neutropenic: disruption of natural anatomical barriers permits the fungus to gain access to the circulation. Most AIDS patients who have developed deep forms of candidosis belong in this second group.

Neutrophil polymorphonuclear (PMN) leucocytes form the earliest and most efficient non-specific mechanism for eliminating *C. albicans* cells that transgress the epithelium of the host. In addition to their non-specific phagocytic function, neutrophils possess receptors which permit them to function as microbicidal effector cells in specific immunological reactions. These receptors augment the process of phagocytosis by assisting with the ingestion of *C. albicans* cells coated with specific IgG or activated C3. Neutrophils can attach to hyphal forms of *C. albicans* and then release factors that damage the fungus, a process that is enhanced with anti-*C. albicans* IgG (Diamond et al., 1978).

Neutrophils possess several microbicidal mechanisms: oxidative mechanisms are most important (Lehrer, 1970; Diamond et al., 1980), but non-oxidative mechanisms are also involved (Lehrer, 1972). Defects in certain of these mechanisms have been associated with the development of deep forms of candidosis (Lehrer and Cline, 1971). In contrast, impaired neutrophil function has seldom been described in patients with chronic mucocutaneous candidosis (Djawari et al., 1977).

It has been observed that in HIV-infected individuals with the pseudomembranous form of oral candidosis, as in other patients with this condition, fungal penetration is restricted to the outermost cornified layer of the epithelium. The intense neutrophilic infiltration that develops in the underlying corium might well be important in preventing the fungus from entering the circulation and causing localized or disseminated deep infection.

Tests on HIV-infected individuals with no other concurrent infection have demonstrated that patients with ARC tend to have lower neutrophil counts than patients who have progressed to AIDS: one recent report noted that 18% of the former group, but none of the latter were neutropenic (Ellis et al., 1988). Impairment of neutrophil chemotactic, phagocytic and microbicidal function has been described in several reports (Lazzarin et al., 1986; Nielsen et al., 1986; Ellis et al., 1988; Murphy et al., 1988). It appears that although neutrophil function is often subnormal in AIDS patients, the defects that occur are not sufficient to prevent the PMN cells from contributing to the containment of *C. albicans*. However, should an overwhelming number of fungal cells be introduced into the circulation, as can occur when vascular catheters become infected, a debilitated individual, such as an AIDS patient, will often succumb to fatal disseminated *C. albicans* infection.

Matthews et al. (1984, 1987) have demonstrated that patients who recover from disseminated candidosis have significant levels of specific serum IgM or IgG to an immunodominant 47kD antigen of *C. albicans* while those who die from this infection have insignificant, fading or no immunoglobulins to this antigen. Immunodominant antigens with similar molecular weights have been described in several other reports (Greenfield and Jones, 1981; Strockbine et al., 1984).

Unlike non-HIV-infected individuals with superficial forms of candidosis, less than half of whom possess serum IgM or IgG against the 47kD antigen (Matthews et al., 1987), 34 AIDS patients with oral or oesophageal candidosis and 19 of 20 ARC patients with oral candidosis were found to have serum IgM to this antigen (Matthews et al., 1988). Increased immunoglobulin production is a characteristic finding in individuals with recent HIV infection owing to activation of existing B-cell populations (Lane and Fauci, 1985). The mechanism for this B-cell activation is unclear, but it could be due to direct viral stimulation (HIV or EBV), or to chronic antigen stimulation. The continued production of anti-*C. albicans* IgM in patients who have progressed to AIDS appears to be a result of chronic antigen stimulation. Matthews et al. (1988) have speculated that specific IgM to the 47kD antigen might be helping to prevent the fungus from disseminating beyond the gastrointestinal tract. This is an interesting suggestion, but one which needs further investigation to establish whether or not anti-*C. albicans* IgM is more than a marker of some other immunological mechanism.

IMMUNOMODULATING EFFECTS OF *C. ALBICANS* IN AIDS PATIENTS

Persistent mucocutaneous infection with *C. albicans* is often associated with depression of cell-mediated immunological reactions to the fungus. Although congenital and acquired T-cell disorders are important in predisposing the host to this infection, it appears that structural components of the fungus itself might also be contributing to the underlying immunosuppression (Domer et al., 1988).

Mannan, a major cell wall component of *C. albicans*, is often released into the circulation of patients with deep forms of candidosis (Weiner and Coats-Stephen, 1979). Mannan has also been detected in serum from several children with chronic mucocutaneous candidosis, and has been held to account for the subnormal *C. albicans*-stimulated T-cell proliferation observed in these patients (Fischer et al., 1978). In a subsequent paper, Fischer et al. (1982) suggested that mannan can impair MN cell function thus blocking its own presentation to T-cells and the consequent T-cell proliferation. Additional support for the suggestion that mannan might be involved in T-cell malfunction is derived from the observation that eradication of *C. albicans* from patients with chronic mucocutaneous candidosis by means of antifungal treatment often leads to remission of the underlying T-cell defect (Paterson et al., 1971; Kirkpatrick and Smith, 1974).

It has still to be established whether mannan (or other antigenic components) derived from *C. albicans* is contributing to the development of AIDS in HIV-infected individuals. It has, however, been observed that there is a much greater likelihood of progression to AIDS in HIV-infected patients with oral candidosis than in similar patients without *C. albicans* infection (Klein et al., 1984). Of 22 patients with low helper/suppressor T-cell ratios and oral candidosis, 13 developed AIDS at a median of 3 months, compared with none of 20 similar patients without candidosis who were followed up for a median period of 12 months. This does suggest that the fungus might be aggravating the underlying host impairment, but further work is needed to confirm this impression.

REFERENCES

Becker, J., Ulrich, P., Kunze, R., Gelderblom, H., Langford, A., and Reichart, P., 1988, Immunohistochemical detection of HIV structural proteins and distribution of T-lymphocytes and Langerhans cells in the oral mucosa of HIV infected patients, Virchows Arch. [A], 412:413.

Belsito, D.V., Sanchez, M.R., Baer, R.L., Valentine, F., and Thorbecke, G.J., 1984, Reduced Langerhans cell Ia antigen and ATPase activity in patients with the acquired immunodeficiency syndrome, N. Engl. J. Med., 310:1279.

Bender, B.S., Davidson, B.L., Kline, R., Brown, C., and Quinn, T.C., 1988, Role of the mononuclear phagocyte system in the immunopathogenesis of human immunodeficiency virus infection and the acquired immunodeficiency syndrome, Rev. Infect. Dis., 10:1142.

Bienenstock, J., and Befus, A.D., 1980, Mucosal immunology, Immunology 41:249.

Budhraja, M., Levendoglu, H., Kocka, F., Mangkornkanok, M., and Sherer, R., 1987, Duodenal mucosal T-cell subpopulation and bacterial cultures in acquired immunodeficiency syndrome, Am. J. Gastroenterol., 82:427.

Carpenter, C.C.J., Mayer, K.H., Fisher, A., Desai, M.B., and Durand, L., 1989, Natural history of acquired immunodeficiency syndrome in women in Rhode Island, Am. J. Med., 86:771.

Diamond, R.D., Krzesicki, R., and Jao, W., 1978, Damage to pseudohyphal forms of *Candida albicans* by neutrophils in the absence of serum *in vitro*, J. Clin. Invest., 61:349.

Diamond, R.D., Clark, R.A., and Haudenschild, C.C., 1980, Damage to *Candida albicans* hyphae and pseudohyphae by the myeloperoxidase system and oxidative products of neutrophil metabolism *in vivo*, J. Clin. Invest., 66:908.

Djawari, D., Bischoff, T., and Hornstein, O.P., 1977, Defect of phagocytosis and intracellular killing of *Candida albicans* by granulocytes in patients with familiar and non-familiar chronic mucocutaneous candidosis, Arch. Dermatol. Res., 260:159.

Domer, J., Elkins, K., Ennist, D., and Baker, P., 1988, Modulation of immune responses by surface polysaccharides of *Candida albicans*, Rev. Infect. Dis., 10(2):S419.

Ellis, M., Gupta, S., Galant, S., Hakim, S., VandeVen, C., Toy, C., and
    Cairo, M.S., 1988, Impaired neutrophil function in patients with
    AIDS or AIDS-related complex: comprehensive evaluation, J. Infect.
    Dis., 158:1268.
Epstein, J.B., Kimura, L.H., Menard, T.W., Truelove, E.L., and Pearsall,
    N.N., 1982, Effects of specific antibodies on the interaction
    between the fungus Candida albicans and human oral mucosa, Arch.
    Oral Biol., 27:469.
Estevez, M.E., Ballart, I.J., Diez, R.A., Planes, N., Scaglione, C., and
    Sen, L., 1986, Early defect of phagocytic cell function in subjects
    at risk for acquired immunodeficiency syndrome, Scand. J. Immunol.,
    24:215.
Fischer, A., Ballett, J.J., and Griscelli, C., 1978, Specific inhibition of
    in vitro Candida induced lymphocyte proliferation by polysaccharide
    antigens present in the serum of patients with chronic mucocutaneous
    candidiasis, J. Clin. Invest., 62:1005.
Fischer, A., Pichat, L., Audinot, M., and Griscelli, C., 1982, Defective
    handling of mannan by monocytes in patients with chronic
    mucocutaneous candidiasis resulting in a specific cellular
    unresponsiveness, Clin. Exp. Immunol., 47:653.
Gough, P.M., Warnock, D.W., Richardson, M.D., Mansell, N.J., and King,
    J.M., 1984, IgA and IgG antibodies to Candida albicans in the
    genital tract secretions of women with or without vaginal
    candidosis, Sabouraudia, 22:265.
Greenfield, R.A., and Jones, J.M., 1981, Purification and characterization
    of a major cytoplasmic antigen of Candida albicans, Infect. Immun.,
    34:469.
Kirkpatrick, C.H., 1984, Host factors in defense against fungal infections,
    Am. J. Med., 77:1.
Kirkpatrick, C.H., and Smith, T.K., 1974, Chronic mucocutaneous
    candidiasis. Immunologic and antibiotic therapy, Ann. Int. Med.,
    80:310.
Klein, R.S., Harris, C.A., Butkus Small, C., Moll, B., Lesser, M., and
    Friedland, G.H., 1984, Oral candidiasis in high risk patients as the
    initial manifestation of the acquired immune deficiency syndrome. N.
    Engl. J. Med., 311:354.
Korting, H.C., Ollert, M., Georgii, A., and Froschl, M., 1988, In vitro
    susceptibilities and biotypes of Candida albicans isolates from the
    oral cavities of patients infected with human immunodeficiency
    virus, J. Clin. Microbiol., 26:2626.
Kotler, D.P., Scholes, J.V., and Tierney, A.R., 1987, Intestinal plasma
    cell alterations in acquired immunodeficiency syndrome, Dig. Dis.
    Sci., 32:129.
Lamm, M.E., Roux, M.E., McWilliams, M., and Phillips-Quagliata, J.M., 1982,
    Differentiation and migration of mucosal plasma cell precursors, in:
    "Recent advances in mucosal immunity", W. Strober, L.A. Hanson, K.W.
    Sell, eds., Raven Press, New York.
Lane, H.C., and Fauci, A.S., 1985, Immunologic abnormalities in the
    acquired immunodeficiency syndrome, Ann. Rev. Immunol., 3:477.
Lazzarin, A., Uberti Foppa, C., Galli, M., Mantovani, A., Poli, G.,
    Franzetti, F., and Novati, R., 1986, Impairment of polymorphonuclear
    leukocyte function in patients with acquired immunodeficiency
    syndrome and with lymphadenopathy syndrome, Clin. Exp. Immunol.,
    65:105.
Lehner, T., Wilton, J.M.A., and Ivanyi, L., 1972, Immunodeficiencies in
    chronic mucocutaneous candidosis, Immunology, 22:775.
Lehrer, R.I., 1970, Measurement of candidacidal activity of specific
    leukocyte types in mixed cell populations. I. Normal,
    myeloperoxidase-deficient and chronic granulomatous disease
    neutrophils, Infect. Immun., 2:42.
Lehrer, R.I., 1972, Functional aspects of a second mechanism of
    candidacidal activity by human neutrophils, J. Clin. Invest.,
    51:2566.

Lehrer, R.I., and Cline, N.J., 1971, Leukocyte candidacidal activity and resistance to systemic candidiasis in patients with cancer, Cancer, 27:1211.

Levine, M.S., Woldenberg, R., Herlinger, H., and Laufer, I., 1987, Opportunistic esophagitis in AIDS: radiographic diagnosis, Radiology, 165:815.

MacFarlane, T.W., and Mason, D.K., 1974, Changes in the oral flora in Sjogren's syndrome, J. Clin. Pathol., 27:416.

Matthews, R.C., Burnie, J.P., and Tabaqchali, S., 1984, Immunoblot analysis of the serological response in systemic candidiasis, Lancet, 2:1415.

Matthews, R.C., Burnie, J.P., and Tabaqchali, S., 1987, Isolation of immunodominant antigens from sera of patients with systemic candidiasis and characterization of serological response to Candida albicans. J. Clin. Microbiol., 25:230.

Matthews, R., Burnie, J., Smith, D., Clark, I., Midgley, J., Conolly, M., and Gazzard, B., 1988, Candida and AIDS: evidence for protective antibody, Lancet, 2:263.

McNabb, P., and Tomasi, T.B., 1981, Host defense mechanisms at mucosal surfaces, Ann. Rev. Microbiol., 35:477.

Melbye, M., Schonheyder, H., Kestens, L., Stenderup, A., Gigase, P.L., Ebbesen, P., and Biggar, R.J., 1985, Carriage of oral Candida albicans associated with a high number of circulating suppressor T lymphocytes, J. Infect. Dis., 152:1356.

Murphy, P.M., Lane, H.C., Fauci, A.S., and Gallin, J.I., 1988, Impairment of neutrophil bactericidal capacity in patients with AIDS, J. Infect. Dis., 158:627.

Nielsen, H., Kharazmi, A., and Faber, V., 1986, Blood monocyte and neutrophil functions in the acquired immune deficiency syndrome, Scand. J. Immunol., 24:291.

Paterson, P.Y., Semo, R., Blumenschein, G., and Swelstad, J., 1971, Mucocutaneous candidiasis, anergy and a plasma inhibitor of cellular immunity: reversal with amphotericin B, Clin. Exp. Immunol., 9:595.

Poli, G., Bottazzi, B., Acero, R., Bersani, L., Rossi, V., Introna, M., Lazzarin, A., Galli, M., and Mantovani, A., 1985, Monocyte function in intravenous drug abusers with lymphadenopathy syndrome and in patients with acquired immunodeficiency syndrome: selective impairment of chemotaxis, Clin. Exp. Immunol., 62:136.

Rhoads, J.L., Wright, D.C., Redfield, R.R., and Burke, D.S., 1987, Chronic vaginal candidiasis in women with human immunodeficiency virus infection, JAMA, 257:3105.

Rodgers, V.D., Fassett, R., and Kagnoff, M.F., 1986, Abnormalities in intestinal mucosal T cells in homosexual populations including those with the lymphadenopathy syndrome and acquired immunodeficiency syndrome, Gastroenterology, 90:552.

Schiodt, M., and Pindborg, J.J., 1987, AIDS and the oral cavity. Epidemiology and clinical oral manifestations of human immune deficiency virus infection: a review, Int. J. Maxillofac. Surg., 16:1.

Schonheyder, H., Melbye, M., Biggar, R.J., Ebbesen, P., Neuland, C.Y., and Stenderup, A., 1984, Oral yeast flora and antibodies to Candida albicans in homosexual men. Mykosen, 27:539.

Schulten, E.A.J.M., ten Kate, R.W., and van der Waal, I., 1989, Oral manifestations of HIV infection in 75 Dutch patients, J. Oral Pathol. Med., 18:42.

Shannon, K., Cowan, M.J., Ball, E., Abrams, D., Volberding, P., and Ammann, A.J., 1985, Impaired mononuclear-cell proliferation in patients with the acquired immune deficiency syndrome results from abnormalities of both T lymphocytes and adherent mononuclear cells, J. Clin. Immunol., 5:239.

Silverman, S., Migliorati, C.A., Lozada-Nur, F., Greenspan, D., and Conant, M., 1986, Oral findings in people with or at high risk for AIDS: a study of 375 homosexual males, J. Am. Dent. Assoc., 112:187.

Smith, P.D., Ohura, K., Masur, H., Lane, H.C., Fauci, A.S., and Wahl, S.M., 1984, Monocyte function in the acquired immune deficiency syndrome: defective chemotaxis, J. Clin. Invest., 74:2121.

Strockbine, N.A., Largen, M.T., Zweibel, S.M., and Buckley, H.R., 1984, Identification and molecular weight characterization of antigens from Candida albicans that are recognized by human sera, Infect.Immun., 43:715.

Tavitian, A., Raufman, J.P., and Rosenthal, L.F., 1986, Oral candidiasis in the acquired immunodeficiency syndrome, Ann. Int. Med., 104:54.

Torssander, J., Morfeldt-Manson, L., Biberfeld, G., Karlsson, A., Putkonen, P.O., and Wasserman, J., 1987, Oral Candida albicans in HIV infection, Scand. J. Infect. Dis., 19:291.

Tsang, P.H., Sei, Y., and Bekesi, J.G., 1987, Isoprinosine-induced modulation of T-helper-cell subsets and antigen-presenting monocytes resulted in improvement of T- and B-lymphocyte functions, in vitro and in ARC and AIDS patients, Clin. Immunol. Immunopathol., 45:166.

Vudhichamnong, K., Walker, D.M., and Ryley, H.C., 1982, The effect of secretory immunoglobulin A on the in-vitro adherence of the yeast Candida albicans to human oral epithelial cells, Arch. Oral Biol., 27:617.

Weiner, M., and Coats-Stephen, M., 1979, Immunodiagnosis of systemic candidiasis: mannan antigenemia detected by radio-immunoassay in experimental and human infections, J. Infect. Dis., 140:989.

Williams, R.C., Gibbons, R.J., 1972, Inhibition of bacterial adherence by secretory immunoglobulin A: a mechanism of antigen disposal, Science, 177:697.

Witkin, S.S., Hirsh, J., Ledger, W.J., 1986, A macrophage defect in women with recurrent-Candida vaginitis and its reversal in vitro by prostaglandin inhibitors, Am. J. Obstet. Gynecol., 155:790.

CONTROVERSIAL ASPECTS OF CANDIDIASIS IN

THE ACQUIRED IMMUNODEFICIENCY SYNDROME

Jack D. Sobel
Division of Infectious Diseases, Department of Internal
Medicine, Wayne State University, Detroit, MI

INTRODUCTION

Fungal infections are widely recognized as common and frequently life-threatening opportunistic infections in patients with the acquired immunodeficiency syndrome (AIDS). *Candida* species and *Cryptococcus neoformans* are responsible for the vast majority of fungal infections in patients with AIDS, and although cryptococcal infections are often fatal, infections due to *Candida* are the most frequent opportunistic infection encountered in AIDS.

Most fungal infections in AIDS are due to reactivation and in this context are not different from the majority of other opportunistic infections. Similarly most *Candida* infections in AIDS patients represent reactivation. Mucosal *Candida* infections, reflect a failure of the patient to clear the offending organisms completely, allowing persistence of the organisms which rapidly reexpress their pathogenic ability as a clinical relapse. Candidiasis in AIDS is a study of repeated relapses at relatively usual but sometimes unique anatomical sites. To date candidiasis in AIDS has rarely been fatal or life-threatening, accordingly *Candida* has not enjoyed significant attention on the part of research or clinical investigators. Nevertheless the sheer numerical frequency of mucosal candidiasis in AIDS, the unusual clinical syndromes and the changing clinical patterns in a disease where therapeutic gains are drastically changing the natural history of AIDS, necessitate a careful review of *Candida* infections in AIDS. This paper will focus on controversial aspects of candidiasis in HIV positive patients and ask critical questions regarding aspects of epidemiology, pathogenesis, diagnosis and treatment (Tables 1 & 2).

OROPHARYNGEAL CANDIDIASIS IN HIV INFECTED INDIVIDUALS

Point prevalence studies reveal approximately 15-55% of normal healthy adults have oropharyngeal colonization with *Candida* species an average figure of approximately 20% (Kozinn and Taschdjian, 1962). HIV infection in homosexual men is associated with increased colonization in asymptomatic

Mycoses in AIDS Patients, Edited by
H. Vanden Bossche *et al.,* Plenum Press, New York, 1990

93

Table 1. Unresolved issues regarding pathogenesis of infections in AIDS

---

1. Host factors facilitating *Candida* mucosal colonization in early HIV infection.
2. Definition of host factors deficient in HIV infection that allow uncontrolled *Candida* mucosal proliferation and mycelial formation.
3. Define the natural CMI* and T lymphocyte protective mechanism that normally exert anti-*Candida* effect.
4. Are *Candida* isolates from HIV infected patients more virulent/less invasive, unusual serotypes etc? DNA typing confirmation needed.
5. Role of *Candida* as cofactor in HIV infection (in facilitating HIV activation/replication) or as immunosuppressor.
6. Why is esophageal candidiasis so often asymptomatic?

---

* CMI: cell-mediated immunity

Table 2. Unresolved clinical issues regarding therapy of *Candida* infections in AIDS

---

1. Should asymptomatic oral and esophageal candidiasis be treated?
2. Indications for esophagoscopy and biopsy in esophageal candidiasis?
3. Can *Candida* induce ulceration in AIDS in absence of *Herpes simplex* and CMV* infection?
4. Can *Candida* cause enteritis/diarrhea in AIDS?
5. Development of resistance with long term use of azoles.

---

* CMV: cytomegalovirus

men of 60-80% (Torssander et al.,1987). The cause and explanation for the increased *Candida* carriage rate in asymptomatic homosexuals is unknown. Moreover increased colonization rate occurs with CD4 cell counts well within normal limits. Thereafter the likelihood of isolating *Candida* increases with stage of HIV infection.

Oropharyngeal candidiasis occurs in approximately 75-95% of subjects with AIDS (Torssander et al., 1987; Klein et al, 1984). The prevalence of upper gastrointestinal tract (GIT) candidiasis is identical in at risk third world communities in Africa and Haiti. The high prevalence of oral thrush and esophageal candidiasis in AIDS patients was already apparent early in the epidemic in 1983. Klein et al. in 1984 first described the significance of oral thrush as a bad prognostic factor, in that more than 50% of patients with oral candidiasis, generalized lymphadenopathy (PGL) and reversed T helper/suppressor cell ratio, developed AIDS over the subsequent 12 months follow-up. Similarly within one year, Murray et al. showed that patients with oral candidiasis in addition to PGL went on to develop AIDS in 46% of cases (Murray et al., 1985). Multiple other retrospective and prospective studies have confirmed the ominous appearance of oral candidiasis as an indicator of incipient AIDS (Kaslow et al., 1987). In a prospective multicenter cohort study, the appearance of oral thrush correlated with advancing immunodeficiency, but whereas *Pneumocystis carinii* infections correlated with a CD4 cell count of <200 cells/mm^3, oropharyngeal candidiasis occurred with only slightly reduced CD4 counts of 400-700/mm^3 (Kaslow et al., 1987). This observation confirms a widely-held clinical view that mucosal *Candida* infection in HIV infected individuals requires a considerably milder reduction in helper T cell population numbers. Oropharyngeal candidiasis is therefore a uniquely sensitive index of mucosal immunity.

The first critical question is what constitutes the normal mucosal anti-*Candida* defense mechanism. Factors thought to be critical in preventing symptomatic *Candida* mucositis are seen in Table 3. Of these 4 poorly studied factors, the issue of how cell-mediated immune mechanisms operate to prevent colonization and the development of symptomatic *Candida* mucosal inflammation have escaped the scrutiny of investigators. Sobel and Opitz (1987) postulate that normal T-helper cells function at the mucosal level to reduce *Candida* numbers colonizing these surfaces and to prevent transformation from the less virulent blastospore to the mycelial phase. They postulated that under normal physiological conditions lymphokines elaborated by surveillance CD4 cells in submucosal and mucosal tissues provide this moderating protective effect and recently Kala-Klein and Witkin (1990) described the effect of γ-interferon in inhibiting *Candida* germ tube formation. The entire subject of anti-*Candida* mucosal defense mechanisms deserves more attention.

Table 3. Mucosal resistance to *Candida*

---

1. Physically intact barrier
2. Normal resident bacterial flora
3. Antimycotic secretions (Salivary)
4. Cell mediated immunity - CD4 cells/lymphocytes

Controversial aspects of oropharyngeal and esophageal candidiasis in AIDS include establishing an explanation for why extensive florid oral and esophageal thrush is so often asymptomatic. This suggests a normal inflammatory reaction to the extensive but superficial yeast population is necessary to induce symptoms prior to the more invasive mucosal phase of the disease. Clinicians experienced in the subject of oral and vaginal candidiasis have long recognized the spectrum of clinical manifestations accompanying *Candida* mucositis, indicating a role of the host inflammatory/immune response in determining the clinical picture. This leads to the next unanswered question viz how does *Candida* induce mucosal inflammation? There is surprisingly little information on this subject.

Attention needs to be directed at whether *Candida* strains associated with candidiasis in AIDS patients differ from those isolated from non-HIV infected individuals. Are the strains in AIDS more virulent in terms of causing mucosal disease but less so in achieving deep mucosal penetration and hence candidemia? Studies at the Pasteur Institute indicate a surprising increased prevalence of serotype B in AIDS patients and similar observations were forthcoming by Brawner and Cutler (1989). Other typing techniques however have not confirmed this difference in strain prevalence (Korting et al., 1988; Matthews et al., 1988) especially when DNA homology techniques were used. Once more this controversial issue demands more study in a larger number of both patients and yeast isolates.

Perhaps the most critical as yet unanswered aspect of recurrent and chronic mucosal candidiasis in AIDS relates to the possible role opportunistic yeast infection could exert as a cofactor in facilitating or activating HIV infection. Numerous investigators have demonstrated that *Candida* in both experimental infection and *in vitro* induces a variety of immunological reactions. These include induction of suppressor T and B lymphocytes subpopulations, the presence of serum inhibitors of lymphocyte function, humoral activation etc. (Matthews et al., 1988; Piccolella et al., 1981; Rogers and Balish, 1980; Domer et al., 1986). Accordingly the potential exists for mucosal *Candida* antigen presence or excess to activate CD4 cells infected with HIV and facilitate viral reactivation or replication or to contribute to host immunosuppression and predispose to opportunistic infections and other auto-immune reactions. These are intriguing issues and if evidence is forthcoming that *Candida* mucosal colonization and mucositis influence the natural history of HIV infection, this may necessitate earlier and more effective therapy aimed at eliminating *Candida* or reducing antigen excess.

Controversial aspects of therapy of upper gastrointestinal tract candidiasis relate not so much to selection of the most effective oral antimycotic agents but to duration of therapy. Only the mildest cases of oral thrush respond to conventional nystatin therapy. Most cases are however effectively treated with clotrimazole troches or oral ketoconazole. More recently both oral systemic fluconazole and itraconazole have been highly effective in both oral and esophageal candidiasis and often in more abbreviated courses. All the azoles are however fungistatic, achieving a complete symptomatic response and negative cultures without eradicating all the offending micro-organisms. Accordingly one should anticipate positive cultures shortly after cessation of therapy and clinical recurrence thereafter, representing true relapse (not reactivation) due to mucosal blastospore persistence. This principle should not be lost when prescribing azole therapy. Given the persistent immunodeficiency, antimycotic therapy is never curative but constitutes highly effective suppressive prophylaxis when used in maintenance regimens. Long term therapy has not as yet been shown to induce azole resistance among *Candida* isolates in HIV infected individuals (O'Connor and Sobel, 1986).

CANDIDEMIA AND SYSTEMIC CANDIDIASIS IN AIDS

It is widely recognized that in spite of the prevalence of oropharyngeal and esophageal candidiasis, deep mucosal invasion, candidemia and disseminated candidiasis is remarkably rare in patients with AIDS. This observation is analogous to that found in children and adults with severe Chronic Mucocutaneous Candidiasis syndrome (Kirkpatrick, 1971). In both conditions, it is postulated that profound defects in CMI result in mucosal inflammation caused by *Candida* but that normal or near normal phagocytic function in the presence of normal opsonins (immunoglobulins and complement) prevent deep penetration by invading *Candida*, hence candidemia and disseminated candidiasis are rare. Polymorphonuclear leukocytes and other mononuclear phagocytic cells provide a resistance barrier to deeper yeast tissue invasion and appear to achieve this goal in AIDS. Matthews et al. (1988) postulated that the high serum titers of antibody directed against the *Candida* 47 KD antigen might also afford protection against candidemia and systemic candidiasis.

Nevertheless there appears to be growing evidence that systemic candidiasis is more common than previously estimated. In unpublished studies, Dr. Jaffee of the CDC (1989) noted several episodes of disseminated candidiasis among the first 30,632 cases of reported AIDS in the USA. This included 114 cases of *Candida pneumonia*, 14 episodes of *Candida sepsis*, 10 with multiple organ involvement, 4 patients with CNS involvement, 3 with bone marrow, 2 cardiac and 1 with retinal and splenic candidiasis. Similarly evidence of disseminated candidiasis was reported in 11/101 (10%) autopsies of adults with AIDS by Wilkes (1988).

Accordingly, disseminated candidiasis may well be more frequent than currently estimated. Given the recent changes in therapy it is also likely that candidemia may become more frequent in the future. This is because more and more patients currently receive IV therapy via long term central implanted venous access sites (Hickman catheters). Moreover patients with more advanced AIDS are more likely to receive broad spectrum antibiotics and parenteral hyperalimenation. Another contributory factor facilitating invasion and dissemination is drug induced neutropenia e.g. by zidovudine or gancyclovir. Although data are still scanty, one study showed that patients with AIDS were five times more likely to develop bacterial or fungal superinfection via Hickman catheters than HIV negative patients with malignancy (Raviglione et al., 1989). Similarly, patients with HIV nephropathy and renal failure were significantly more likely to develop peritoneal dialysis catheter infection and *Candida* peritonitis than matched controls (Dressler et al., 1989). It seems prudent to assume that although candidemia and disseminated candidiasis are currently relatively uncommon, their prevalence may be on the increase and it behoves the clinician not to exclude *Candida* as a cause of sepsis or systemic disease in persons with AIDS.

CANDIDA INFECTIONS IN HIV POSITIVE WOMEN

Vulvovaginal candidiasis (VVC)

Given the enormous prevalence of oral and esophageal candidiasis one might anticipate a similar frequency of symptomatic candidal vaginitis in women with ARC and AIDS. During the first few years of the AIDS epidemic, although women constituted a tiny percentage of patients with AIDS, this author searched in vain for HIV positive women with candidal vaginitis. The first report of candidal vaginitis in women with HIV infection appeared finally in 1987 indicating both the presence and chronic incalcitrant nature of VVC (Rhoads et al., 1987). Seven of twenty nine HIV infected women (24%) had chronic VVC as their presenting complaint, which invariably predated the onset of oral thrush. All patients had a transient response to topical therapy only. The seven women with VVC all subsequently developed oral thrush, six of whom developed frank AIDS within 30 months. None of the twenty two HIV positive women without mucosal candidiasis developed AIDS during the period of follow-up. This small study suggests that candidal vaginitis in HIV positive patients does not have the same predictive value as oral thrush with regard to incipient AIDS. Perhaps the most interesting aspect of this study is the fact that three-quarters of the women did not develop vaginal candidiasis in spite of repeated bouts of antibiotic therapy. This may reflect the lack of long term follow-up in the study. In contrast, candidal vaginitis was documented in 17/24 (70%) of women with AIDS in a study in Rhode Island, USA. In the latter study *Candida* vaginitis and oropharyngitis were equally prevalent (Carpenter et al., 1989).

It is apparent that additional studies are needed on the frequency and significance of VVC in HIV infected women. While the new onset of repeated bouts of candidal vaginitis in a woman at risk of HIV infection should suggest the need for HIV testing, it should also be emphasized that thousands of women without HIV infection suffer from recurrent VVC and should not be terrified into suspecting HIV infection.

Mucocutaneous candidiasis infections in HIV positive women

To date there has been surprisingly little published on the clinical syndromes that are characteristic of HIV infection and AIDS in women. In one review of 24 women with AIDS, candidal esophagitis was the commonest AIDS defining event which is in sharp contrast to males where the dominant initial opportunistic infection is *Pneumocystis carinii* pneumonia (Carpenter et al., 1989). Mucocutaneous candidiasis was evident in 23/24 (90%) of the women and suggests a gender related increased susceptibility to *Candida* infection. Clearly additional studies are needed to confirm this observation recognizing the importance of female reproductive hormones or host susceptibility to *Candida* infection as well as directly influencing yeast physiology and virulence potential.

RESPIRATORY TRACT CANDIDIASIS IN AIDS

As a general axiom, clinicians have long been perplexed by the rarity of candidal involvement of the lower respiratory tract mucosa in severely immunocompromised non-HIV positive patients. Pneumonia due to *Candida* is similarly rare and usually diagnosed at autopsy as a reflection of a hematogenous process rather than due to bronchogenic spread. The description of mucosal candidiasis involving the larynx, trachea, and bronchi in patients with AIDS is therefore unique and an indication of the severity of depletion of anti-*Candida* mucosal defense mechanisms in AIDS. In 1987, involvement of the respiratory pathway and lungs was recognized by the CDC as an AIDS-defining syndrome similar to *Candida* esophagitis. Many of the AIDS patients with extensive involvement of the upper and lower respiratory tract, characterized by mucosal plaques varying degrees of mucositis, are asymptomatic. The trachea-bronchial findings, (which are often very impressive), are often found coincidently at the time of bronchoscopy for unexplained pneumonia caused by non-*Candida* pathogens.

Pulmonary candidiasis has been described in 3% of AIDS patients as determined by autopsy. Thus although the entire respiratory tract appears intrinsically highly resistant to *Candida* mucositis and invasion, in AIDS patients even this organ is vulnerable to *Candida* infection. Even the profound *Candida* specific immunodeficiency accompanying chronic mucocutaneous candidiasis is not associated with respiratory tract involvement.

CANDIDA GASTROINTESTINAL INVOLVEMENT

While the most frequent sites of GIT involvement by *Candida* in AIDS remain the oropharynx and esophagus, the role of *Candida* in producing gastric and intestinal disease is less well understood. Clinical studies performed over many years indicate that *Candida* usually invades existing gastric lesions both benign and malignant as well as incision sites producing, "thrush-like" plaques on endoscopy (Odds, 1988). Long before the appearance of AIDS, *Candida* was thought to be associated with microscopic and macroscopic ulceration and invasion of the gastric mucosa in patients with hematologic malignancy (Kozinn and Taschdjian, 1962; Odds, 1988; Eras et al., 1972). In patients with AIDS, gastritis and gastric ulceration (in the absence of surgery and neoplasia) are common and are multifactorial in origin. CMV and to a lesser extent HSV infection induce mucosal inflammation and ulceration. Any study of the pathogenic role of *Candida* in the stomach must of necessity define the role of concomitant viral infection in causing or contributing to gastric disease. Accordingly whether *Candida sp.* are capable of independently causing gastritis, gastric ulceration and gastric invasion remains controversial. Nevertheless mucosal infiltration and invasion by *Candida* can occur in the presence of preexisting and concomitant disease.

Enteric involvement of both small and large bowel, similarly remains controversial. Numerous autopsy studies of patients with disseminated malignancy usually lymphohemotological in nature, have demonstrated invasive *Candida* lesions of small intestine and colon, although much less frequently than esophagitis (Odds, 1988; Eras et al., 1972). Lesions include pseudomembrane formation consisting of massive hyphae formation, isolated and multiple ulcerations with rare perforations and peritonitis. Most of these studies did not include viral cultures but histologically often the only pathogen visualized and invading the intestinal wall was *Candida*.

The possibility that *Candida sp.* may cause diarrhoea especially in AIDS patients remains controversial. Diarrhoea is a frequent, chronic and extremely debilitating clinical syndrome of complex etiology in patients with AIDS. Not infrequently the only pathogen cultured in the stool of such patients is *Candida* and the question of culpability is then entertained. Positive stool cultures of *Candida* is hardly indicative of a causal relationship. In contrast, it has been claimed that direct microscopic examination of fecal smears permits differentiation between the saprophytic and the pathogenic phase of intestinal *Candida* (Kozinn and Traschdjian, 1962). The latter phase is characterized by the presence of large numbers of mycelia in direct smears. Proponents of the theory that *Candida* causes diarrhoea in debilitated patients quote anecdotes of favorable therapeutic response to oral therapy with oral nystatin (Kozinn and Traschdjian, 1962), although most of these studies dealt with infants only. Currently in spite of frequent and often massive *Candida* colonization of the GIT of AIDS patients, the role of these organisms in causing enteritis and diarrhoea remains speculative and prospective controlled studies are required both to establish a causal relationship, mechanism and to evaluate under blinded controlled circumstances the effect of antifungal intervention therapy.

REFERENCES

Brawner, D.L., and Cutler, J.E., 1988, Oral *Candida abicans* isolates from nonhospitalized normal carriers, immunocompetent hospitalized patients and immunocompromised patients with or without AIDS, J. Clin. Microbiol., 27:1335.
Carpenter, C.C.J., Mayer, K.H., Fisher A., Desai, M.B., and Durand, L., 1989, Natural history of Acquired Immunodeficiency Syndrome in women in Rhode Island, Am. J. Med., 86:771.
Domer, J.E., Stashak, P.W., Elkins, K., Prescott, B., Caldes, G., and Baker, P.J., 1986, Separation of immunomodulatory effects of mannan from *Candida albicans* into stimulatory and suppressive components, Cell Immunol., 101:403.
Dressler, R., Peters, A.T., and Lynn, R.I., 1989, Pseudomonal and candidal peritonitis as a complication of continuous ambulatory peritoneal dialysis in Human Immunodeficiency Virus-infected patients, Am. J. Med., 86:787.

Eras, P., Goldstein, M.J., and Sherlock, P., 1972, *Candida* infection of the gastrointestinal tract, Medicine, 51:367.

Jaffee, H., 1989, Centers for Diseases Control, Atlanta, GA, U.S.A. Personal communication to J.E. Edwards, Jr.

Kala-Klein, A., and Witkin, S.S., 1990, Prostagladin E2 enhances and gamma interferon inhibits germ tube formation in *Candida albicans*, Infect. Immun., in press.

Kaslow, R.A., Phair, J.P., Friedman, H.B., Lyter, D., Solomon, R.E., Dudley, J., Polk, B.F., and Blackwelder, W., 1987, Infection with the Human Immunodeficiency Virus: Clinical manifestations and their relationship to immune deficiency, Ann. Intern. Med., 107:474.

Kirkpatrick, C.H., Rich, R.R., and Bennett, J.E., 1971, Chronic mucocutaneous candidiasis. Model building in cellular immunity, Ann. Intern. Med., 74:955.

Klein, R.S., Harris, C.A., Small, C.B., Moll, B., Lesser, M., and Friedland, G.H., 1984, Oral candidiasis in high-risk patients as the initial manifestation of the acquired immunodeficiency syndrome, N. Eng. J. Med., 311:354.

Korting, H.C., Ollert, M., Georgii, A., and Froochl, M., 1988, *In vitro* susceptibilities and biotypes of *Candida albicans* isolates from the oral cavities of patients infected with Human Immunodeficiency Virus, J. Clin. Microbiol., 26:2626.

Kozinn, P.J., and Taschdjian, C.W., 1962, Enteric Candidiasis. Diagnosis and clinical considerations, Paediatr., 71.

Matthews, R., Burnie, J., and Smith, D., Clark, I., Midgley, J., Conolly, M., and Gazzard, B., 1988, *Candida* and AIDS: Evidence for protective antibody, Lancet, ii:263.

Murray, H.W., Hillman, J.K., Rubin, B.Y., Kelly, C.D., Jacobs, J.L., Tyler, L.W., Donelly, D.M., Carriero, S.M., Godbold, J.H., and Roberts, R.B., 1985, Patients at risk for AIDS-related opportunistic infections: Clinical manifestations and impaired gamma interferon production, N. Eng. J. Med., 313:1504.

O'Connor, M.I., and Sobel, J.D., 1986, Epidemiology of recurrent vulvovaginal candidiasis: identification and strain differential of *Candida albicans*, J. Infect. Dis., 154:358.

Odds, F.C., 1988, "*Candida* and Candidosis - A review and bibliography", Baillière Tindall, London.

Piccolella, E., Lombardi, G., and Morelli, R., 1981, Generation of suppressor cells in the response of human lymphocytes to a polysaccharide from *Candida albicans*, J. Immunol., 126:2151.

Raviglione, M.C., Batton, R., Pablos-Mendez, A., Aceves-Casillas, P., Mullen, M.P., and Taranta, A., 1989, Infections associated with Hickman Catheters in patients with Acquired Immunodeficiency Syndrome, Am. J. Med., 86:780.

Rhoads, J.L., Wright, D.C., Redfield, R.R., and Burke, D.S., 1987, Chronic vaginal candidiasis in women with human immunodeficiency syndrome, J.A.M.A., 257:3105.

Rogers, T.J., and Balish, E., 1980, Immunity to *Candida albicans*, Microbiol. Rev., 44:660.

Sobel, J.D., and Opitz, M., 1987, Immunomodulation of germ tube formation in *Candida albicans*. Annual Meeting of the Interscience Conference Antimicrob Agents Chemother. Abstr. 581.

Torssander, J., Morfeld-Manson, L., Biberfeld, G., Karlsson, A., Putkonen, P.O., and Wasserman, J., 1987, Oral *Candida albicans* in HIV infection, Scand. J. Infect. Dis., 19:291.

Wilkes, M.S., Fortin, A.H., Felix, J.C., Godwin, T.A., and Thompson, W.G., 1988, Valve of necropsy in acquired immunodeficiency syndrome, Lancet, 2:85.

Cryptococcosis

ECOLOGY OF *CRYPTOCOCCUS NEOFORMANS* AND PREVALENCE OF ITS TWO VARIETIES IN

AIDS AND NON-AIDS ASSOCIATED CRYPTOCOCCOSIS

K.J. Kwon-Chung[1], A. Varma[1], and D.H. Howard[2]

[1]Laboratory of Clinical Investigation, National Institute of
Allergy and Infectious Diseases, NIH, Bethesda, MD, and
[2]Department of Microbiology and Immunology UCLA School of
Medicine, Los Angeles, CA.

INTRODUCTION

Cryptococcus neoformans is a yeast-like fungus that causes
cryptococcosis in man and animals. The fungus produces basidiomycetous
teleomorphs belonging to the genus *Filobasidiella* of Filobasidiaceae in
Heterobasidiomycetes. *Cryptococcus neoformans* is subdivided into two
varieties: *C. neoformans* var. *neoformans* and *C. neoformans* var. *gattii*. The
ecological niche of *C. neoformans* var. *neoformans* is well established,
while that of *C. neoformans* var. *gattii* is still unknown.

The occurrence of cryptococcosis has increased markedly over the last
several years since AIDS became the leading predisposing factor for
cryptococcosis. The prevalence of the two varieties of *C. neoformans* in
some geographic areas has also been changed since the advent of AIDS. In
this paper we review the ecology of *C. neoformans* and study the frequency
of the two varieties in AIDS- and non-AIDS-associated cryptococcosis from
the two cities, Los Angeles and São Paulo, previously known to be prevalent
for the var. *gattii*. The biotypes of the serotype A isolates from AIDS- and
non-AIDS-associated cryptococcosis which occurred in southern California
were also compared with those of Zairian AIDS cultures.

Ecology of *Cryptococcus neoformans*

Cryptococcus neoformans was first isolated from peach juice by
Sanfelice (1894) in Italy. During the following 60 years, *C. neoformans* was
known only from clinical specimens of man or animals. In 1955, Emmons first
established the ecological association of *C. neoformans* with pigeon excreta
and pigeon nests in the Maryland-Virginia-District of Columbia area
(Emmons, 1955). Two years later, this association was confirmed by Kao and
Schwartz (1957) in Ohio. Since then, weathered pigeon droppings and soil
contaminated with droppings have been constant environmental sources of
C. neoformans throughout the world (Bauwens et al., 1986; Bennett et al.,
1977; Swinne et al., 1989). Other environmental sources reported for
C. neoformans include guanos of at least five different kinds of birds and
various substrates (Table 1). All the isolates reported from these sources,
regardless of the geographic origin, have been *C. neoformans* var.
neoformans serotype A or D (Bennett et al., 1977). The isolates of D
serotype have been more commonly isolated from Europe than any other part
of the world (Bennett et al., 1977). Since pigeon droppings are the best

Mycoses in AIDS Patients, Edited by
H. Vanden Bossche *et al.,* Plenum Press, New York, 1990

Table 1. *Cryptococcus neoformans* from natural sources

C. neoformans var. *neoformans*	Avian Guano	Other Sources
	Pigeons	Rotting vegetables
	Canaries	Fruits and fruit juices
	Parrots	Wood
	Budgerigars	Dairy products
	Swallows	Soil
	Munia birds	
C. neoformans var. *gattii*	_____	_____

known and the most consistent natural reservoir, it has been assumed that exposure to pigeon excreta is the most likely way for man and animals to develop cryptococcosis; however, there have been only a few incidences of pigeon-dropping-associated cryptococcosis, and the role of pigeons in the disease is still unclear. The lack of well documented focal epidemics in the areas heavily contaminated by *C. neoformans*-ridden pigeon droppings, and the lack of the var. *gattii* in the droppings suggest the existence of other sources besides avian guano. Recently, Swinne et al. (1989) reported the isolation of *C. neoformans* from houses of AIDS associated cryptococcosis patients in Bujumbra. The samples tested were soil and pigeon droppings collected from the outside, and dust from the inside of the patients' houses. Samples from the environment of seven of twenty patients yielded *C. neoformans*. Since no epidemiological marker other than the serotype of the isolates was used, it is difficult to assess the relationship between the environmental isolates and the patient isolates. Epidemiological tools sensitive enough to differentiate individual isolates are needed to solve such problems. Electrophoretic karyotype may be used as an effective tool for the epidemiological study, since the karyotype pattern varies widely even among isolates of the same serotype (Polacheck and Lebens, 1989; Perfect et al., 1989). In our experience with 16 isolates, no two had identical karyotypes.

The ecological niche of *Filobasidiella neoformans* (perfect state of *C. neoformans*) is not known. It is suspected that the natural reservoir of *F. neoformans* is different from that of *C. neoformans* for the following reasons: 1) *Filobasidium floriforme*, a fungus having a close phylogenetic relationship with *F. neoformans*, was isolated from dead florets of the large plume grasses, *Erianthus giganteus* or *Nibiscus siriaca* (Olive, 1968; Kwon-Chung, 1987; Kwon-Chung, 1984b). The anamorph of *F. floriforme*, *Cryptococcus albidus*, however, has been most frequently isolated from clinical specimens and less commonly from other substrates such as wine, dew-rotted flax, soil, leaves, animal excreta, polluted water, and dairy products (Rodrigueuz de Miranda, 1984). 2) Another basidiomycete, *Tremella* (Jelly Fungi), also closely related phylogenetically to *F. neoformans* (Kwon-Chung, 1984a) is found as a saprophyte on dead tree branches while its haplophase (Yeast belonging to the genus *Cryptococcus*) is isolated from various other substrates besides dead tree branches. 3) The strains of *C. neoformans* isolated from pigeon droppings, pigeon nests, and soil in the United States were mostly serotype A and they have all been of the mating type α (Kwon-Chung and Bennett, 1978). If the primary ecological niche of *F. neoformans* is avian droppings, both α and a mating types should be found from these sources.

Epidemiology of the two varieties of *C. neoformans*

Prior to the AIDS epidemic, cryptococcosis was most commonly reported from patients with underlying diseases such as Hodgkins disease, leukemia, or other kinds of cancer, sarcoid and rheumatoid arthritis (steroid therapy). Since the early 1980s, AIDS has become the leading predisposing factor for cryptococcosis (Dismukes, 1988; Chuck and Sande, 1989; Kovacs et al., 1985). In the United States, at least 50% of the total cryptococcosis cases reported annually are from AIDS patients, ranking as the fourth most life-threatening infection in these patients. According to one study in 106 AIDS patients with cryptococcal infection, cryptococcosis was the first manifestation of their AIDS in 45 per cent of the patients (Chuck and Sande, 1989).

The epidemiological studies of the two varieties conducted prior to the AIDS epidemic indicated that the infection caused by the isolates of the *neoformans* variety is world-wide in distribution. The infection caused by the isolates of var. *gattii*, however, was prevalent only in tropical and subtropical regions and was rarely found in regions with cold climates (Kwon-Chung and Bennett, 1984). Since the advent of AIDS, however, the overall frequency of infection due to var. *gattii* in tropical and subtropical regions has diminished drastically. This is because cryptococcosis occurs most commonly in AIDS patients and almost all AIDS patients with cryptococcosis are due to the var. *neoformans* regardless of the geographic location. There have been only three isolates of var. *gattii* reported from AIDS patients. They were one each from Zaire (Kapend'a et al., 1987), Canada (St. Germain et al., 1988), and Los Angeles, California (Clancy et al., in press). The Canadian patient was in Mexico for six months before he was diagnosed as having AIDS (St. Germain et al., 1988). It is likely that he was exposed to *C. neoformans* var. *gattii* in Mexico where that variety has been known to be prevalent (Kwon-Chung and Bennettt, 1984).

The most drastic shift in the frequency of the two varieties has been observed in the central part of Africa. As shown in Table 2, only one of the 13 isolates originating from Cameroon, Ivory Coast, Kenya, and Zaire between 1951 to 1969 was var. *neoformans*, indicating that the prevalence of var. *gattii* was 92% during that period of time (Kwon-Chung and Bennett, 1984; Swinne, 1986). Contrary to these results, 40 isolates obtained from Zaire between 1970 to 1985 were all var. *neoformans* (Swinne, 1986). During

Table 2. Prevalence of the two varieties of *C. neoformans* among the clinical isolates from central Africa between 1951 - 1985[*]

1951-1969	var. *neoformans*	var. *gattii*
Cameroon	0	2
Ivory Coast	0	1
Kenya	0	2
Zaire	1	7
1970 - 1985 from Zaire		
1970	2	0
1977	1	0
1978	2	0
1980	1	0
1981	3	0
1982	6	0
1983	12	0
1984	10	0
1985	3	0

[*] Data from Kwon-Chung and Bennett (1984) and from Swinne et al. (1986).

this period, the number of cryptococcosis cases increased drastically, and all the isolates obtained since 1980 were from suspected or proven AIDS cases (Swinne, 1986). By examining the data presented in Table 2, it is tempting to hypothesize that AIDS was present in Zaire as early as 1970 when only *C. neoformans* var. *neoformans* started appearing among the human isolates. With these observations, Swinne et al. in 1986, suggested that *C. neoformans* var. *gattii* is disappearing from central Africa. Two years later, however, the isolation of *C. neoformans* var. *gattii* from AIDS patients was first reported from Zaire (Kapend'a et al., 1987). There is no recent information concerning the prevalence of the two varieties among cryptococcosis patients without AIDS in central Africa, and it is premature to conclude that var *gattii* is disappearing from that region.

Prevalence of *C. neoformans* var. *gattii* among AIDS and non-AIDS patients in the same geographical areas

Our previous study on southern Californian isolates indicated that the predominance of *C. neoformans* var. *neoformans* among AIDS isolates from that region is not due to the disappearance of var. *gattii* from the environment (Kwon-Chung et al., 1988). We have expanded the study by including more recent isolates from southern California and isolates from São Paulo, Brazil. These two regions were previously known as the high prevalence area for *C. neoformans* var. *gattii*. Table 3 shows the prevalence of the two varieties among the non-AIDS and AIDS associated isolates obtained before and after 1980 in southern California. All isolates of the var. *neoformans* were serotype A. The isolates of var. *gattii* obtained before 1980 were either serotype B or C, while those obtained since 1980 were all serotype B. The results indicate that var. *gattii* is causing infection in non-AIDS patients at the same rate before and after the advent of AIDS in that region. Similar results were obtained from the Brazilian isolates. Table 4 shows that the prevalence of var. *neoformans* and var. *gattii* remained the same among the isolates from non-AIDS patients before and after 1980.

In both southern California and Brazil, however, AIDS patients were infected almost always by the isolates of var. *neoformans* (98-100%). These results support our hypothesis (Kwon-Chung et al., 1988) that the AIDS patients are somehow more predisposed to infection by the var. *neoformans* than the var. *gattii*. This can be due to more frequent exposure to var. *neoformans* than to var. *gattii* by these patients. The question of exposure will remain unanswered until the ecological niche of var. *gattii* is discovered.

Table 3. *C. neoformans* received from Los Angeles, California

Date of Isolation	Patients	*C. neoformans* var. *neoformans* (%)	*C. neoformans* var. *gattii* (%)	Total
Before 1980	Non-AIDS	43 (59)	30 (41)	73
	AIDS	0 (0)	0 (0)	0
Since 1980	Non-AIDS	11 (58)	8 (44)	19
	AIDS	49 (98)	1 (2)	50
TOTAL		88 (70)	39 (30)	126

Table 4. *C. neoformans* received from São Paulo, Brazil

Date of Isolation	Patients	*C. neoformans* var. *neoformans* (%)	*C. neoformans* var. *gattii* (%)	Total
Before 1980	Non-AIDS	20 (65)	11 (35)	31
	AIDS	0 (0)	0 (0)	0
Since 1980	Non-AIDS	3 (60)	2 (40)	5
	AIDS	25 (100)	0 (0)	25
TOTAL		48 (79)	13 (21)	61

DNA polymorphism among the isolates of *C. neoformans* var. *neoformans* serotype A from AIDS and non-AIDS patients

The predominance of the var. *neoformans* among the isolates from AIDS patients in southern California and Brazil is puzzling, since the var. *gattii* still is prevalent among the isolates from non-AIDS-associated cryptococcosis cases in those regions. The high incidence of cryptococcosis due to serotype A in AIDS patients may be due to infection with unique or more virulent isolates of serotype A strains. Alternately, the AIDS patients may be infected by var. *gattii* at a frequency similar to non-AIDS patients but var. *gattii*, while in an AIDS host, may become var. *neoformans* by switching antigenic and biochemical properties. This hypothesis can be tested by molecular biotyping methods based on DNA polymorphism.

We have compared 14 isolates of serotype A from non-AIDS patients with 14 AIDS isolates of serotype A, obtained from Los Angeles. Fourteen isolates of serotype A and one isolate of serotype B from AIDS patients in Zaire were also compared with those of Californian isolates. The DNA was isolated by the method described previously (Varma and Kwon-Chung, in press), digested with endonucleases SalI or XbaI and was electrophoresed on 0.8% agarose gels. The DNA from the gels was transferred to nitrocellulose filters (Southern, 1975) and the filters were hybridized with three different probes. The probes used were an XA6 fragment of mtDNA from a serotype A isolate cloned into the *E. coli* vector, PUC 8, an SC3 fragment of mtDNA from a serotype C isolate cloned into the same vector PUC 8, and an EcoRI-HindIII fragment of the cytochrome oxidase gene (subunit I) from *Saccharomyces cerevisae* in YEP358 obtained from Dr. A. Tzagoloff. The XA6 fragment was previously shown to be a var. *neoformans*-specific mtDNA sequence isolated from NIH 68, a standard culture used to produce serotype A antiserum (Bennett et al., 1977; Dismukes, 1988). The SC3 fragment was isolated from mtDNA of the type culture of *F. neoformans* var. *bacillispora* (NIH 191), the teleomorph of *C. neoformans* var. *gattii*, and it distinguishes the two varieties when used as a hybridization probe (Varma and Kwon-Chung, in press).

The hybridization patterns of the XA6 fragment with the DNA of AIDS and non-AIDS isolates obtained from southern California show no homology to DNA of serotype B and C standard cultures (Fig. 1). Hybridization was observed to a 1.1 kb fragment of the standard cultures of *C. neoformans* var. *neoformans* A and D serotypes and all isolates of serotype A, non-AIDS cultures. The hybridization pattern of XA6 with the DNA from AIDS cultures showed essentially the same picture, except that two of the fourteen isolates showed different patterns: one did not show any bands and the

AIDS (CALIFORNIA)

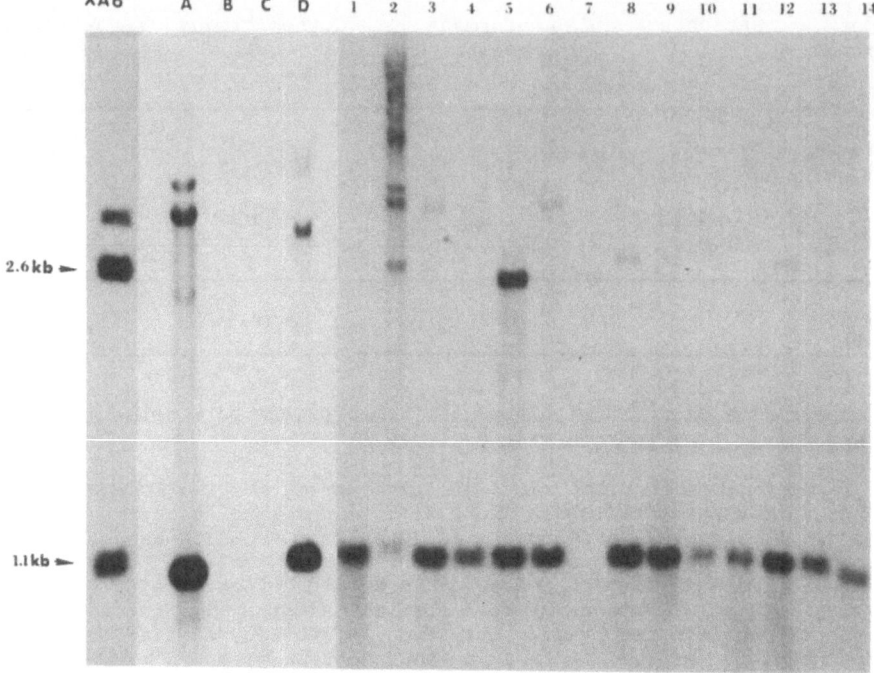

NON-AIDS (CALIFORNIA)

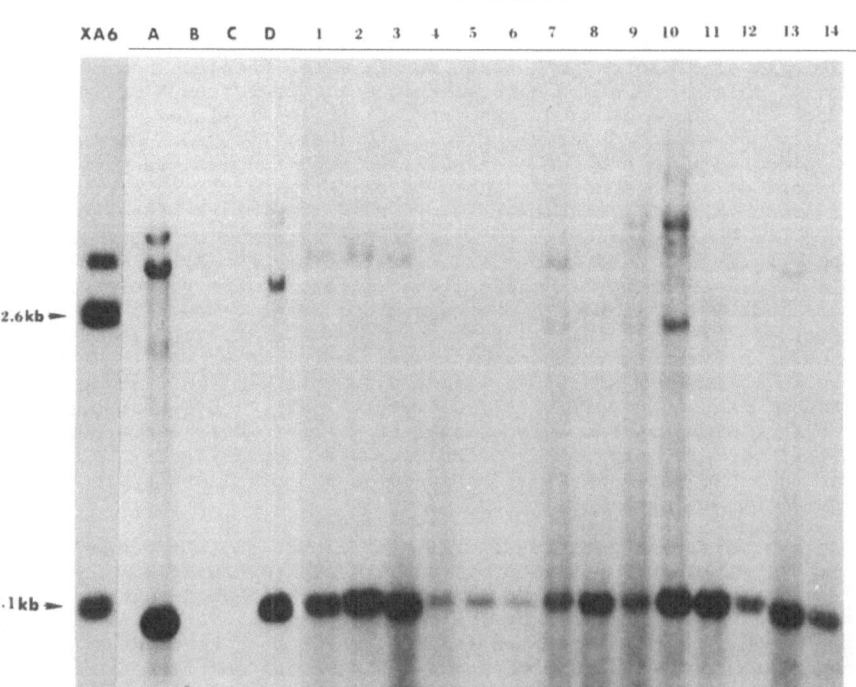

Fig.1. Hybridization patterns of a mtDNA sequence XA6 (see text) with XbaI digested total DNA of *C. neoformans* var. *neoformans* isolates from AIDS-and non-AIDS-associated cryptococcosis in southern California. ABCD indicate the DNA from standard cultures of four serotypes.

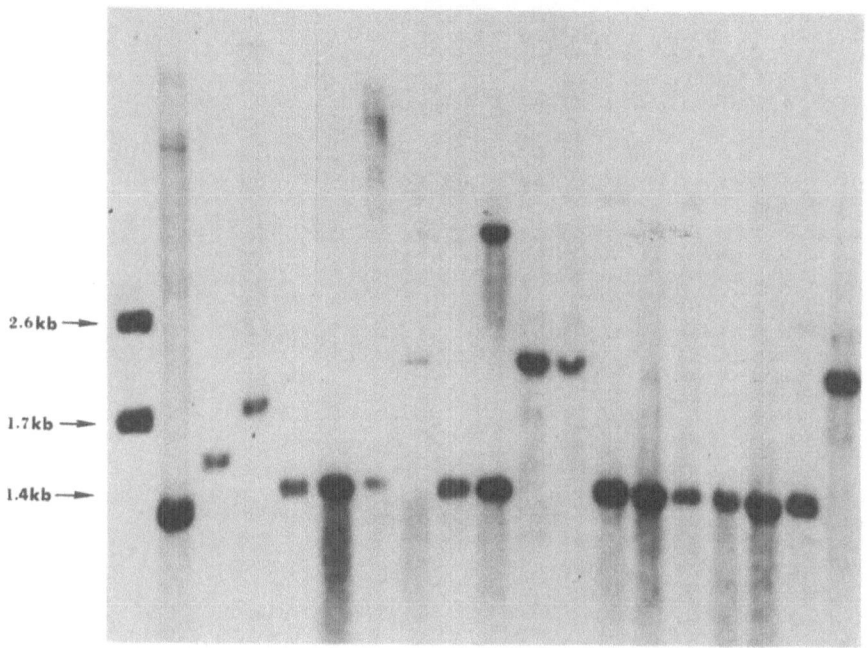

AIDS (CALIFORNIA)

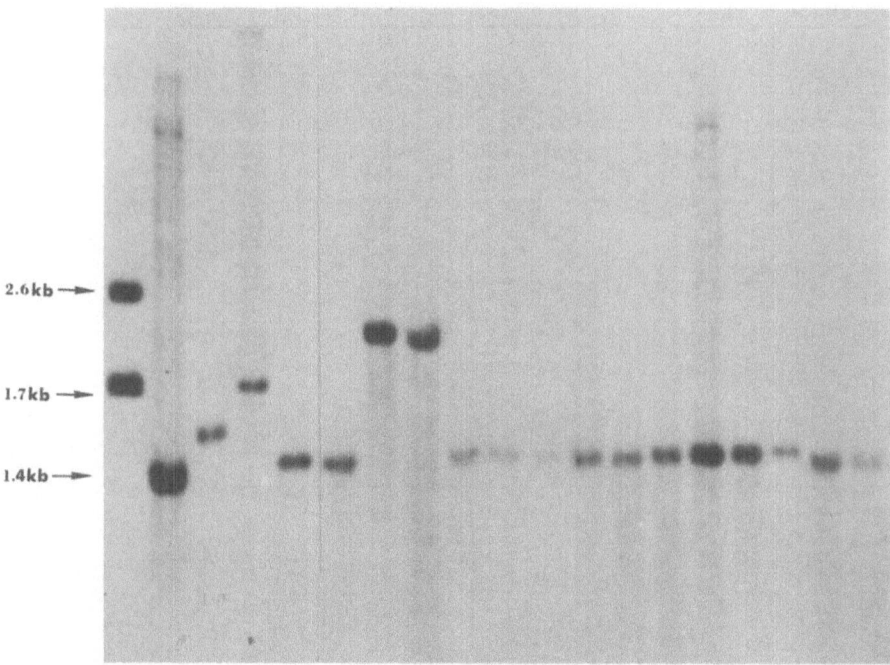

NON-AIDS (CALIFORNIA)

Fig.2. Hybridization patterns of the mt DNA sequence, SC3, with the total
DNA (SalI digested) from the same cultures used in Fig.1.

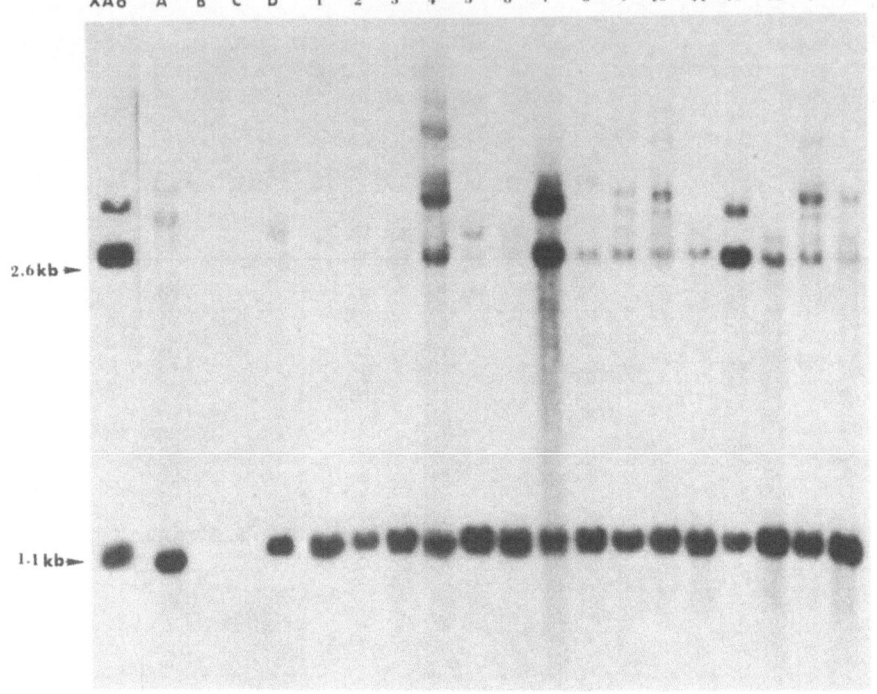

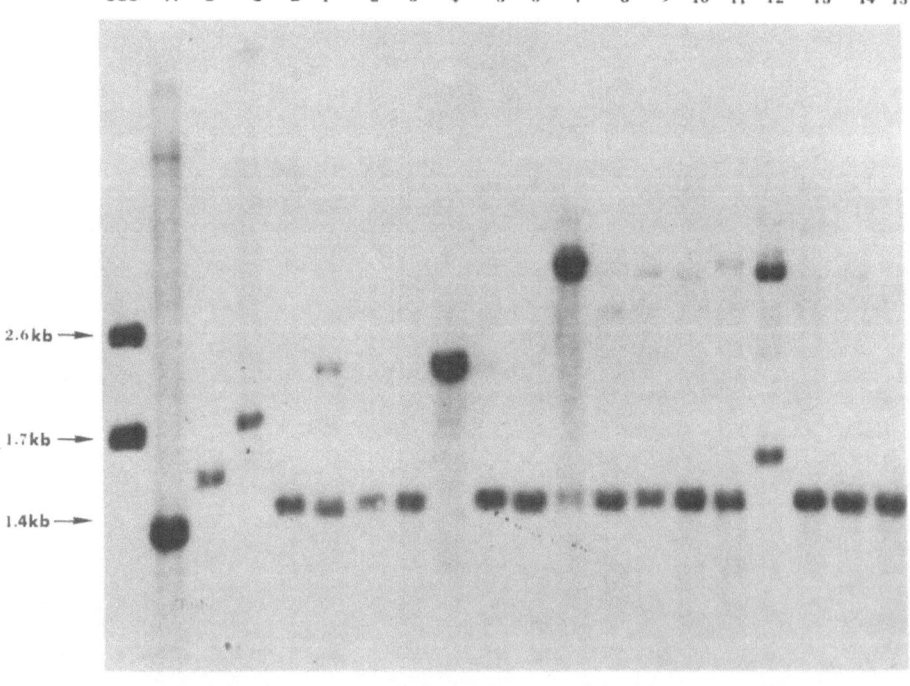

Fig.3. Hybridization patterns of XA6 and SC3 with total DNA from 15 isolates of C. *neoformans* var. *neoformans* from Zairian AIDS patients.

other had an additional band of 2.6 kb. With respect to the hybridization patterns of the 1.7 kb SC3 fragment (Fig. 2) with the DNA from AIDS and non-AIDS isolates, two of 14 non-AIDS isolates showed the 2.3 kb band, while the remaining 12 isolates showed a 1.4 kb band. The standard cultures of serotype A and D showed a 1.4 kb band, while the isolates of B and C serotypes showed 1.5 kb and 1.7 kb bands, respectively. The hybridization patterns of SC3 with DNA from AIDS cultures were essentially similar to those from non-AIDS cultures except one isolate (lane 5) which showed two bands, one of 1.4 kb and the other of over 4 kb.

Among the hybridization patterns of XA6 with the AIDS cultures from Zaire (Fig. 3), all 15 isolates showed a 1.1 kb band as was the case in Californian isolates of serotype A; however, over 50% of the isolates showed multiple bands of 2.6 kb or larger which were not seen among Californian isolates. One isolate of *C. neoformans* var. *gattii* (from AIDS) included in this figure (Lane 12) was not distinguishable from some of the serotype A isolates (Lanes 4 and 7). The isolate, however, was distinguishable from the serotype A isolates by having a 1.5 kb band typical of serotype B when hybridized with SC3 (Fig. 3). The hybridization pattern of this isolate was different, however, from the standard culture of serotype B by having an additional band of >3 kb. These results indicate that DNA polymorphism exists even among isolates of the same serotypes.

Heterogeneity of the isolates revealed by probing the DNA with cytochrome oxidase gene of *Saccharomyces cerevisiae*.

The DNA of 28 isolates from southern California (14 from AIDS, 14 from non-AIDS) and 15 isolated from Zaire (all AIDS) were probed with an EcoRI-HindIII fragment of the cytochrome oxidase gene from *S. cerevisiae* to detect polymorphism. The cytochrome oxidase gene is highly conserved among eukaryotic microorganisms (Fox, 1979). Isolates from AIDS and non-AIDS patients from California showed three patterns. Eight of 14 AIDS isolates and 9 of 14 non-AIDS isolates from southern California had identical hybridization patterns, a 4.4 kb band, indicating that this type is the

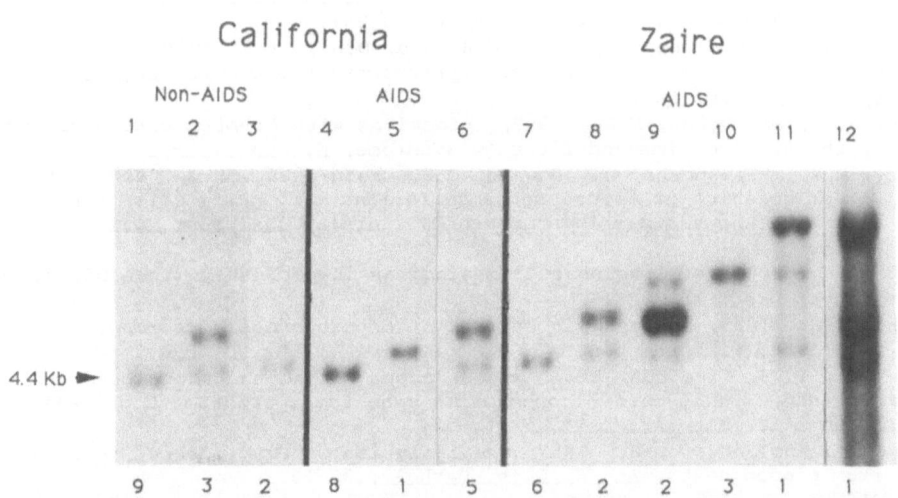

Fig.4. DNA polymorphism found among AIDS-and non-AIDS-associated isolates of *C. neoformans* var. *neoformans* from California and Zaire. The probe for hybridization was cytochrome oxidase gene (subunit I) from *Saccharomyces cerevisiae*. The number of isolates are indicated in the lower part of the figure.

most common isolate in both groups. It is apparent from these results that AIDS patients are not necessarily predisposed to infection by a unique virulent serotype A strain. The AIDS isolates from Zaire showed 6 different patterns, two of which, were the same as those found among southern Californian isolates. The most common patterns found among the isolates of these two geographical areas were the same (Lanes 1, 4, and 7 of Fig. 4). Nearly 50% of Zairian isolates, however, had the patterns not found among the Californian isolates. These results indicate that the Zairian isolates are genetically more diverse than those of Californian AIDS or non-AIDS isolates.

SUMMARY

The ecological niche of *C. neoformans* var. *gattii* remains unknown, while avian guanos, especially that of pigeons, are the major environmental source of *C. neoformans* var. *neoformans*. With a few exceptions, cryptcoccosis occurring in AIDS patients is caused by *C. neoformans* var. *neoformans*. The lack of *C. neoformans* var. *gattii* among AIDS-associated cryptococcosis is not due to the disappearance of the variety from the environment. *C. neoformans* var. *gattii* is still prevalent among non-AIDS associated cryptococcosis in tropical and subtropical regions. The biotypes of the var. *neoformans* isolates from AIDS-associated and non AIDS-associated cryptococcosis from southern California were essentially the same. This indicates that the predominance of var. *neoformans* serotype A among the AIDS-associated cryptococcosis is not due to the presence of unique virulent strains to which AIDS patients are selectively predisposed. The AIDS isolates from Zaire were genetically more diverse than the Californian AIDS isolates.

ACKNOWLEDGEMENTS

We thank Ms. Amina Bhatia for her valuable technical assistance.

REFERENCES

Bauwens, L., Swinne, D., De Vroey, C., and De Meurichy, W., 1986, Isolation of *Cryptococcus neoformans* var. *neoformans* in the aviaries of the Antwerp Zoological Gardens, Mykosen, 29:291.

Bennett, J.E., Kwon-Chung, K.J., and Howard, D.H., 1977, Epidemiologic differences among serotypes of *Cryptococcus neoformans*. Am. J. Epidemiol., 105:582.

Chuck, S.L., and Sande, M.A., 1989, Infections with *Cryptococcus neoformans* in the acquired immunodeficiency syndrome, N. Eng. J. Med., 321:794.

Clancy, M.N., Fleishmann, J., Howard, D.H., Kwon-Chung, K.J., and Shimizu, R.Y., Isolation of *Cryptococcus neoformans* var. *gattii* from a patient with AIDS who resides in southern California, J. Infect. Dis., in press.

Dismukes, W.E., 1988, Cryptococcal meningitis in patients with AIDS, J. Infect. Dis., 157:628.

Emmons, C.W., 1955, Saprophytic sources of *Cryptococcus neoformans* associated with the pigeon (Columbia livia.), Am. J. Hyg., 62:227.

Fox, T.D., 1979, Five TGA "Stop" codons occur within the translated sequence of the yeast mitochondrial gene for cytochrome C oxidase. Subunit II, Nat. Acad. Sci., 76:6534.

Kao, C.J., and Schwarz, J., 1957, The isolation of *Cryptococcus neoformans* from pigeon nests, Am. J. Clin. Pathol., 27:652.

Kapend'a, K., Komichelo, K., Swinne, D., and Vandepitte, J., 1987, Meningitis due to *Cryptococcus neoformans* diagnostic medium for *Cryptoccocus neoformans* var. *gattii*, Eur. J. Clin. Microbiol., 6:320.

Kovacs, J.A., Kovacs, A.A., Polis, M., Wright, W.C., Gill, V.J., Tuazon, C.U., Gelmann, E.P., Lane, H.C., Longfield, R., and Overturf, G., 1985, Cryptococcosis in the acquired immunodeficiency syndrome, Ann. Int. Med., 103:533.

Kwon-Chung, K.J., 1987, Filobasidiaceae - A. taxonomic survey, in: "The Expanding Realm of Yeast-like Fungi", G.S. de Hoog, M.Th. Smith, and A.C.M. Weijman, eds., Studies in Mycology 30:75.

Kwon-Chung, K.J., 1984a, Filobasidiaceae, the yeast-like genera, in: "The Yeasts, a taxonomic study", N.J.W. Kreger-van Rij, ed., Elsevier Science Publishers, Amsterdam.

Kwon-Chung, K.J., 1984b, Filobasidium Olive, in: "The Yeasts, a taxonomic study", N.J.W. Kreger-van Rij, ed., Elsevier Science Publishers, Amsterdam.

Kwon-Chung, K.J., and Bennett, J.E., 1978, Distribution of α and a mating types of Cryptococcus neoformans among natural and clinical isolates, Am. J. Epidemiol., 108:337.

Kwon-Chung, K.J., and Bennett, J.E., 1984, Epidemiologic differences between the two varieties of Cryptococcus neoformans, Am. J. Epidemiol., 120:123.

Kwon-Chung, K.J., Varma, A.K., and Howard, D.H., 1988, Ecology and epidemiology of Cryptococcus neoformans: A recent study of isolates in the United States. Proceedings of the Xth Congress of International Society of Human and Animal Mycology, Barcelona, 107.

Olive, L.S., 1968, An unusual heterobasidiomycetes with Tilletia like basidia, J. Elisha Mitchell Sci.Soc., 84:261.

Pal, N., 1989, Cryptococcus neoformans var. neoformans and Munia birds, Mycoses, 32:250.

Perfect, J.R., Magee, B.B., and Magee, P.T., 1989, Separation of chromosomes of Cryptococcus neoformans by pulsed field gel electrophoresis, Infect. Immun. 57:2624.

Polacheck, I., and Lebens, G.A., 1989, Electrophoretic karyotype of the pathogenic yeast, Cryptococcus neoformans, J. Gen. Microbiol., 135:65.

Rodriquez de Miranda, 1984, Cryptococcus Kutzing emend. Phaff et Spencer, in: "The Yeasts, a taxonomic study", N.J.W. Kreger- van Rij, ed., Elsevier Science Publishers, Amsterdam.

Sanfelice, F., 1894, Contributo alla morphologia e biologia dei blastomiceti che si sviluppano nei succi di alcuni frutti, Ann. d'ig., 4:463.

Southern, E., 1975, Detection of specific sequences among DNA fragments separated by gel electrophoresis, J. Mol. Biol., 98:503.

St. Germain, G., Noel, G., and Kwon-Chung, K.J., 1988, Disseminated cryptococcosis due to Cryptococcus neoformans var. gattii in a Canadian patient with AIDS, Eur. J. Clin. Microbiol., 7:587.

Swinne, D., Nkurikiyinfura, J.B., and Muyembe, T.L., 1986, Clinical isolates of Cryptococcus neoformans from Zaire, Eur. J. Clin. Microbiol., 5:50.

Swinne, D., Deppner, M., Laroche, R., Floch, J.J., and Kadende, P., 1989, Isolation of Cryptococcus neoformans from houses of AIDS associated cryptococcosis patients in Bujumbura (Burundi), AIDS, 3:389.

Varma, A., and Kwon-Chung, K.J., Restriction fragment polymorphism in mitochondrial DNA of Cryptococcus neoformans, J. Gen. Microbiol., in press.

CLINICAL ASPECTS OF CRYPTOCOCCOSIS IN PATIENTS WITH AIDS

Jozef Vandepitte

Dept. of Bacteriology, St-Raphael University Hospital
B-3000 Leuven, Belgium

ABSTRACT

Cryptococcosis, generally presenting as meningitis, is the most life threatening fungal infection in patients with AIDS. Its prevalence varies from 5% in Western Europe to 20% in Tropical Africa. Besides this dramatically increased frequency, cryptococcosis in AIDS patients differs in several respects from the mycosis as observed in non-AIDS patients. The paucity of meningeal signs and the low cell count and protein level in the cerebrospinal fluid tend to delay the diagnosis. A high index of suspicion and systematic screening of all suspected HIV positive patients for serum cryptococcal antigen may detect the disease at an earlier stage and improve survival.

INTRODUCTION

Cryptococcosis is a systemic mycosis, affecting animals and man, and caused by the encapsulated yeast, *Cryptococcus neoformans*. The ubiquitous nature of this fungus is in sharp contrast with the rarity of the sporadic infection in man. The number of cases estimated to occur in the pre-AIDS era in the U.S.A. and the U.K. was respectively 300 and 10 per year (Diamond, 1965; Mackenzie and White, Comm. Dis. Rep. 88/20, p.3).

Human cryptococcosis almost invariably starts as a mild or a subclinical pulmonary infection. This primary form may remain localized and undergo spontaneous resolution, or the yeast may disseminate through the bloodstream and reach other organs, most notably the central nervous system. Cryptococcal meningitis is the most important clinical manifestation of the disease. In the absence of treatment, it is uniformly fatal. Impaired T-cell-mediated immunity is the major predisposing factor in patients with the disseminated form: predisposition includes malignant lymphoma, solid neoplasms, sarcoidosis, collagen disease, corticosteroid or other immunosuppressive treatment in organ transplant patients. It is less often recognized however, that in more than 50% of cases, no predisposing factor is recognized, although subtle abnormalities in cellular immunity have been described in patients with an apparently normal host status (Schimpf and Bennett, 1975). The absence of underlying immune defect is particularly striking in tropical countries where the majority of cases of cryptococcosis arise in patients with a previously normal host status (Pillay and Simjec, 1976; Gould and Gould, 1985; Pathmanathan and Soon, 1982; Renoirte et al., 1967).

Since the early eighties, patients with AIDS have emerged as a major risk group for cryptococcosis, which in the U.S.A. is the fourth most common life threatening infection complicating AIDS, after cytomegalovirus, *Pneumocystis carinii* and *Mycobacterium avium-intracellulare*.

Cryptococcus is, after *Candida*, the second most common fungal infection complicating AIDS. In tropical Africa, where laboratory facilities are often rudimentary, cryptococcal meningitis is, together with tuberculosis, the only serious opportunistic infection that can reliably be diagnosed in patients with AIDS. It was the sudden increase of patients admitted for cryptococcosis at the Kinshasa general hospital in 1980-81 that attracted the attention of the medical world on the emergence of an AIDS epidemic in this central African city (Lamey and Melameka, 1982).

The prevalence of cryptococcosis in patients with AIDS varies from country to country and also depends on selection, follow-up and survival of the patients. In the U.S.A., cryptococcosis has been reported in approximately 7% of patients with AIDS (Kovacs et al., 1985; Zuger et al., 1986; Eng et al., 1986; Chuck and Sande, 1989), with a much higher frequency (20.3%) in blacks (Atkinson et al., 1989).

In Western Europe, somewhat lower prevalences have been recorded: 5.4% in Belgium (Taelman et al., 1988), 3.2% in the U.K. (Leading article, 1988), 6.1% in the Netherlands (Teunissen and Zanen, 1987), 5% in Germany (Weinke and Pohle, 1988), and 5.8% in France (De Closets et al., 1989). By contrast, despite problems with laboratory diagnosis, cryptococcosis is a more frequent complication of AIDS in tropical Africa. This is particularly striking for African patients investigated in Europe. Among African patients, observed in Belgium, 28/137 (20.4%) were diagnosed with cryptococcosis, against only 5/93 (5.4%) among Europeans, a highly significant (P<0.01) difference (Taelman et al., 1988). Such marked geographic differences may be related to different degrees of domestic exposure to the environmental reservoir of the yeast (Swinne et al., 1986; 1989).

Besides the dramatically increased frequency, cryptococcosis in AIDS patients differs from the disease in the classical risk groups in several respects: mycological, epidemiological, clinical, diagnostic and prognostic. One of the most puzzling differences, the rarity of the biovar *gattii* of *C. neoformans* in Africans with AIDS, will not be discussed here.

EPIDEMIOLOGY

In tropical Africa, where cryptococcosis used to affect uncompromised hosts, most cases were observed in young patients. In a series of 15 cases observed at the Kinshasa University Hospital in the period 1960-77, the average age was only 17.4 years. In the following decade, which showed a steep rise of cryptococcosis in AIDS patients, the average age rose to 35.6 years (36.8 in males, 33.0 in females) corresponding to the most promiscuous part of the urban population (Vandepitte and Kayembe, unpublished data). The male/female ratio, which was 1.8 before the AIDS era, remained almost unchanged at 2. The young age of cryptococcosis patients before 1978 refutes the hypothesis that these cases represented the earliest sporadic manifestation of the emerging AIDS epidemic.

In the U.S.A. there used to be a threefold preponderance of males over females, but since its association with AIDS cryptococcosis is almost exclusively found in males. In the series published by Chuck and Sande (1989) there was only one woman among 106 patients with AIDS and cryptococcal infection and their mean age was 38 years (range 23-65). Only few cases of paediatric AIDS and cryptococcosis have been described (Pippard et al., 1986).

CLINICAL MANIFESTATIONS

Cryptococcosis in AIDS patients has been detected at various times during the clinical evolution of the syndrome. It may be the initial manifestation of AIDS: in 41% of the 133 patients described by Kovacs et al., 1985, and Chuck and Sande, 1989; or it may occur at the same time or subsequent to other opportunistic infections or tumours.

Central Nervous System

The most common clinical presentation of cryptococcosis in AIDS, as in other immunodeficient patients, is meningitis or meningo-encephalitis. This was the case in 81% of 159 patients reviewed in three reports from the U.S.A. (Kovacs et al., 1985; Zuger et al., 1986; Chuck and Sande, 1989). The onset of meningitis in patients with AIDS often presents in a subtle, insidious manner with a paucity of meningeal signs, reflecting the lack of immune response. In other patients however, with advanced AIDS, the course may be fulminant and resemble acute encephalitis. On the average, and despite some quantitative differences, the clinical presentation of cryptococcal meningitis in AIDS and in non-AIDS patients is rather similar, as is illustrated in the following table (Table 1), adopted from Patterson and Andriole (1989) and Chuck and Sande (1989).

High fever and headache are very common but specific meningeal signs, such as neck stiffness, are seen in less than half of the patients. Focal neurological findings (seizures) are rare. Both in AIDS and in non-AIDS patients, cryptococcal meningitis may remain clinically silent for a variable time and may only be revealed by the fortuitous discovery of the yeast or its antigen in the C.S.F. Every patient with extraneural cryptococcosis should therefore be evaluated for meningeal involvement. Although a nonspecific clinical presentation, with predominance of fever and headache, has also been reported in African patients (Testa et al., 1989), other observers stress the fact that patients with cryptococcosis tend to be seen at a later stage of the disease often with coma or mental confusion. Also typical for African patients is the association with profound weight loss and pulmonary tuberculosis, suggesting a diagnosis of tuberculous meningitis (Lamey and Melameka, 1982; Laroche et al., 1986).

Cryptococcoma, a focal lesion of the C.N.S., is rare in AIDS patients (Zuger et al., 1986). There are a few reports of cryptococcal meningitis with simultaneous cerebral toxoplasmosis (Bahls and Sumi, 1986).

Table 1. Clinical presentation of cryptococcal meningitis

Sign or Symptom	% Affected		
	non-AIDS	AIDS	
Headache	87[a]	81[b]	73[c]
Fever	60	88	65
Nausea,Vomiting	53	38	42
Mental Status Changes	52	19	17
Meningeal Signs	50	31	27
Visual Changes, Photophobia	33	19	18
Seizures	15	8	4
No Signs or Symptoms	10	12	

[a]Patterson and Andriole, 1989; [b]Zuger et al., 1986; [c]Chuck and Sande , 1989

Respiratory Tract

Although the respiratory tract is the classical portal of entry of cryptococcal infection, primary pulmonary involvement is generally asymptomatic in the normal host. In immunocompromised subjects, the pulmonary infection tends to give rise to dissemination with meningitis. The situation in patients with AIDS is not clear. According to Kovacs et al. (1985), pneumonitis was the initial manifestation of cryptococcosis in only one of 27 patients with AIDS. This is contradicted by another American report describing pulmonary involvement as the initial manifestation of cryptococcosis in 5/11 patients with AIDS. Patients presented with fever, cough, dyspnoea, and pleuritic chest pain. Chest X-ray findings varied from localized and diffuse infiltration, to adenopathy and pleural effusion (Wasser and Talavera, 1987). In two further American series, culture-documented pulmonary involvement was demonstrated in a quarter (Zuger et al., 1986) to half (Gal et al., 1986) of patients with AIDS and cryptococcal meningitis . The diagnosis of cryptococcal pneumonitis in AIDS patients is complicated by the high frequency of pulmonary infection due to *P. carinii*, which may lead the clinician to ascribe pulmonary infiltration to this organism. In African patients, the diagnosis of pulmonary cryptococcosis is rendered difficult by the absence of laboratory confirmation and the multiplicity of other causes of pulmonary infection, especially tuberculosis. However, the presence of cough in 35% of Burundese patients with AIDS and cryptococcal meningitis, is compatible with coexistent fungal infection of the lung (Laroche et al., 1986).

Cryptococcaemia

According to Gal et al. (1987) cryptococci could be cultivated from the blood in 14/27 patients with AIDS and cryptococcosis, examined at various times during their clinical course. Other sources quote positive rates of 18 to 50% in AIDS patients with cryptococcal meningitis (Patterson and Andriole, 1989). Positive blood cultures seem to be more frequent than in non-AIDS patients with cryptococcal meningitis (Perfect et al., 1983).

Urinary Tract

Symptomatic pyelonephritis is a rare manifestation of cryptococcal infection and may even precede meningitis (Randall et al., 1968). Renal involvement must however be common in disseminated infection but remains most often asymptomatic. According to Butler et al. (1964), the urine harbours cryptococci in approximately 35 per cent of patients with non-AIDS cryptococcal meningitis. The importance of this site in AIDS patients with cryptococcosis has still to be assessed, but the microscopic and culture examination of urine sediment in suspected patients, seems to be a useful diagnostic approach, particularly at the primary health level.

Prostatic infection is a common cause of persistent urinary tract infection after apparently effective therapy for cryptococcal meningitis in patients with AIDS (Larsen et al., 1989).

Other Sites

A wide variety of other extraneural cryptococcal infections has been reported in patients with AIDS. Such localizations may precede other evidence of infection or occur, often in combination, in patients with documented meningitis. Skin lesions are a rather frequent manifestation of disseminated cryptococcosis. They are often located on the head but may also be seen on the trunk or the extremities. Different clinical types have been described, including molluscum contagiosum-like lesions (Miller, 1988; Concus et al., 1988), pustular or maculopapular rash (Cornevale et al., 1988), or herpetiform lesions (Borton and Wintroube, 1984). Typical yeasts can be visualized or cultivated from skin exudate or biopsy, offering a readily accessible source for early diagnosis.

Several other cryptococcal localizations have been described in patients with AIDS including myocarditis (Lafont et al., 1987), pericarditis (Brivet et al., 1987), local or generalized lymphadenopathy (Witt et al., 1987; Scalfano et al., 1988), arthritis (Ricciardi et al., 1986), pleural effusion without pulmonary involvement (Newman et al., 1987), prostatitis (Lief and Sarfazi, 1986), mucosal ulceration in the mouth (Lynch and Naftolin, 1987; Glick et al., 1987) and peritonitis (Kovacs et al., 1985). Multiple sites of disseminated infection may be disclosed at autopsy such as liver, spleen, bone marrow, kidney, adrenals, digestive tract, eyes, etc...

DIAGNOSTIC FEATURES

Cerebrospinal Fluid (C.S.F.)

Examination of the C.S.F. of AIDS patients with cryptococcal meningitis reveals the same general findings as those seen in non-AIDS patients with, however, some notable exceptions (Patterson and Andriole, 1989). There are no differences for the high opening pressure, hypoglycorrhachia, and elevated protein (Table 2). By contrast there is a striking lack of cellular response and in one series (Kovacs et al., 1985), 65% of patients had fewer than 5 leukocytes per mm . Rates of positive culture are close to 100% and the yeast cells are more abundant and more often seen in a direct India ink preparation. This heavy fungal load is reflected in the high cryptococcal antigen titers detected by commercial latex kits in 90-100% of cases.

Other Specimens

Due to the disseminated nature of the disease and the greater load of organisms, direct examination and culture of other specimens will be more often positive in untreated AIDS than in other patients. In the series of Chuck and Sande 68% of patients had at least one positive culture from an extrameningeal source. Blood, lymphnode aspirate, skin and mucous membrane lesions, urine sediment and biopsy material should always be examined. For

Table 2. Laboratory features in AIDS and non-AIDS patients with cryptococcal meningitis[a]

	non-AIDS	AIDS	
CSF High Opening Pressure	65%	62%	
Low Glucose	75%	65%	
Elevated Protein	90%	64%	
Pleocytosis	Present	Mild or Absent	
<20 Leukocytes/mm^3	30%	69%	
India Ink Positive	60%	80%	
Number of Yeasts	+	+++	
Small Capsule Yeast	Rare	++	
Culture Positive	95%	95%	
Antigen Positive	>90%	90-100%	(High
Blood Culture Positive	28%	50%	Titer)
Serum Antigen Positive	>50%	100%	

[a]Compiled from several sources

119

respiratory tract secretions and urine, appropriate selective and differential culture media should be used, particularly to differentiate *C. neoformans* from other opportunistic yeasts. Transbronchial biopsy constitutes a better specimen for the demonstration of cryptococci than expectorated sputum, bronchoalveolar lavage fluid, or bronchoscopic aspirate.

PROGNOSTIC FEATURES

Patients with AIDS and cryptococcosis can now be treated with antifungals with a success rate similar to that for other immunodeficient subjects (Zuger et al., 1986). However, there is a much higher relapse rate and it is now generally admitted that suppressive treatment should be maintained for the rest of life. Poor prognostic factors in patients with AIDS are not always the same as those applicable to non-AIDS patients (Diamond and Bennett, 1974). The following factors appeared predictive of rapid fatal outcome in a series of 34 AIDS patients treated for cryptococcosis by Zuger et al. (1986): a positive India ink preparation, high initial C.S.F. and serum cryptococcal antigen titers. The correlation between serum antigen titers and outcome was not confirmed by Gal et al. (1987). In another series (Chuck and Sande, 1989) a positive culture from a nonmeningeal site was significantly associated with a shorter survival, Other factors such as the absence of C.S.F. pleocytosis and a positive culture were not significant.

In a survey of HIV positive Zairian patients with confirmed cryptococcal meningitis, the time elapsed between the appearance of symptoms and the confirmation of diagnosis was inversely proportional to the survival time (Kayembe-Kalambayi and Desmet, 1989). Systematic screening of all suspected HIV positive patients for cryptococcal antigen in serum seems therefore a reasonable approach to the early diagnosis of cryptococcosis (Masci and Nicholas, 1989, Desmet et al., 1989).

CONCLUSIONS

Cryptococcosis is one of the most frequent and most devastating complications of AIDS, particularly in Africa. In spite of the persisting immune defect and the ultimately fatal prognosis of the underlying syndrome, early antifungal therapy is remarkably efficient. If the prognosis of cryptococcal infection in AIDS patients is to be further improved, it would seem reasonable to establish the diagnosis at the earliest moment. Greater awareness of the varied clinical presentation, more aggressive laboratory investigation, and a programme of systematic screening for serum cryptococcal antigen in all HIV positive subjects or those with compatible symptoms, may be required for optimal therapeutic response.

REFERENCES

Atkinson, W., Traxler, S., and McFarland, L., 1989, Extrapulmonary cryptococcal disease as a manifestation of Aids in Louisiana, 5th Intern. Conf. on AIDS, Montreal, B.519.
Bahls, F., and Sumi, S.M., 1986, Cryptococcal meningitis and cerebral toxoplasmosis in a patient with acquired immunodeficiency syndrome, J. Neurol. Neurosurg. Psychiatry, 49:328.
Borton, L.K., and Wintroub, B.U., 1984, Disseminated cryptococcosis presenting as herpetiform lesions in a homosexual man with acquired immunodeficiency syndrome, J. Am. Acad. Dermatol., 10:387.
Brivet, F., Livartowski, J., Herve, P., Rain, B., and Dormont, J., 1987, Pericardial cryptococcal disease in acquired immune deficiency syndrome, Am. J. Med., 82:1273.

Butler, W.T., Alling, D.W., Spickard, A., and Utz, J.P., 1964, Diagnostic and prognostic value of clinical and laboratory findings in cryptococcal meningitis, N. Engl. J. Med., 270:59.

Chuck, S.L., and Sande, M.A., 1989, Infection with *Cryptococcus neoformans* in the acquired immunodeficiency syndrome, N. Engl. J. Med., 321:794

Concus, A.P., Helfand, R.F., Imber, M.J., Lerner, E.A., and Sharpe, R.J., 1988, Cutaneous cryptococcosis mimicking molluscum contagiosum in a patient with AIDS, J. Infect. Dis., 158:897.

Cornevale, G., Lombardi, L., Sacchi, P., Pan, A., Filia, C., and Maserati, R., 1988, Disseminated cryptococcosis presenting as cutaneous disease in AIDS patients, 4th Intern. Conf. on AIDS, Stockholm, Abstr. 7624.

De Closets, F., Barrabes, A., and Cotty, F., 1989, Mycoses et parasitoses opportunistes au cours du SIDA, Bull. Soc. Fr. Mycol. Med., 28:97.

Desmet, P., Kayembe, K.D., and De Vroey, C., 1989, The value of cryptococcal antigen screening among HIV-positive AIDS patients in Kinshasa, Zaire, AIDS, 3:77.

Diamond, R.D., 1985, *Cryptococcus neoformans*, in: "Principles and Practice of Infectious Diseases", G.L. Mandell, R.C. Douglas, J.E. Bennett, eds., John Wiley, New York.

Diamond, R.D., and Bennett, J.E., 1974, Prognostic factors in cryptococcal meningitis, Ann. Intern. Med., 80:176.

Eng, R.H.K., Bishburg, E., Smith, S.M., and Kapila, R., 1986, Cryptococcal infections in patients with acquired immune deficiency syndrome, Am. J. Med., 81:19.

Gal, A.A., Evans, S., and Meyer, P.R., 1987, The clinical laboratory evaluation of cryptococcal infections in the acquired immunodeficiency syndrome, Diagn. Microbiol. Infect. Dis., 7:249.

Glick, M., Cohan, S.G., Cheney, R.T., Crooks, G.W., and Greenberg, M.S., 1987, Oral manifestations of disseminated *Cryptococcus neoformans* in a patient with acquired immunodeficiency syndrome, Oral Surg., 64:454.

Gould, P.R., and Gould, I.M., 1985, Cryptococcosis in Zimbabwe, Trans. R. Soc. Trop. Med. Hyg., 79:67.

Kayembe-Kalambayi, M.D., and Desmet, P., 1989, Prognostic factors of cryptococcal meningitis in AIDS patients, 5th Intern. Conf. on AIDS, Montreal, W.B.P.4.

Kovacs, J.A., Kovacs, A.A., Polis, M., Wright, W.C., Gill, V.J., Tuazon, C.U., Gelmann, E.P., Lane, H.C., Longfield, R., Overturf, G., Macher, A.M., Fauci, A.S., Parrillo, J.E., Bennett, J.E. and Masur, H., 1985, Cryptococcosis in the acquired immunodeficiency syndrome, Ann. Intern. Med., 103:533.

Lafont, A., Wolff, M., Marche, C., Clair, B., and Regnier, B., 1987, Overwhelming myocarditis due to *Cryptococcus neoformans* in an AIDS patient, Lancet, ii:1145.

Lamey, B., and Melameka, N., 1982, Aspects cliniques et epidemiologiques de la cryptococcose a Kinshasa. A propos de 15 cas personnels, Med. Trop., 42:507.

Laroche, R., Hategekimana, T., Ndabaneze, E., Kadende, P., Petat, E., and Aubry, P., 1986, La cryptococcose au Burundi en 1985, Med. Trop., 46:249.

Larsen, R.A., Bozzette, S., McCutchan, J.A., Chiu, J., Leal, M.A., and Richman, D.D., 1989, Persistent *Cryptococcus neoformans* infection of the prostate after successful treatment of meningitis, Ann. Intern. Med., 111:125.

Leading Article, 1988, Cryptococcosis and AIDS, Lancet, i:1434.

Lief, M., and Sarfarazi, F., 1986, Prostatic cryptococcosis in the acquired immune deficiency syndrome, Urology, 28:318.

Lynch, D.P., and Naftolin, L.Z., 1987, Oral *Cryptococcus neoformans* infection in AIDS, Oral Surg., 64:449.

Masci, J., Pierone, G., and Nicholas, P., 1989, Serum cryptococcal antigen screening in the early diagnosis of cryptococcal infection in patients with HIV infection, 5th Intern. Conf. on AIDS, Montreal, W.B.P.6.

Miller, S.J., 1988, Cutaneous cryptococcasis resembling molluscum contagiosum in a patient with acquired immunodeficiency syndrome drome, *Cutis*, 41:411.

Newman, T.G., Soni, A., Acaron, S., and Huang, C.T., 1987, Pleural cryptococcosis in the acquired immune deficiency syndrome, *Chest*, 91:459.

Pathamanathan, R., and Soon, S-H.T., 1982, Cryptococcosis in the University Hospital, Kuala Lumpur and review of published cases, *Trans. R. Soc. Trop. Med. Hyg.*, 76:21.

Patterson, T.F., and Andriole, V.T., 1989, Current concepts in cryptococcosis, *Eur. J. Clin. Microbiol. Infect. Dis.*, 8:457.

Perfect, J.R., Durack, D.T., and Gallis, H.A., 1983, Cryptococcemia, *Medicine*, 62:98.

Pillay, N., and Simjee, A.E., 1976, Cryptococcal meningitis. Our experience in 24 black patients, *S. Afr. Med. J.*, 50:1604.

Pippard, M.J., Dalgleish, A., Gibson, P., Malkovsky, M., and Webster, A.D.B., 1986, Acquired immunodeficiency with disseminated cryptococcosis, *Arch. Dis. Child*, , 61:289.

Randall, R.E., Stacy, W.K., Toone, E.C., Prout, G.R., Madge, G.E., and Shadomy, H.J., 1968, Cryptococcal pyelonephritis, *N. Engl. J. Med.*, 279:60.

Renoirte, R., Michaux, J.L., Gatti, F., Vanbreuseghem, R., Bastin, J.P., Drexler, L., Maertens, K., and Renoirte-Montjoie, A.M., 1967, Nouveaux cas d'histoplasmose africaine et de cryptococcose observés en République démocratique du Congo, *Bull. Acad. Roy. Med. Belg.*, 7:465.

Ricciardi, D.D., Sepkowitz, D.V., Berkowitz, L.B., Bienenstock, H., and Maslow, M., 1986, Cryptococcal arthritis in a patient with acquired immune deficiency syndrome. Case report and review of the literature, *J. Rheumatol.*, 13:455.

Scalfano, F.P. Jr., Prichard, J.G., Lamki, N., Athey, P.A., and Graves, R.C., 1988, Abdominal cryptococcoma in AIDS: a case report. *J. Comput. Tomogr.*, 12:237.

Schimpff, S.C., and Bennett, J.E., 1975, Abnormalities is cell-mediated immunity in patients with *Cryptococcus neoformans* infection, *J. Allergy Clin. Immunol.*, 55:430.

Swinne, D., Deppner, M., Laroche, R., Floch, J.-J., and Kadende, P., 1989, Isolation of *Cryptococcus neoformans* from houses of AIDS-associated cryptococcosis patients in Bujumbura (Burundi), *AIDS*, 3:389.

Swinne, D., Kayembe, K., and Niyimi, M., 1986, Isolation of *Cryptococcus neoformans* var. *neoformans* in Kinshasa, Zaire, *Ann. Soc. Belge Med. Trop.*, 66:57.

Taelman, H., Clumeck, N., Sonnet, J., and Desmyter, J., 1988, 4th Intern. Conf. on AIDS, Stockholm, Abstr. 5550.

Testa, J., Vohito, M.D., and Georges, A.J., 1988, Cryptococcose et SIDA a Bangui, *Bull. Soc. Fr. Mycol. Med.*, 17:117.

Teunissen, A.W.J., and Zanen, H.C., 1987, Dertig patienten met een "verborgen" ziekte: cryptokokken-meningitis, *Nederl. Tschr. Geneesk.*, 131:1123.

Wasser, L., and Talavera, W., 1987, Pulmonary cryptococcosis in AIDS, *Chest*, 92:692.

Weinke, T., and Pohle, H.D., 1988, Cryptococcosis and AIDS, *Lancet*, ii:221.

Witt, D., MacKay, D., Schwann, L., Goldstein, D., and Gold, T., 1987, Acquired-immune deficiency syndrome presenting as bone marrow and mediastinal cryptococcosis, *Am. J. Med.*, 82:149.

Zuger, A., Louie, E., Holzmann, R.S., Simberkoff, M.S., and Rachal, J.J., 1986, Cryptococcal disease in patients with the acquired immunodeficiency syndrome. Diagnostic features and outcome of treatment, *Ann. Intern. Med.*, 104: 234.

IMMUNOLOGICAL ASPECTS OF CRYPTOCOCCOSIS IN AIDS PATIENTS

Johannes Müller

Mycology Section, Institute for Medical Microbiology and
Hygiene, University Of Freiburg i. Br., Fed. R. Germany

INTRODUCTION

Cryptococcosis shows several peculiarities compared with other
infectious diseases. This concerns all aspects of this type of mycosis with
emphasis of its immunobiology.

Cryptococcus neoformans (Cr. n.) cells are widely present in nature,
particularly in excrements of birds. The fungal cells released into the
soil or dust dessicate and are able to shrink to a dimension which allows
aspiration into the respiratory airways of man. No reliable information is
available on the rate of subclinical infection of man worldwide which would
allow conclusions on the incidence of exposure of humans to *Cr. n.* cells or
to the basidiospores of *Filobasidiella,* its perfect state.

It is generally accepted that the aspiration of *Cr. n.* leads first to
a primary infection of the lungs and that this event can be asymptomatic.
The incidence of asymptomatic infections as well as the ratio between
asymptomatic infections and clinical manifestations is unknown. Therefore,
the question arises: How potent is the host's ability to control the
infections via innate or via adaptive immunity? And further: is *Cr. n.* a
primary or an opportunistic pathogen?

Recent epidemiological and clinical data on cryptococcosis in AIDS
patients show that the incidence in this group is more than 100-fold higher
than in non-AIDS patients. This reflects a much higher general exposure of
man to *Cr. n.* than previously assumed, since no special exposure of AIDS
patients to this fungus is recognizable. The incidence of cryptococcosis in
AIDS patients is reported to be at minimum 7%. The incidence of
cryptococcosis in non-AIDS patients according to own observations as well
as to reports published is 2 cases per million population per year, i.e.
0.0002% (Kappe and Müller, 1986a; Dupont, 1986; Müller et al., 1988;
Vandepitte, 1990). If we assume that each AIDS patient exposed to *Cr. n.*
also develops clinical cryptococcosis the general exposure of man must be
calculated to a rate of several percent. The relevant conclusion - drawn
from the rarity of cryptococcosis - is that more than 99% of all primary
infections due to *Cr. n.* must be controlled effectively by the
immunocompetent host.

Mycoses in AIDS Patients, Edited by
H. Vanden Bossche *et al.,* Plenum Press, New York, 1990

This is a mere conclusion, not an observation. *Cr. n.* is very rarely isolated from clinical specimens taken from the respiratory tract. It is noteworthy to acknowledge that the primary infection does not induce humoral antibodies in the host as an expression of the preceding infection, and no affirmative skin reaction can be practised to support epidemiological data. The effective elimination of *Cr. n.* from immunocompetent hosts allows to designate cryptococcosis as an opportunistic infection. The mechanisms of this elimination are effectuated by neutrophilic granulocytes and alveolar macrophages both being the first control barrier in respiratory infections. What is the relationship between these cells of host defense and the fungus?

Cr. n. is characterized by a huge capsule consisting mainly of glucuronoxylomannan and galactoxylomannan – both substances being regarded as the essential virulence factor (Drouhet et al., 1950; Drouhet and Segretain, 1951; Murphy et al., 1988; Kozel, 1989). The capsule impairs the phagocytosis of *Cr. n.* cells. Capsular substance is widely spread in the body fluids, thus enabling the diagnosis of cryptococcosis via the demonstration of circulating antigen by agglutinating antibody-coated latex particles. While the antigen titers in non–AIDS patients are mainly below 1:1000, antigen titers as high as 1:100.000 to > 1:1.000.000 have been observed in AIDS patients with cryptococcosis. This is – apart from the severe symptomatology – the most impressive fact which leads to the view that AIDS obviously offers conditions of a particularly insufficient control of this disease.

The human organism does not possess hydrolytic enzymes sufficiently active in breaking down cryptococcal capsular substance for subsequent rapid elimination from the host. The half-life time of this elimination process is 24 h in comparison with a 2 h half-life time of *Candida* mannan (Kappe and Müller, 1986b).

Figure 1 demonstrates a cultured *Cr. n.* cell with its electrontransparent solid cell wall (CW) and the exterior capsule (CAP). Incubation of the cell with ferritin-labelled anti-*Cryptococcus* serum makes the spongy texture of the capsular substances visible, and the distribution pattern of the labelling ferritin (F) shows the free access of the offered antibodies to a layer of remarkable depth. A basic capsular layer with a network structure, however, is not reacting with the antibodies (Kuttin et al., 1984a).

Cr. n. cells having been harvested via peritoneal lavage from non-immunized rabbits infected 2 h before show a very electron-dense capsular structure and, as expected, no antibodies of the non-vaccinated host animal being present (Figure 2). The density of the capsule represents an opsonization *in vivo* based on unspecific acute phase proteins which allows phagocytosis of *Cr. n.* in the immunocompetent host.

Alveolar macrophages undergo an AIDS-typical suppression supported by a feed-back mechanism derived from specific cell-mediated immunity. *Cryptococcus* cells phagocytized by neutrophilic granulocytes (to a minor extent by macrophages) are killed by mechanisms mainly depending on peroxidase activity (Diamond et al., 1972) or by non-oxidative mechanisms (Miller and Kohn, 1983). This is obviously the way of controlling the early invasion of *Cr. n.* cells, a process which is clinically asymptomatic and not reflected by serologically detectable antibodies (Kuttin et al., 1984a).

ANTIBODY–MEDIATED IMMUNITY

The immunocompetent host is able to form antibodies against the *Cr. n.* capsular substance. There are, however, some particularities in this

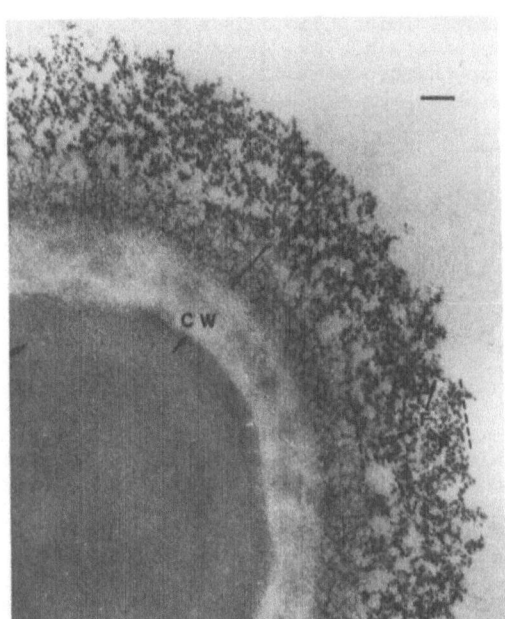

Figure. 1.   Electronmicrograph of a *Cryptococcus neoformans* culture cell
incubated with ferritin-labelled anti-*Cr. n.* immunoglobulin.
The capsule (CAP) is ferritin-labelled (F) down to its deep
layers except a basic capsular layer with a network structure
being in direct contact with the cell wall (CW). Bar represents
100 nm. From Kuttin et al., 1984a.

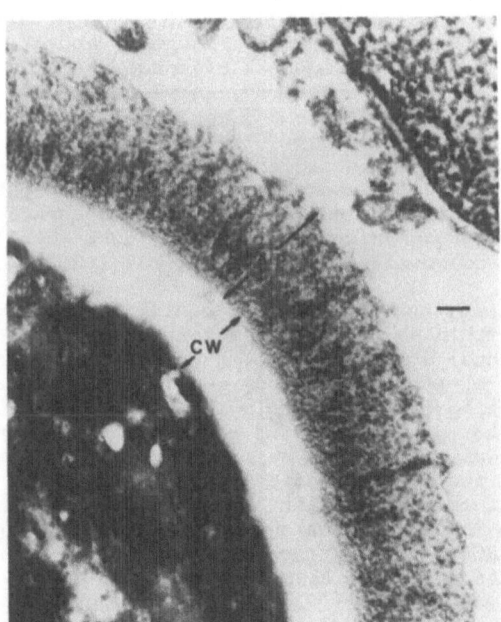

Figure 2.   Electronmicrographs of a *Cryptococcus neoformans* cell harvested
via a peritoneal lavage from a non-vaccinated rabbit 2 h after
intraperitoneal infection, and incubated with ferritin-labelled
anti-rabbit immunoglobulin: the higher electrondensity of the
capsule (CAP) represents an opsonization *in vivo* based on
unspecific acute phase proteins. The absence of ferritin
demonstrates that no specific opsonization has occurred. Bar
represents 100 nm. From Kuttin et al., 1984a.

specific branch of humoral immunity. It is a general experience that experimental immunization against *Cr. n.* capsular antigen is difficult: Animal assays show that the production of anti-cryptococcal antibodies is controlled by genetic factors – the progeny of good responders are productive responders as well. Otherwise one can easily fall a victim to immunological tolerance with too high inoculation doses during immunization which is due to the long persistence of capsular antigen in the circulation.

Figure 3 A demonstrates the aspect of *Cr. n.* cells having been harvested via peritoneal lavage from a rabbit vaccinated with ethanol-inactivated *Cr. n.* cells, 2 h after infection. The capsule is increased in size and – compared with the previous electronmicrographs (Figures 1 and 2) very densely packed: Ferritin-labelled anti-cryptococcal antibodies (F), in pre-embedding incubation experiments, can react only with surface structures indicating the presence of non-completed antigenic determinants, but the antibodies offered are not able to infiltrate into the deeper capsular layer due to its density. The higher magnification (Figure 3 B) demonstrates the release of *Cryptococcus* capsular substance from the capsule surface into the environment of the infection focus (Kuttin et al., 1984a and b).

The same situation with *Cr. n.* cells harvested from vaccinated rabbits, but incubated with ferritin-labelled anti-rabbit immunoglobulin, shows the presence of host antibodies on the surface of the capsule (Figure 4 A and B). One can assume that – as in the assay before – the ferritin-labelled antibodies (F) cannot invade the deeper capsular layers thus preventing the reagent from demonstrating host antibodies there. In summary the surface of the capsule is partially completed by host antibodies, but free antigen determinants are likewise still present (Kuttin et al., 1984a and b).

Cr. n. cells opsonized specifically in this manner are readily phagocytized by neutrophil granulocytes (Figure 5, above) and macrophages (Figure 5, below). There is experimental evidence that also complement factors are fixed to the antigen-antibody-complexes of the capsule, however, no lytic activity of complement against *Cr. n.* cells can be observed (Kuttin et al., 1984a and b). The progressive course of cryptococcosis in non-AIDS patients can, therefore, at least partially be explained by the imbalance between the rapidity of propagation of the fungal cells and their elimination via phagocytosis especially in compartments with low phagocytic activity, e.g. in cerebral tissues, although unspecific and specific opsonization may function (Waldorf, 1989).

A particular inhibition of B-cell productivity may, however, be effectuated by the high mass of free capsular antigen present in cryptococcosis patients: When the B-cell's Fc-receptor is cross-linked to its antigen receptor by antigen-antibody complexes, a signal is delivered to the B-cell inhibiting it from entering the antibody production phase. Henderson et al. (1986) demonstrated in both animal and human studies that cryptococcal antigenemia can induce suppression of the formation of anticryptococcal IgM and IgG antibodies. This is a further possibility of explaining the modest response in antibody production with cryptococcosis patients in general. That this downregulation in antibody production leads to withstanding the attack of neutrophils and to survival has been made evident in experiments performed by Miller and Kohn (1983).

In AIDS patients the specific lack of T-helper cells leads to an insufficient stimulation of B-cells which is serologically reflected by the well-known AIDS-specific IgM-deficiency. In consequence the capsular antigen material produced is not completed by antibodies, and its elimination is, therefore, impaired to a great extent which results in those high titers of circulating antigen observed in AIDS patients.

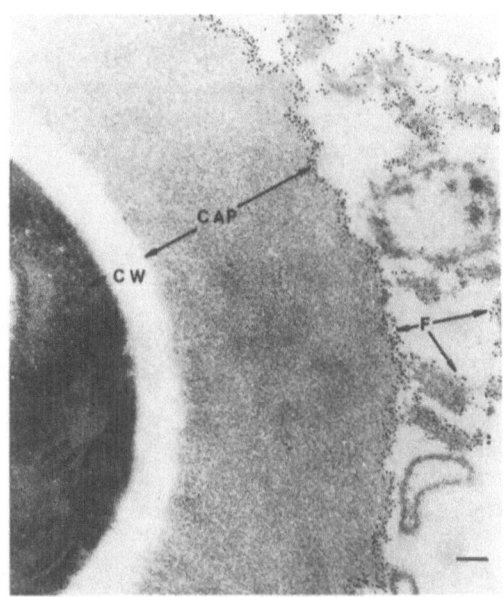

Figure 3 A. Electronmicrograph of a *Cryptococcus neoformans* cell harvested from a rabbit vaccinated with ethanol-inactivated *Cr. n.* cells, via a peritoneal lavage 2 h after infection and incubated with ferritin-labelled anti-*Cr. n.* immunoglobulin: The capsule (CAP) is increased in size and very densely packed. In pre-embedding incubation experiments the ferritin-labelled anti-cryptococcal antibodies can react only with surface structures indicating the presence of non-complexed antigenic determinants. These antibodies are not able to infiltrate into the deeper capsular layers due to its density. CW—Cell wall, F = ferritin. Bar represents 100 nm. From Kuttin et al., 1984a.

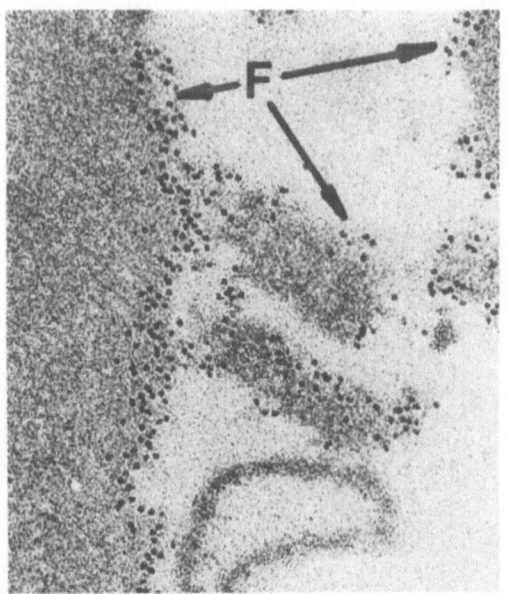

Figure 3 B. The same cell as in Figure 3 A: The higher magnification demonstrates the release of *Cr. n.* capsular antigenic substances into the environment of the infection focus. F = ferritin. Bar represents 100 nm. From Kuttin et al., 1984a.

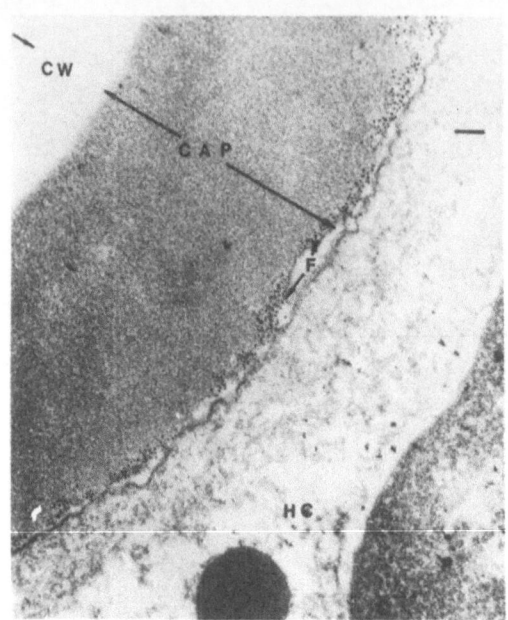

Figure 4 A. Electronmicrograph of *a Cryptococcus neoformans* cell harvested
from a rabbit vaccinated with ethanol-inactivated *Cr. n.* cells,
via a peritoneal lavage 2 h after infection and incubated with
ferritin-labelled anti-rabbit immunoglobulin: The ferritin
molecules (F) represent the sites in which antibodies of the
host have reacted with antigen determinants of the capsule
(CAP). As in Figure 3 one must assume that in these pre-
embedding experiments the ferritin-labelled antibodies cannot
invade the deeper capsular layer, thus preventing the reagent
from demonstrating host antibodies-there. CW = cryptococcal
cell wall, HC = peritoneal host cell. Bar represents 100 nm.
From Kuttin et al., 1984a.

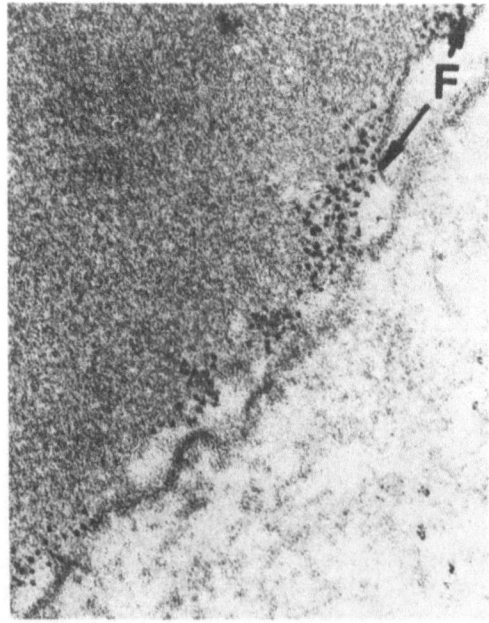

Figure 4 B. Higher magnification of Figure 4 A.

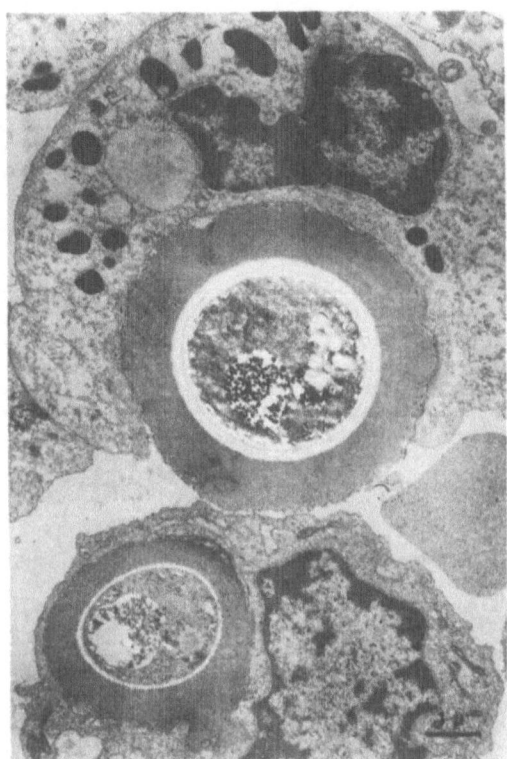

Figure 5.   Electronmicrograph of *Cryptococcus neoformans* cells harvested
             from a rabbit vaccinated with ethanol-inactivated *Cr. n.* cells,
             via a peritoneal lavage 2 h after intraperitoneal infection.
             the specific opsonization of the *Cryptococcus* cells is
             demonstrated by the densely packed capsules and the presence of
             ferritin on the capsular surface due to the treatment with
             ferritin-labelled anti-rabbit immunoglobulin (not recognizable
             because of the low magnification). The *Cryptococcus* cell above
             is being phagocytized by a neutrophilic granulocyte, the fungus
             cell below by a macrophage. Bar represents 1 μ. From Kuttin et
             al., 1984a.

In the case of successful therapy of cryptococcosis the propagation of
Cr. n. cells is restrained slowly recognizable in the fall of antigen
titers and in the conversion to antibody possitivity – a signal of good
prognosis for recovery. This development, however, is not observed in
cryptococcosis with AIDS patients.

CELL-MEDIATED IMMUNITY

Cell-mediated immunity (CMI) against *Cr. n.* is certainly the most
important protective mechanism, and in recent years most immunological
studies on cryptococcosis have focused on the CMI response and its
modulation, excellently reviewed by Murphy (1989).

In murine cryptococcosis Lim et al. (1980) found an initial increase and fall in number of *Cr. n.* cells followed by an increase in circulating antigen. The subsequent decrease of antigen was correlated with a peak of delayed-type hypersensitivity (DTH) against cryptococcal antigen. This delayed type hypersensitivity is transferable to immunologically naive mice. Several authors have, therefore, tried to use skin testing in humans with cryptococcal antigen preparations as a tool for raising diagnostic or epidemiologic information, but all these data are difficult to interpret, since different antigen preparations and dosages were applied and also cross-reactions with other fungal pathogens were observed.

Anticryptococcal CMI responses in humans also have been studied by measuring the proliferation of blood lymphocytes after stimulation with *Cryptococcus* cells of cryptococcal antigen. These investigations established that lymphocytes from cryptococcosis patients do not display as strong a lymphocyte transformation response as do cells from cured cryptococcosis patients. The anti-cryptococcal CMI response in mice recognizes a mannan protein constituent and not the capsular polysaccharides of *Cr. n.* as it was shown by Murphy et al. (1988). In general, our knowledge of up-regulation of the anticryptococcal CMI response is limited.

In contrast, our understanding of the mechanisms responsible for suppression of cell-mediated immunity response in cryptococcosis is better. There is experimental evidence that immunocompetent animals (mice and rats) display suppressed CMI response after infection with *Cr. n.* This is also obtainable by infection of avirulent *Cr. n.* strains, but not with cryptococcal antigen preparations, and the suppression was assessed by *in vitro* response to T- and B-cell mitogens and *in vivo* by cell-mediated immune responses to sheep red blood cells (Blackstock and Hall, 1981, 1984; Blackstock et al., 1987; Robinson et al., 1982). Other authors (Masih et al., 1986; Sotomayor et al., 1987) could demonstrate that this suppression phenomenon is transferable to immunologically naive animals, and the cells involved were shown to be nylon wool-nonadherent cells which points to a non-specific immunosuppression.

Cr. n.-specific immunosuppression has been investigated intensely by Murphy et al. (reviewed by Murphy, 1989) as well as by Blackstock et al. (1987). T-suppressor cells can be induced in normal mice by an intravenous infection of a culture filtrate antigen of *Cr. n.* or by infection of sera from *Cr. n.*-infected mice with high titers for cryptococcal antigen. This suppression is mediated by a complicated cascade of effector cells, as Murphy and coworkers could demonstrate (Murphy, 1989). First-order T-suppressor cells are induced in lymphenodes, these Ts1 cells mediate their effect through a soluble factor which induces second-order T-suppressor cells in the spleens, and they suppress the induction of T-lymphocytes responsible for delayed-type hypersensitivity. Again a soluble factor released by Ts2 cells induces third-order T-suppressor cells which control the expression of T-cells responsible for delayed-type hypersensitivity. Further experiments proved that these phenomena are *Cr. n.*-specific.

Blackstock et al. (1987) demonstrated the induction of a suppressive lymphokine which inhibits phagocytizing abilities of macrophages. These observations link the suppressed specific cell-mediated immunity back to the non-specific cell-mediated defense mechanisms.

In the AIDS patient the characterizing lack of CD4-T-helper cells is aggravated by the induction of CD8-suppressor cells, by this way further decreasing the CD4/CD8-T-cell ratio. This may explain the high incidence and the severe symptomatology of cryptococcosis in AIDS patients, since the early control mechanisms of innate immunity as well as the late phenomena of adaptive humoral and cell-mediated immunity are downregulated synergistically by mechanisms based on immunological particularities of both cryptococcosis and AIDS.

SUMMARY

Cryptococcosis is a fungal disease in which immunological downregulation mechanisms are proven or at least probable:

1. Phagocytosis of *Cr. n.* effectuated by cells of the non-specific host defense, i.e. neutrophils and macrophages, is inhibited by the cryptococcal capsule. Alveolar macrophages undergo an AIDS-typical suppression supported by a feed-back mechanism derived from specific cell-mediated immunity. Unspecific opsonization of *Cr. n.* cells – probably highly effective in immunocompetent hosts – is not a sufficient clearance mechanism in AIDS patients.

2. The presence of cryptococcal capsular antigen in excess in the host leads to a downregulation of the antibody production thus impairing specific opsonization. This phenomenon is aggravated by the AIDS-specific lack of antibody formation.

3. Mechanisms of downregulation can further be seen in blocking specific T-cells and cytotoxic cells by free antigen or immunecomplexes.

4. Cryptococcal antigens of different types induce suppressor T-cells causing blockades in cell-mediated immunity via a cascade of effector cells and soluble factors.

The AIDS patient enters cryptococcosis with a stage comparable to a *Cr. n.*-infected non-AIDS patient whose cell-mediated immunity is already downregulated by the fungal disease itself. Both these phenomena mycosis-induced downregulation of cell-mediated immunity as well as AIDS-specific lack of T-helper cells and its consequences – act synergistically in the same direction thus resulting in that fulminant course of cyptococcosis typical for the patient suffering from the acquired immunedeficiency syndrome.

REFERENCES

Blackstock, R., and Hall, N.K., 1981, Immunosuppression by avirulent, pseudohyphal forms of *Cryptococcus neoformans*, Mycopathologia, 80:96.

Blackstock, R., and Hall, N.K., 1984, Nonspecific immune suppression by *Cryptococcus neoformans* infection, *Mycopathologia*, 86:36.

Blackstock, R., McCormack, J.M., and Hall, N.K., 1987, Induction of a macrophage suppressive lymphokine by soluble cryptococcal antigens and its association with models of immunologic tolerance, Infect. Immun., 66:233.

Diamond, R.D., Root, R.K., and Bennett, J.E., 1972, Factors influencing killing of *Cryptococcus neoformans* by human leukocytes *in vitro*, J. Infect. Dis., 126:267.

Drouhet, E., and Segretain, G., 1961, Inhibition de la migration leucocytaire *in vitro* par un polyoside capsulaire de *Torulopsfs (Cryptococcus) neoformans*, Ann. Microbiol., (Inst. Pasteur) 81B:674.

Drouhet, E., Segretain, G., and Aubert, J.-P., 1960, Polyoside capsulaire d'un champignon pathogène, *Torulopsis neoformans*. Relation avec la virulence, Ann. Microbiol., (Inst. Pasteur) 79B:891.

Dupont, B., (Rapp.), 1986, La cryptococcose en France en 1986. Résultats d'une enquete épidémiologique, Bull. Soc. Franc. Mycol. Méd., 16:46.

Henderson, D.K., Kan, V.L., and Bennett, J.E., 1986, Tolerance to cryptococcal polysaccharide in cured cryptococcosis patients: failure of antibody secretion *in vitro*, Clin. Exp. Immunol., 66:639.

Kappe, R., and Müller, J., 1986a, Mycoses of the central nervous system in Freiburg i. Br. (Germany) and surrounding area in 1981, Mykosen, 29:64

Kappe, R., and Müller, J., 1986b, *Candida* mannan elimination following intravenous application in rabbits and mice, 22nd Ann. Meeting British Soc. Mycopathol., Cardiff.

Kozel, T.R., 1989, Antigenic structures of *Cyptococcus neoformans*. Capsular polysaccharides, in: "Immunology of fungal diseases", E. Kurstak (ed.), Marcel Dekker Inc., New York and Basel.

Kuttin, E.S., Jaeger, R., and Müller, J., 1984a, Licht- und elektronenmikroskopische Untersuchungen an Peritoneallavagen von nicht-immunisierten und immunisierten *Cryptococcus neoformans*-infizierten Versuchstieren, 18. Ann. Meeting German-Speaking Mycol. Soc., Bremen.

Kuttin, E.S., Müller, J., Jaeger, R., and Scheidecker, I., 1984b, Etude de complexes immuns au microscope électronique dans des lavages péritonéales d'animaux non-immunisés et immunisés, inrectés par *Cryptococcus neoformans*, Réunion Soc. Franc. Mycol. Méd, Paris.

Lim, T.S., Murphy, J.W., and Cauley, L.K., 1980, Host-etiological agent interactions in intranasally and intraperitoneally induced cryptococcosis in mice, Infect. Immun., 29:633.

Masih, D.T., Rubinstein, H.R., Sotomayor, C.E., Ferro, M.E., and Riera, C.M., 1986, Non-specific immunosuppression in experimental cryptococcosis in rats, Mycopathologia, 94:79.

Miller, G.P.G., and Kohn, S., 1983, Antibody dependant leukocyte killing of *Cryptococcus neoformans*, J. Immunol., 131:1466.

Müller, J., and Kappe, R., 1989, Immunochemistry of fungal antigens at the electronmicroscopic level, in: "Fungal antigens", E. Drouhet et al. (eds.), Plenum Press, New York, London.

Müller, Kappe, R., Kubitza, D., Fessler, R., and Scheidecker, I., 1988, The incidence of deep-seated mycoses in Freiburg i. Br. (Federal Republic of Germany), Mycoses, 31 (Supplement 1):9.

Murphy, J.W., 1989, Immunoregulation in cryptococcosis, in: Immunology of fungal diseases, E. Kurstak (ed.), Marcel Dekker Inc., New York and Basel.

Murphy, J.W., Mosley, R.L., Chernia, R., Reyes, G., Kozel, T., and Reiss, E., 1988, Serological, electrophoretic, and biological properties of *Cryptococcus neoformans* antigens, Infect. Immun., 66:424.

Robinson, B.E., Hall, N.K., Bulmer, G.S., and Blackstock, R., 1982, Suppression of responses to cryptococcal antigen in murine cryptococcosis, Mycopathologia, 80:167.

Sotomayor, C.E., Rubinstein, H.R., Riera, C.M., and Masih, D.T., 1987, Immunosuppression in experimental cryptococcosis in rats. Induction of afferent T suppressor cells to a nonrelated antigen, J. Med. Vet. Mycol., 26:67.

Vandepitte, J.M., 1990, Clinical aspects of cryptococcosis in AIDS patients, in: "Mycoses in AIDS patients", H. Vanden Bossche et al. (eds), Plenum Press, New York, London.

Waldori, A.R., 1989, Pulmonary defense mechanisms against opportunistic fungal pathogens, in: "Immunology of fungal diseases", E. Kurstak (ed.), Marcel Dekker Inc., New York and Basel.

Dermatomycoses and Rare Mycoses in AIDS Patients

DERMATOPHYTES AND *PITYROSPORUM* IN AIDS PATIENTS:

ECOLOGY AND EPIDEMIOLOGY

Ch. De Vroey* and M. Song**

* Laboratory for Mycology, Institute of Tropical Medicine,
   Antwerp, Belgium
** Department of Dermatology, University of Brussels,
   Brussels, Belgium

INTRODUCTION

Various skin disorders are claimed to be commonly associated with
human immunodeficiency virus (HIV) infection. Actually several of these
skin diseases, including dermatomycoses are not more prevalent in HIV
infected individuals as compared to the general population (Fisher and
Warner, 1987; Valle, 1987; Goodman et al., 1987; Sindrup et al., 1987).
The difference, as will be shown by other participants at this symposium,
lies in the unusual presentation and the severity of the clinical
manifestations.

Several differences exist in the field of ecology and epidemiology
between dermatophytes-dermatophytoses and *Pityrosporum*-pityrosporoses. This
is well illustrated by the difference in the sources of infection : the
dermatophytoses originate from environmental sources whereas the patient
himself is the source of *Pityrosporum*.

Dermatophytes and Dermatophytoses

Dermatophytes, are divided for epidemiological reasons into 3 groups,
namely, geophilic, zoophilic and anthropophilic species. Infections caused
by geophilic dermatophytes are of saprophytic origin. Infections caused by
zoophilic or anthropophilic species are on the contrary contagious, since
they orginate from a contact with infective particles of parasitic origin
which are usually transmitted indirectly from one host to another. The
practical implications of this epidemiological classification has been
reviewed by one of us (De Vroey, 1985).

Besides exposure to sources of ringworm infection there are a number
of known and unknown host-related innate and immunological factors
predisposing to infection. The role of T lymphocyte mediated responses is
reviewed by Hay (this book). HIV infected patients have altered clinical
response to superficial infections, including dermatophytosis, however,
there is little evidence that they occur more frequently in this group of
patients.

For example, in a recent study on the prevalence and clinical spectrum
of skin diseases in 100 consecutive HIV-infected patients, dermatophytoses

Mycoses in AIDS Patients, Edited by
H. Vanden Bossche *et al.*, Plenum Press, New York, 1990

were not more prevalent than in a control population (Coldiron and Bergstresser, 1989).

In another study the prevalence of dermatophytoses was not significantly higher in a group of HIV+ (37.3%) and HIV- (31.8%) homosexual males. However the prevalence of ringworm infection was much more important in a homosexual male as compared to a control group of heterosexual males, where the prevalence was 8.6%. This could be due to a higher exposure to sources of dermatophytoses (Torssander et al., 1988).

Nevertheless, another study seems to indicate that the prevalence of dermatophytosis is up to 4 times higher in HIV+ individuals (Goodman et al., 1987). It should however be mentioned that this relates to tinea pedis and/or nail infections whereas the number of cases of tinea cruris and tinea corporis reported in this study is unusually low.

In other studies the higher prevalence of tinea pedis in HIV+ individuals could only be assessed on clinical impressions but not on mycological criteria (Valle, 1987). In the same study the occurence of inguinal lesions suggestive of dermatophyte infection could not be correlated with HIV seropositivity !

Likewise, onychomycosis of the toe nails is claimed to be associated with HIV infection. However, this statement is usually based alone on clinical criteria, and cases of "yellow toe nail changes" are without proof considered as a manifestation of fungal infection. The statement recently made by Morfeldt-Manson et al. (1989) that in their patients the yellow nail changes are caused by dermatophytes should be corroborated. Recently, several cases of "white nails" in ARC/AIDS patients due to *Trichophyton rubrum* have been reported (Weismann et al.,1988)

Pityrosporum and Pityrosporoses

Pityrosporum ovale: ecological aspects. The polymorphic, lipophilic yeast *P. ovale* is a member of the normal skin microflora, mainly of the scalp, the face and the trunk. It is very simple to obtain cultures of this episaprophytic yeast with the use of contact plates (RODAC) (Vermander and De Vroey, 1985). We have used this method to compare the quantity of *Pityrosporum* on the skin of HIV+ and HIV- caucasian males. Table 1 summarises the characteristics of the patient population.

Table 1. Details of patient population studied*

Patients	Total No.	Heterosexual No.	Age (years)	Homosexual No.	Age (years)
HIV+	33[a]	12	36 (23-53)	21	39 (24-62)
HIV-	17	13	34 (23-54)	4	27 (21-30)
Total	50	25		25	

[a] CDC class. syst. - Gr2 : 5
      - Gr3 : 12
      - Gr4 : 3
      AIDS : 13

* The patients were seen at the Sexual Transmitted Diseases and the AIDS Consultation of the Institute of Tropical Medicine, Antwerp, between 27.07.89 and 27.10.89. Thanks are due to the consultant clinicians: D. Avonts, R. Colebunders, J. Goeman, M. Peeters, M. Vandenbruaene.

A direct sample was taken from the intrascapular area and the forehead with contact plates (RODAC) containing Dixon's medium (*) enriched with 0.5 mg/ml chloramphenicol. The number of colonies per plate (CFU/± 16 cm^2) were recorded after 5-10 days incubation at 37°C.

The results obtained from the intrascapular area are shown in Table 2: surprisingly, more positive cultures were obtained from HIV- individuals (16/17) compared with HIV+ individuals (24/33). The mean number of CFU was also much higher in the HIV- group than in the HIV+ group (87 versus 31).

Table 2. *P.ovale*: Isolations from intrascapular area

No. CFU/plate	HIV+ (33)	HIV- (17)
0	9	1
1- 20	19	6
21- 40	0	1
41- 60	3	0
61- 80	0	1
81- 100	0	3
101- 150	0	1
151- 200	1	3
250- 300	1	0
Total positive	24	16
Mean colony count	31	87

If the number of the individuals with CFU counts higher than the mean CFU, i.e. 41-60, in each group are compared (Table 2), another difference appears. Five out of 33 HIV+ patients and 8 out of 17 HIV- patients have more than 40 CFU obtained from their intrascapular area ($p < 0.02$).

It should be noticed that there is no significant difference between homosexual and heterosexual males, nor between the different HIV+ groups.

Four HIV+ patients had oral antifungal therapy with azoles. If they are omitted the difference between HIV+ and HIV- remains.

Interestingly, as can be seen from Table 3, the 11 patients treated with AZT seem to harbour a more 'normal' *P. ovale* population, since *P. ovale* is isolated from the intrascapular area in 9 of them.

The following results were obtained from the forehead:

P. ovale was isolated from 15 out of 33 HIV+ patients and from 10 out of the 17 HIV-. The number of colonies was is much lower than from the intrascapular area.

A noteworthy finding was that, the mean counts in the HIV+ group was 9 colonies/plate and in the HIV- group about 28 colonies/plate.

(*)Dixon's medium: malt extract agar (Oxoid)   60 g
ox bile (Oxoid)   20 g
tween 40   10 ml
glycerol mono-oleate   2.5 g
distilled water   1000 ml

Table 3. *P. ovale* : Isolations from intrascapular area in AZT treated patients

No. CFU/plate	No.(11)
0	2
1-10	5
11-20	2
21-40	0
41-60	1
>250	1
Total positive	9
Mean colony count	40

Using another method Hakansson et al. (1988) also made a quantitative evaluation of the number of *P. ovale* yeast cells on the skin of HIV+ vs HIV- homosexual males. They could not find a correlation between the immune status and the number of yeast cells.
Our results seem to indicate that the number of *Pityrosporum* yeast cells on the skin of HIV+ patients is lower than in HIV- individuals. Further studies are needed to corroborate these findings.

P. ovale: pathological implications

P. ovale is not only the etiologic agent of pityriasis versicolor: it has also been implicated in the pathogenesis of several other skin disorders.

The host related factors which are responsible for the change of this harmless commensal to cause pityriasis versicolor and other 'pityrosporoses' have still to be understood.

Pityriasis versicolor does not apparently occur more frequently in HIV-infected patients than in the general population (Hughes, 1988).

Several studies, usually based of the efficacy of certain antifungal drugs, tend to demonstrate an association between *P. ovale* and seborrhoeic dermatitis (S.D.) (Shuster, 1984; Faergemann, 1986; Ford, 1984). However, for others (Leyden, 1986; Massone et al., 1988; Cachao et al. 1989) this association remains questionable since antifungal treatment, for example econazole shampoo, is not effective in all patients, since the disease may apparently occur in the absence of *P. ovale,* or since *P. ovale* may be still found after clinical cure !

Several authors suggest that *P. ovale* is abundant on the skin of patients with S.D., but since this condition is not usually associated with seborrhoea (Burton and Pye, 1983) it becomes difficult to correlate the higher number of *P. ovale* yeast cells with a more abundant presence of lipids. According to Bergbrant and Faergemann (1989), however, there is no significant difference in the number of *P. ovale* or serum antibodies to this yeast in patient with S.D. as compared to healthy individuals. They also showed that there is no difference in number of yeast cells in lesions compared to healthy skin in the patient group. Since these authors are convinced that *P. ovale* is always associated with S.D. they conclude that an abnormal reaction to *P.ovale* causes the inflammation, and that the number of yeast cells is of minor importance.

The prevalence of S.D. is important in HIV infected individuals being present in up to 83% in AIDS, 42% in ARC and 20% in HIV+ patients with PGL without AIDS (Eisenstat and Wormser, 1984; Mathes and Douglas, 1985; Farthing et al., 1985, Valle, 1987; Matis et al., 1987). However, in other studies similar figures have been observed in ARC and AIDS patients (Senaldi et al., 1987; Wishner et al., 1987; Coldiron and Bergstresser, 1989).

In our group of 33 HIV+ patients, S.D. was observed in 10, (6 of the AIDS group) but in none of the 17 HIV- individuals.

The etiological role of *P. ovale* in HIV+/AIDS associated S.D. remains questionable. Groisser et al. (1989), in a study of 10 AIDS patients with S.D. have noticed that the number of *Pityrosporum* yeast cells per keratinocyte decreases after ketoconazole treatment.
Our results show at least that HIV+ patients do not harbor more *Pityrosporum* on their skin, although we have not really collected specimens from the most affected sites.
That S.D. in HIV infected patients may differ from the same condition observed in other patients is further demonstrated by Soeprono et al. (1986) who observed that the pathological changes in S.D. associated with HIV-infection differ from those of the 'classical' S.D. These authors preferred to use the terminology 'seborrhoeic-like dermatitis'.
S.D. may, as proposed by Burton and Pye (1983), be defined as a "dermatitis of the sebaceous areas occuring in heterogenous groups of patients". It is therefore better to consider S.D. as a syndrome i.e. a clinical manifestation with different etio-pathogenesis. Further studies are needed to understand the factors involved in various types of patients particularly in HIV infected individuals.

In conclusion, the prevalence of dermatophyte and *Pityrosporum* infections is not higher in HIV+ asymptomatic, ARC or AIDS patients. This could further be deduced from the fact that we hardly ever received any request for mycological investigations for this group of patients except for oral and oesophageal candidosis and cryptococcal meningitis.

REFERENCES

Bergbrant, I.M., and Faergemann, J., 1989, Seborrhoeic dermatitis and *Pityrosporum ovale* : a cultural and immunological study, Acta Dermato. Venereol., 69:332.
Burton , W., and Pye, R.J., 1983, Seborrhoea is not a feature of seborrhoeic dermatitis, Br. Med. J., 286:1169.
Cachao, P., Sequeira, H., Cabrita, J., and Rodrigo, F.G., 1989, Scalp seborrhoeic dermatitis and *Pityrosporum ovale*. A clinical trial with an econazole shampoo, in: "Abstracts 1st Congress European Academy of Dermatology and Venereology".
Coldiron, B.M., and Bergstresser, P.R., 1989, Prevalence and clinical spectrum of skin disease in patients infected with human immunodeficiency virus, Arch. Dermatol., 125:357.
De Vroey, Ch., 1985, Epidemiology of ringworm (Dermatophytosis), Semin. in Dermatol., 4:185.
Eisenstat, B.A., and Wormser, G.P., 1984, Seborrhoeic dermatitis and butterfly rash in AIDS (letter), N. Engl. J. Med., 311:189.
Faergemann, J., 1986, Seborrhoeic dermatitis and Pityrosporum orbiculare: treatment of seborrhoeic dermatitis of the scalp with miconazole-hydrocortisone (Daktacort®), miconazole and hydrocortisone, Br. J. Dermatol., 114:695.
Farthing, C.F., Staughton, R.C.D., and Rowland Payne, C.M.E., 1985, Skin disease in homosexual patients with acquired immune deficiency syndrome (AIDS) and lesser forms of human T cell leukaemia virus (HTLV III) disease, Clin. Exp. Dermatol., 10:3.

Fisher, B.K., and Warner, L.C., 1987, Cutaneous manifestations of the acquired immunodeficiency syndrome, *Int. J. Dermatol.*, 26:615.

Ford, G.P., Farr, P.M., Ive, F.A., and Shuster, S., 1984, The response of seborrhoeic dermatitis to ketoconazole, *Br. J. Dermatol.*, 111:603.

Goodman, D.S., Teplitz, E.D., Wishner, A., Klein, R.S., Burk, P.G., and Hershenbaum, E., 1987, Prevalence of cutaneous disease in patients with acquired immunodeficiency syndrome (AIDS) or AIDS-related complex, *J. Am. Acad. Dermatol.*, 17:210.

Groisser, D., Bottone, E.J., and Lebwohl, M., 1989, Association of *Pityrosporum orbiculare* (*Malassezia furfur*) with seborrhoeic dermatitis in patients with acquired immuno deficiency syndrome (AIDS), *J. Am. Acad. Dermatol.*, 20:770.

Häkansson, C., Faergemann, J., and Löwhagen, G.B., 1988, Studies on the lipophilic yeast *Pityrosporum ovale* in HIV-seropositive and HIV-seronegative homosexual men, *Acta Derm. Venereol.*, 68:422.

Leyden, J.J., 1986, The pathogenic role of microbes in seborrhoeic dermatitis, *Arch. Dermatol.*, 122:16-17.

Massone, L., Borghi, S., Pestarino, A., Piccini, R., Solari, G., Casini Lemmi, M, and Isola, V., 1988, Seborrhoeic dermatitis in otherwise healthy patients and in patients with lymphadenopathy syndrome/AIDS-related complex : treatment with 1% bifonazole cream, *Chemiotherapia*, 7:109.

Mathes, B.M., and Douglass, M.C., 1985, Seborrhoeic dermatitis in patients with acquired immuno deficiency syndrome, *J. Am. Acad. Dermatol.*, 13:947.

Matis, W.L., Triana, A., Shapiro, R., Eldred, L., Polk, B.F., and Hood, A.F., 1987, Dermatologic findings associated with human immuno deficiency virus infection, *J. Am. Acad. Dermatol.*, 17:746.

Morfeldt-Manson, L., Julander, I., and Nilsson, B., 1989, Dermatitis of the face, yellow toe nail changes, hairy leukoplakia and oral candidiasis are clinical indicators of progression to AIDS/opportunistic infection in patients with HIV infection, *Scand. J. Infect. Dis.*, 21:497.

Senaldi, G., Di Perri, G., Di Silverio, A., and Minoli, L., 1987, Seborrhoeic dermatitis : an early manifestation in AIDS, *Clin. Exp. Dermatol.*, 12:72.

Shuster S., 1984, The aetiology of dandruff and the mode of action of therapeutic agents, *Br. J. Dermatol.*, 111:695.

Sindrup, J.H., Lisby, G., Weismann, K., and Lange Wantzin, G., 1987, Skin manifestations in AIDS, HIV infection and AIDS-related complex, *Int. J. Dermatol.*, 26:267.

Skinner, R.B., Zanolli, M.D., Noah, P.W., and Rosenbert, E.W., 1986, "Seborrhoeic dermatitis and acquired immuno deficiency syndrome", *J. Am. Acad. Dermatol.*, 14:147.

Soeprono, F.F., Schinella, R.A., Cockerell, C.J., and Comite, S.L., 1986, Seborrhoeic-like dermatitis of acquired immuno deficiency syndrome, A clinicopathologic study, *J. Am. Acad. Dermatol.*, 14:242.

Torssander, J., Karlsson, A., Morfeldt-Manson, L., Putkonen, P.O., and Wasserman, J., 1988, Dermatophytosis and HIV infection, *Acta Derm. Venereol.*, 68:53.

Valle, S.L., 1987, Dermatologic findings related to human immunodeficiency virus infection in high-risk individuals, *J. Am. Acad. Dermatol.*, 17:951.

Vermander, F., and De Vroey, Ch., 1985, Dénombrement de *Pityrosporum ovale* par empreinte directe, *Bull. Soc. Franc. Mycol. Méd.*, 14:23.

Weismann, K., Knudsen, E.A., and Pedersen, C., 1987, White nails in AIDS/ARC due to *Trichophyton rubrum* infection, *Clin. Exp. Dermatol.*, 13:24.

Wishner, A.J., Teplitz, E.D., and Goodman, D.S., 1987, *Pityrosporum*, ketoconazole and seborrhoeic dermatitis, *J. Am. Acad. Dermatol.*, 17:140.

CLINICAL ASPECTS OF DERMATOMYCOSES IN AIDS PATIENTS

R.J. Hay

Department of Dermatology, United Medical and Dental Schools
Guy's Hospital, London SE1 9RT

The acquired immunodeficiency syndrome (AIDS) is a modern plague of
unparalleled proportions caused by human immunodeficiency viruses types 1
and 2 (HIV 1 & 2) and affecting predominantly young sexually active
individuals (Pinching, 1986). It also affects those receiving contaminated
blood products and intravenous drug abusers. The major clinical marker of
HIV infection is the development of tumours and opportunistic infections
both of which are associated with defective T lymphocyte function. The
infections range from *Pneumocystis carinii* pneumonia to mycobacterial
infection. Amongst the fungal complications of AIDS, oropharyngeal
candidosis is the commonest infection but cryptococcal meningitis is also
particularly important occurring in between 4-11% of affected patients.

While the deep mycoses have made a significant contribution to the
mortality and morbidity of the disease, superficial fungal infections are
very common in this group of patients. The commonest of these mycoses in
AIDS patients is superficial candidosis but dermatophyte infections and
seborrhoeic dermatitis are also found both in those with the established
syndrome and AIDS-related complex (ARC). The cutaneous manifestations of
AIDS, have been the subject of several recent reviews (Farthing, 1986;
Goodman et al., 1987).

AIDS AND IMMUNITY IN SUPERFICIAL MYCOSES

Resistance in superficial mycoses depends on both innate and acquired
immunological factors (Sohnle, 1989). Innate immunity has been the subject
of considerably less research although there are a number of different and
intriguing mechanisms which may affect the behaviour of superficial
pathogens such as the dermatophytes. Serum is inhibitory to the growth of
dermatophytes and certain other fungi including *Torulopsis (Candida)
glabrata*. This has been linked to the presence of unsaturated transferrin
which by a direct interaction with the organisms rather than, or possibly
in addition to, competition for iron prevents adequate growth (Artis et
al., 1983). Medium chain length fatty acids (FA's) such as those with C7-
13 residues are also inhibitory to dermatophytes *in vitro*. Different chain
length FA's suppress the growth of *Pityrosporum* yeasts (C4-8). The presence
of organisms invading the epidermal surface is also affected by the rate of
epidermal replication. Increased cell turnover can be demonstrated by
increased uptake of tritiated thymidine into basal cells.

Mycoses in AIDS Patients, Edited by
H. Vanden Bossche *et al.,* Plenum Press, New York, 1990

It is not clear how this reaction is regulated although with *Candida* infections the time course is more suggestive of a mechanisms dependent on the activation of T lymphocytes (Sohnle et al., 1976). Regulation of epidermal turnover by cytokines such as the interleukins IL-2 and IL-4 is well recognised and although a similar mechanism operative in superficial mycoses has not been proven it is possible that it also may play a role in the development of infection. The important issue here is whether this is dependent on an immunologically mediated reaction or whether fungi can directly stimulate epidermal growth by triggering the generation of inflammatory mediators such as leukotrienes to which T lymphocyte activation contributes an amplification arc or whether this is solely dependent on T cell activation. Likewise there is clearly another interaction between innate and acquired mechanisms in the recruitment of polymorphonuclear leukocytes (PMN's). In histological sections of superficial mycosis particularly candidosis and inflammatory dermatophyte infections there are usually large numbers of PMNs. These are known to be able to destroy both *Candida* and dermatophytes, in the latter instance by generating respiratory burst enzymes (Calderon and Hay, 1987). Macrophages are less active in this respect but both macrophages and PMNs can be stimulated by release of gamma interferon - in the control of *Candida* at least. It is therefore difficult to separate immune from non-immune mechanisms of skin defense.

Of the purely immunological mechanisms the development of T lymphocyte mediated responses are thought to be most important. In experimentally infected mice, T lymphocytes with the Ly2-phenotype have been shown to be largely responsible for the transfer of immunity to naive animals Calderon and Hay, 1984). Defective T cell mediated antigen specific blastogenesis has been associated with chronic infection and the clearance of ringworm in infected humans correlates with the emergence of delayed type hypersensitivity to intradermal trichophytin. The effector mechanisms are less well understood but two potential regulatory pathways include the amplification of eicosanoid elicited inflammation and epidermal hyperproliferation via cytokines such as IL-4 and tumour necrosis factor.

In AIDS the major immune defect is in both the numbers and function of helper lymphocytes bearing the CD4 (helper subset) marker. In addition other parameters of immune capacity from PMN counts to IgE mediated responses may be affected. In the latter instance the proliferation of B lymphocytes is associated with hyperglobulinaemia. These changes will affect the integrity of the immune response in the skin by interfering with a number of important functions such as the expression of HLA (Class 2) antigens by keratinocytes, the production of cytokines by T cells which aid recruitment of other inflammatory cells or regulate epidermal growth as well as the functioning of Langerhans cells. Natural killer (NK) activity is also reduced in AIDS. It follows that patients with AIDS will present with altered clinical responses to superficial fungal infection although, in the case of exogenous infections such as dermatophytosis, the incidence of disease in this population will still depend on frequency of exposure.

DERMATOPHYTOSIS

Dermatophyte infections are seen in patients with AIDS and are said to be common. However there is little evidence that they occur more frequently than in comparable groups In one survey the incidence of dermatophytosis was no different in homosexual males with or without AIDS (37.3 v. 31.8%) although infection was commoner in both groups than in the general populace (Torssander et al., 1988). The main site of infection was the toe clefts and in 70% it was caused by *Trichophyton rubrum*. Mixed infection was only seen in the homosexual patients. Those with dermatophytosis had had more sexual partners over the previous 12 months than those without infection. There was no correlation between CD4/CD8 ratios and the appearance of dermatophytosis. Sixteen of 46 HIV positive individuals had nail infections but the fungal involvement of the foot was not significantly different in clinical expression to that seen in HIV negative individuals. The situation is similar to that seen with renal transplant recipients where dermatophyte infections are no commoner, but may show clinical distinguishing features and are unusually resistant to therapy. Tinea pedis, cruris and corporis are all seen in AIDS and in many cases have a classical appearance. However in some patients the characteristic inflammatory rim is absent, the only signs of the infection being minimal scaling and the appearance of pustular or papular lesions around hair follicles. It has been suggested that despite the reservations expressed above, the development of extensive or clinically atypical dermatophyte infections in AIDS patients is more frequent in patients with established AIDS or persistent generalised lymphadenopathy, compared to HIV positive but asymptomatic individuals (Muhlemann et al, 1986). This includes a number of different clinical variants. For instance tinea incognito affecting the face has been described by several investigators (Pernicario and Peters, 1986). Facial dermatophytosis may also be mistaken for seborrhoeic dermatitis in AIDS patients where it involves the centrofacial region. Nail infections may also occur in HIV infected individuals. Recently a characteristic form of rapidly spreading proximal subungual onychomycosis has been associated with AIDS (Weissman et al., 1988) and the clinical appearances are sufficiently unusual to arouse suspicion. Here the infection spreads rapidly under and through the nail until the plate is completely opaque.

The response to therapy is very variable but, where appropriate, treatment with a topical antifungal such as an imidazole agent would be the first choice. In many cases it is necessary to use systemic therapy with griseofulvin or itraconazole, particularly in the non-inflammatory form which appear to be more recalcitrant. Relapse is very common. There is no evidence, at present, of an increase in adverse reactions to griseofulvin or other antifungals such as itraconazole used in dermatophytosis in patients with AIDS.

CANDIDOSIS

Oral and perianal candidosis are well recognised in AIDS (Farthing, 1986). Both tend to be persistent and may require oral therapy in many cases. There is often an association with underlying herpes simplex (HSV 1 & 2) infection, which requires treatment to prevent recurrence of the fungal disease. Oesophageal candidosis is another important complication seen in this group (Tavitian et al., 1986). There is a high correlation between the presence of oral and oesophageal candidosis, but the latter may be asymptomatic. Oral candidosis is common in patients with established AIDS but is less so in those presenting with Kaposi's sarcoma alone. The clinical appearances range from small white plaques to moderate mucosal hypertrophy similar to that seen in individuals with chronic mucocutaneous candidosis, although chronic atrophic changes are not seen. There are often widespread plaques which bleed easily and may involve the tongue and pharynx as well as the buccal mucosa. Eating may become distinctly uncomfortable. Invasion of adjacent skin surfaces is common and perleche or angular cheilitis is often seen where there is active oral candidosis. Candidosis elsewhere on the skin may also be found, and intertrigal infection of the groins in males is common. Vaginal infections were originally reported less frequently, but as the disease has spread into the heterosexual population some surveys have shown it to be commoner than oral candidosis (Carpenter et al., 1989). The symptoms and signs do not differ significantly from those seen in otherwise healthy subjects. While the latter survey suggested that the infection was normally responsive some investigators have claimed that vaginal infection in women with AIDS may be refractory to therapy. Persistent perianal candidosis may also be seen and this may be associated with HSV infection. There have now been reports of an increased incidence of *Candida* onychomycosis, including paronychia, in AIDS patients. This is of interest given the prominence of nail plate invasion in *Candida* infections seen in patients with chronic mucutaneous candidosis who in many ways closely resemble AIDS patients in their pattern of susceptibility to cutaneous infections.

As with dermatophytosis it is important to initiate therapy with the simplest approach using topical antifungals. However, many patients with oral disease may require ketoconazole, itraconazole or fluconazole, which is rapidly active in AIDS candidosis. Presumed ketoconazole resistance has been reported in some patients with oropharyngeal infections (Tavitian et al., 1986).

PITYROSPORUM IMFECTIONS

Infections caused by or conditions associated with *Pityrosporum* yeasts include pityriasis versicolor, *Pityrosporum* folliculitis and seborrhoeic dermatitis. Pityriasis versicolor has been seen in AIDS patients but is not particularly common and the clinical features do not really differ from those seen normally. Seborrhoeic dermatitis has long been associated with the presence of *Pityrosporum* yeasts on the skin surface. With the observation that this skin condition responds well to ketoconazole as well as other azoles interest in the relationship between the two has revived. It has been found in some groups to be one of the earliest clinical markers of HIV infection occurring in a high proportion of those with Walter Reed scale types 1 and 2 infections (Berger et al., 1988). Widespread seborrhoeic dermatitis has also been found in those with advanced symptomatic AIDS (Alessi et al., 1988). Overall assessments of its prevalence in AIDS patients have ranged from 30-80% (Groisser et al., 1989). Seborrhoeic dermatitis is common in both AIDS and ARC (Goodman et al., 1987). .It is often very florid in these patients affecting both the face and upper part of the trunk. Pustular lesions are particularly common. The onset of seborrhoeic dermatitis in this group is often very rapid and the rash spreads extensively (Mathes and Douglass, 1985). It may affect areas not often involved such as the groins. In some patients the development of psoriasis has been recorded in patients first presenting with seborrhoeic dermatitis. A correlation between the numbers of yeasts adhering to a single keratinocyte and the severity of the condition has been suggested (Groisser et al., 1989). The pathological mechanisms which may explain the prevalence of this condition in AIDS are not understood. There is no evidence of contact sensitization to *Pityrosporum* antigen although antibody titres against these organisms in non AIDS patients may be increased. *Pityrosporum* folliculitis is also very common in AIDS patients (Midgely and Hay, 1987). The fault may therefore lie in a defect of immunoregulation of antibody reponses. It has been suggested that the seborrhoeic dermatitis in AIDS differs histologically to that seen in other groups, with large numbers of plasma cells being found in the dermal infiltrate (Soeprono et al., 1986). The rash responds well to topically applied azole antifungal creams such as clotrimazole/hydrocortisone as well as topical corticosteroids, but relapse is to be expected.

OTHER PORMS OF SUPERFICIAL MYCOSES IN AIDS PATIENTS

In addition to the common superficial mycoses. other rarer forms of fungal skin infection may occur in AIDS patients. These have ranged from cutaneous alternariosis to skin lesions of disseminated systemic mycoses such as histoplasmosis and cryptococcosis. Increased skin surface carriage of *Trichosporon* species in the perianal area has been described in HIV positive patients without the resulting infection, white piedra. One further complication which may occur in AIDS patients is that skin lesions may have a dual aetiology (Gretzula and Penneys, 1987). This is seen in particular in the coexistence of *Candida* and herpes simplex viruses and may result in a bizarre and atypical clinical appearance. Skin biopsy may be necessary in such cases to establish the true nature of the infection.

REFERENCES

Alessi, E., Cusini, M., and Zerboni, R., 1988, Mucocutaneous manifestions in patients infected with human immunodeficiency virus, J. Am. Acad. Dermatol., 19:290.

Artis, W.M., Patrisky, E., Rastinejard, F., and Duncan, R.L., 1983, Fungistatic mechanism of human transferrin for *Rhizopus oryzae* and *Trichophyton mentagrophytes*; alternative to simple iron depletion, Infect. Imm., 41:1269.

Berger, R.S., Stoker, M.F., Hobbs, E.R., Hayes, J.J., and Boswell, R.W., 1988, Cutaneous manifestations of early human immunodeficiency virus exposure, J. Am. Acad. Dermatol., 19:298.

Calderon, R., and Hay, R.J., 1984, Cell-mediated immunity in experimental murine dermatophytosis II. Adoptive transfer of immunity to dermatophyte infection by lymphoid cells from donors with acute or chronic infections, Immunology, 53:465.

Calderon, R., and Hay, R.J., 1987, Fungicidal activity of human neutrophils and monocytes on dermatophyte fungi, *Trichophyton quinckeanum* and *T. rubrum*, Immunology, 612:289.

Carpenter, C.S., Mayer, K.H., Fisher, A., Desai, M.B., and Durand, L., 1989, Natural history of acquired immunodeficiency syndrome in women in Rhode Island, Am. J. Med., 86:771.

Farthing, C., 1986, Non-malignant cutaneous disease in AIDS and related conditions, in 'AIDS and HIV Infection', Clin. Immunol. Allergy, 6:559.

Goodman, D.S., Teplitz, E.D., Wishner, A., Klein, R.S., Burk, P.G., and Herschenbaum, E., 1987, Prevalence of cutaneous disease in patients with acquired immunodeficiency syndrome (AIDS) or AIDS-related complex, J. Am. Acad. Dermatol., 17:210.

Gretzula, D.O., and Penneys, N.S., 1987, Complex viral and fungal skin lesions of patients with acquired immunodeficiency syndrome, J. Am. Acad. Dermatol., 16:1154.

Groisser, D., Bottone, E.J., and Lebwohl, M., 1989, Association of *Pityrosporum orbiculare* (*Malassezia furfur*) with seborrheic dermatitis in patients with acquired immunodeficiency (AIDS), J. Am. Acad. Dermatol., 20:770.

Mathes, B.M., and Douglass, M.C., 1985, Seborrheic dermatitis in patients with acquired immunodeficiency syndrome, J. Am. Acad. Dermatol., 13:947.

Midgley, G., and Hay, R.J., ,1988, Serological responses to *Pityrosporum* (*Malassezia*) in seborrhoeic dermatitis demonstrated by ELISA and Western blotting, Bull. Soc. Fr. Mycol. Med., 17:267.

Muhlemann, M.F., Anderson, M.G., Paradinas, F.J., Key, P.R., Dawson, S.S., Evans, B.A., Murray Lyon, I.M., and Cream, J.J., 1986, Early warning skin signs in AIDS and persistent generalised lymphadenopathy, Br. J. Dermatol., 114:419.

Perniciaro, C., and Peters, M.S., 1986, Tinea faciale mimicking seborrheic dermatitis in a patient with AIDS, N. Engl. J. Med., 314:315.

Pinching, A.J., 1986, The spectrum of human immunodeficiency virus (HIV) infection; routes infection, natural history, prevention and treatment.in AIDS and HIV infection, Clin. Immunol. Allergy, 6:467.

Soeprono, F.F., Schinella, R.A., Cockerell, C.J., and Comite, S.L., 1986, Seborrheic-like dermatitis of acquired immunodeficiency syndrome. A clinicopathologic study. J. Am. Acad. Dermatol., 14:242.

Sohnle, P.G., Frank, M.M., and Kirkpatrick, C.H., 1976, Mechanisms involved in elimination of organisms from experimental cutaneous *Candida albicans* infections in guinea pigs, J. Immunol., 117:523.

Sohnle, P.G., 1989, Dermatophytosis, in: "Immunology of Fungal Diseases", R.A. Cox, ed., CRC Press Inc, Florida.

Tavitian, A., Raufman, J.P., and Rosenthal, L.E., 1986, Oral candidosis as a marker of oesophageal candidiasis in AIDS, Ann. Intern. Med., 104:54.

Tavitian, A., Raufman, J.R., Rosenthal, L.E., Webber, C.A., and
    Dincsoy,H.P., 1986, Ketoconazole resistant candida oesophagitis in
    patients with acquired immunodeficiency syndrome, Gastroenterology,
    90:443.
Torssander, J., Karlsson, A., Morfeldt-Manson, L., Putkonen, P., and
    Wasserman, J., 1988, Dermatophytosis and HIV infection – a study in
    homosexual men, Acta Dermatovenereol., 68:53.
Weismann, K., Knudsen, E.A., and Pedersen, C., 1988, White nails in
    AIDS/ARC due to Trichophyton rubrum infection, Clin. Exp. Dermatol.,
    13:24.

UNUSUAL MYCOSES IN AIDS PATIENTS

Maria Anna Viviani and Anna Maria Tortorano

Istituto di Igiene
Universita di Milano, Italy

Among the unusual fungal infections that have been reported from time to time in HIV positive patients, some are caused by dimorphic fungi and are characterized by their presentation in a disseminated form (Table 1). These unusual mycoses may be considered as specifically related to the T-lymphocyte depletion induced by the human immunodeficiency virus (HIV). Two of them, penicilliosis marneffei and sporotrichosis, deserve special attention.

Out of the 30 cases of penicilliosis marneffei reported since 1959, 6, diagnosed within the last 2 years (Table 2), have occurred in HIV positive patients native of or visitors to rural areas in South East Asian countries such as Thailand, Vietnam, Hong Kong and Guangxi Zhuang in Southern China, where the fungus is endemic. Among these 6 patients only one was native to the endemic area, a 43 year old woman who had undergone gynecological surgery and blood transfusion 5 years before she developed disseminated penicilliosis. The remaining patients had been travelling in South East Asia. A 7th case could have been added to the group. This case was reported 4 years ago (So et al., 1985). It concerned a 53 year old Chinese sailor from Hong Kong whose clinical history satisfies some of the criteria for the diagnosis of AIDS. However HIV serology was not performed.

Table 1. Unusual mycoses caused by dimorphic fungi

Disseminated mycoses	No*	Reference
Penicilliosis marneffei	6	see Table 2
Sporotrichosis	4	see Table 4
Paracoccidioidomycosis	3	Goldani et al., 1989
		Pedro et al., 1989**
Blastomycosis	2	Chiu et al., 1988**
Histoplasmosis duboisii	1	Peeters et al., 1987

* Number of cases reported up to June 1989
** Personal communication

Table 2. Penicilliosis marneffei in HIV positive patients

Pt. Sex/ Age (years)	Country of origin	Risk factor for HIV infection	P. marneffei present in*	Successful treatment with**	References
m/46	England	homosexual	l,b,sk,bm,u	AMB+KETO	Peto et al.,1988
m/40	France	homosexual	l,b,pf	AMB+5FC	Ancelle et al., 1988
m/33	U.S.A.	homosexual	l,b,sk,bm,st	AMB	Piehl et al., 1988
m/40	France	homosexual	l	KETO	Romana et al., 1989
f/43	Thailand	blood transfusion	l,b,sk,bm,ln	?	Sathapathaya-vongs et al., 1989
m/33	Italy	iv drug user	l,b,sk	AMB+5FC	Viviani et al., in preparation

* l=lung; b=blood; sk=skin nodule; bm=bone marrow; ln=lymph node;
pf=pleural fluid; st=stools; u=urine
**AMB=intravenous amphotericin B; KETO=ketoconazole; 5FC=flucytosine
Pt.=patient

The clinical manifestation of penicilliosis marneffei reflected the diffuse invasive process of the disease. The infection was disseminated in all cases except one (Romana et al., 1989) where the diagnosis was made in the early stages, by isolation of the fungus from bronchoalveolar lavage. At the time of diagnosis there was no radiological evidence of pulmonary involvement. Predominant clinical signs and symptoms were high fever, weakness, weight loss, cough, lymphoadenopathy, hepatomegaly, leukocytosis, and anemia. These symptoms are not specific. Patients with AIDS present similar manifestations caused by HIV infection as well as by other coexisting opportunistic infections or lymphoma. The clinician who is not aware of penicilliosis marneffei may confuse it with tuberculosis or other systemic infections. Similarly, the infection may be misdiagnosed by the microbiologist and the pathologist. The microbiologist, not being aware that *Penicillium marneffei* is a pathogen, may discard it as a contaminant. And the pathologist may overlook the fungus in tissue sections because of its poor staining with haematoxylin-eosin or may confuse the yeast-like cells within the macrophages/histiocytes with *Histoplasma capsulatum* var. *capsulatum*, especially if the characteristic septation of the yeast-like cells is rare. In our patient the internal septations were extremely uncommon (Figure 1). However, blood cultures grew a great number of yeast cells as did cultures of the skin biopsy (Figure 2 and 3). The mycosis may not be misdiagnosed, if the clinician bears in mind that an HIV positive patient, who has visited an endemic area, may have become infected by *P.marneffei*. The mycologist and pathologist, who are involved in the diagnosis, have to be kept informed about these anamnestic data.

Table 3. *In vitro* sensitivity of 10 clinical and rodent isolates of *Penicillium marneffei**

Antifungal	MIC (μg/ml)	MFC (μg/ml)
Amphotericin B	0.19-0.78	0.78-3.12
Flucytosine	0.19-0.78	0.78-12.5
Miconazole	0.19	0.19
Ketoconazole	0.19-0.39	0.19-0.78
Itraconazole	≤0.19	≤0.19

* Sekhon et al., 1989, personal communication

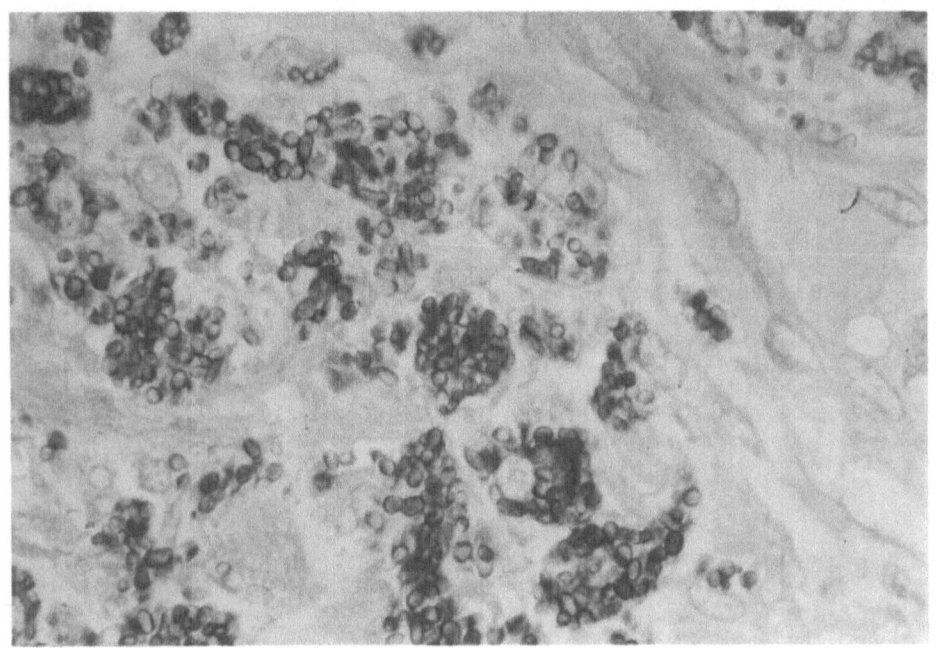

Fig. 1. Tissue section of a skin nodule showing unicellular yeast-like cells of *Penicillium marneffei* within and outside macrophages (Gomori-Grocott's methenamine silver stain).

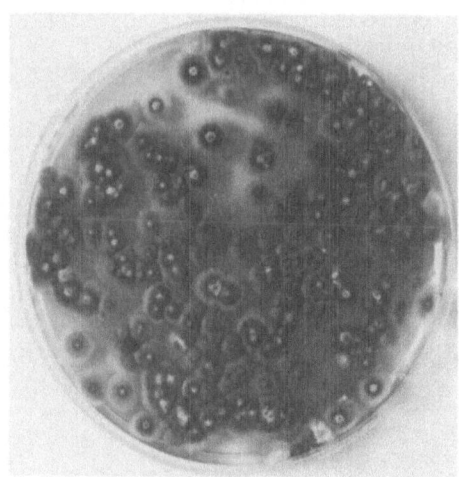

Fig. 2. Colonies of *P. marneffei* grown from blood culture on Sabouraud dextrose agar after 7 days at 37°C.

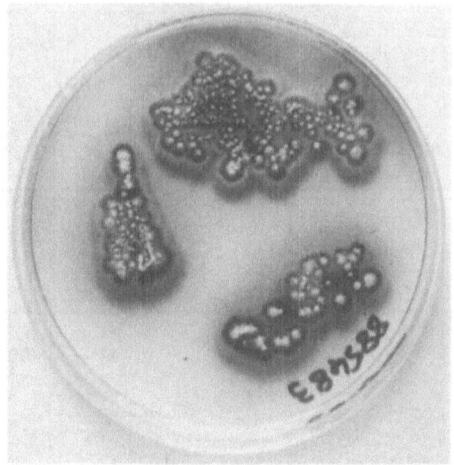

Figure 3: Colonies of *P. marneffei* grown from skin biopsy on Sabouraud dextrose agar after 7 days at 37°C.

Treatment of this fungal infection does not present a real problem. Amphotericin B, flucytosine, miconazole, ketoconazole, and itraconazole were tested *in vitro* against 10 clinical and rodent isolates of *P. marneffei* (Table 3). All were found to be active, in particular, the azoles. Despite the numerous deaths reported in the past, penicilliosis marneffei affecting AIDS patients has been successfully treated with antifungal drugs (Table 2). Treatment of the patient from Thailand was not reported. The need for a long-term antifungal treatment has not yet been established. Nevertheless, it should be noted that the patient from the U.S.A., whose amphotericin B therapy was discontinued after resolution of symptoms, experienced a relapse within 2 months. Unfortunately, it was not reported whether the patient was also mycologically cured when the drug was discontinued. Our patient, after amphotericin B plus flucytosine primary therapy, received itraconazole as maintenance treatment. At death, necropsy did not reveal any evidence of *P. marneffei* infection.

The other dimorphic fungus that has been reported to cause unusual disseminated infections in HIV positive patients is *Sporothrix schenckii*. The disseminated form of sporotrichosis is rare, but it correlates well with T-cell depletion characteristic of the acquired immunodeficiency syndrome. It is well known that normal cellular mediated immunity keeps this infection restricted to the skin.

Four cases of disseminated sporotrichosis have been reported up to now in AIDS patients (Table 4). No relation to one of the predisposing factors of HIV infection was recognized. The disease began with an insidious onset of fatigue, weight loss and low grade fever. Later on the infection was characterized by widespread enlargement of the erythematous cutaneous papules or nodules that ulcerated. The etiology of these cutaneous lesions may be easily identified if proper diagnostic procedures are adopted. The skin biopsy, besides being histologically examined, should always be cultured. Histologic sections may yield very few yeast cells and these cells can be brought into evidence by special stains, such as periodic acid Schiff or methenamine silver. Cultures, in this case, are more often diagnostic. It is also possible that histologic sections of the nodules reveal numerous, round, oval or cigar-shaped budding yeasts within histiocytes and macrophages in the dermal abscess, as in the case reported by Fitzpatrick and Eubanks (1988). In their patient, nearly 20 small erythematous to violaceous papules of the trunk and extremities were present following an initial cutaneous ulcerated plaque at the right cheek. The fungus may also be cultured from blood and bronchial secretions. Colonies from cultures of bronchial secretions may remain glabrous and white, mimicking *Geotrichum spp.*, and it may happen that they are discarded as contaminants.

Table 4. Disseminated sporotrichosis in HIV positive patients

Pt. Sex/Age (years)	Risk factor for HIV infection	Involved organ or tissue	Reference
m/34	iv drug user	skin, lung, liver spleen, joint fluid	Lipstein-Kresh et al., 1985
f/71	factor VIII infusion	skin (multiple lesions)	Bibler et al., 1986
m/30	homosexual	skin (multiple lesions), eye	Kurosawa et al., 1988
m/43	homosexual	skin (multiple lesions), lung	Fitzpatrick and Eubanks, 1988

Table 5. Cutaneous phaeohyphomycosis

Patient Sex/Age (years)	Risk factor for HIV infection	Isolated fungus	Reference
f/7	blood transfused	*Alternaria sp.*	Levy-Klotz et al., 1985
m/31	homosexual	*A. alternata*	Wiest et al., 1987
m/28	homosexual	*Curvularia sp.* *Bipolaris sp.*	Duvic et al., 1987

Intravenous amphotericin B has been till now the treatment of choice in case of disseminated sporotrichosis. Itraconazole, which is highly effective in lymphocutaneous sporotrichosis cases (Grant and Clissold, 1989), has not yet been utilized to our knowledge in the treatment of the disseminated form.

Differential diagnosis of cutaneous lesions in patients with AIDS should take into consideration phaeohyphomycosis. Three cases have been reported, 2 in homosexuals and 1 in a blood transfused patient (Table 5). In all the patients their cutaneous phaeohyphomycosis was characterized by a unique lesion that was ulcerated in 2 of them. This was in contrast to the cutaneous pattern of sporotrichosis in AIDS.

Several other unusual mycoses have been reported in AIDS patients, however they are related to predisposing factors other than T-lymphocyte depletion. Treatment of the HIV infection and of the coexisting opportunistic infections or malignancies causes further deterioration of the immunological defenses of the patient, enhancing mycoses such as zygomycosis, phaeohyphomycosis, aspergillosis, fusariosis, trichosporonosis, saccharomycosis (Cuadrado et al., 1988; Mostaza et al., 1989; Smith et al., 1989; Colon et al., 1988, personal communication; Del Palacio Hernanz et al., 1989a; Holmberg and Meyer, 1986; Del Palacio-Hernanz et al., 1989b; Leaf and Simberkoff, 1989; Sethi & Mandell, 1988; Tawfick et al., 1989).

Also some of the factors that predispose to HIV infection may predispose to unusual mycoses. Fungal encephalitis has been underscored as occurring in intravenous drug abusers (Wetli et al., 1984; Micozzi and Wetli, 1985). The injury to the central nervous systems microvasculature that is induced by drugs predisposes to opportunistic infections. Phaeohyphomycosis and zygomycosis of the central nervous system have been reported in intravenous drug addicts with AIDS (Colon et al., 1988, personal communication; Del Palacio-Hernanz et al., 1989a; Cuadrado et al., 1988; Mostaza et al., 1989).

Finally, homosexual behaviour may increase the carrier rate of *Trichosporon beigelii* in the anal region. An increase from 2% to about 16% was found in two different investigations (Torssander et al. 1985; Stenderup et al., 1986). This colonization may imply a potential risk of developing invasive trichosporonosis of endogenous origin. One case has been recently described in AIDS (Leaf and Simberkoff, 1989).

It is likely that in the near future other unusual opportunistic mycoses will be seen with increasing frequency in AIDS patients. Physicians should be prepared to consider any fungus, including the rarely encountered ones, as a potential etiologic agent and to suspect penicilliosis marneffei in any HIV positive patient who lived in, or visited, the South Eastern Asian areas endemic for that disease.

Lastly we recommend that two unusual mycoses, namely disseminated penicilliosis marneffei and disseminated sporotrichosis, be listed among the fungal infections that are indicative of AIDS.

REFERENCES

Ancelle, T., Dupouy-Camet, J., Pujol, F., Nassif, X., Ferradini, L., and Lapierre, J., 1988, Un cas de penicilliose disseminée à *Penicillium marneffei* chez un malade atteint de SIDA, Bull. Soc. Fr. Mycol. Med., 17:73.

Bibler, M.R., Luber, H.J., Glueck, H.I., and Estes, S.A., 1986, Disseminated sporotrichosis in a patient with HIV infection after treatment for acquired factor VIII inhibitor, J.A.M.A., 256:3125.

Cuadrado, L.M., Guerrero, A., Lopez Garcia Asenjo, J.A., Martin, F., Palau, E., and Garcia Urra, D., 1988, Cerebral mucormycosis in two cases of acquired immunodeficiency syndrome, Arch. Neurol., 45:109.

Del Palacio-Hernanz, A., Moore, M.K., Campbell, C.K., Del Palacio-Perez-Medel, A., and Del Castillo-Cantero, R., 1989a, Infection of the central nervous system by *Rhinocladiella atrovirens* in a patient with acquired immunodeficiency syndrome, J. Med. Veter. Mycol., 27:127.

Del Palacio-Hernanz, A., Vera Casado, A., Fernandez Lopez, A., Herrero Quiros, O., and Moreno Palancar, P., 1989b, Infeccion oportunista pulmonar por *Fusarium* moniliforme en paciente con SIDA, Rev. Iber. Micologia, 6:144.

Duvic, M., Lowe, L., Rios, A., MacDonald, E., and Vance, P., 1987, Superficial phaeohyphomycosis of the scrotum in a patient with the acquired immunodeficiency syndrome, Arch. Dermatol., 123:1597.

Fitzpatrick, J.E., and Eubanks, S., 1988, Acquired immunodeficiency syndrome presenting as disseminated cutaneous sporotrichosis. Int. J. Dermatol., 27:406.

Goldani, L.Z., Martinez, R., Landell, G.A.M., Machado, A.A., and Continho, V., 1989, Paracoccidioidomycosis in a patient with acquired immunodeficiency syndrome, Mycopathol., 105:71.

Grant, S.M., and Clissold, S.P., 1989, Itraconazole. A review of its pharmacodynamic and pharmacokinetic properties and therapeutic use in superficial and systemic mycoses, Drug, 37:310.

Holmberg, K., and Meyer, R.D., 1986, Fungal infections in patients with AIDS and AIDS-related Complex, Scand. J. Infect. Dis., 18:179.

Kurosawa, A., Pollock, S.C., Collins, M.P., Kraff, C.R., and Tso, M.O.M., 1988, *Sporothrix schenckii* endophthalmitis in a patient with human immunodeficiency virus infection, Arch. Ophthalmol., 106:376.

Leaf, H.L., and Simberkoff, M.S., 1989, Invasive trichosporonosis in a patient with acquired immunodeficiency syndrome, J. Infect. Dis., 160:356.

Levy-Klotz, B., Badillet, G., Cavelier-Balloy, B., Chemaly, Ph., Leverger, G., and Civatte, J., 1985, Alternariose cutanée au cours d'un SIDA, Ann. Dermatol. Venereol., 112:739.

Lipstein-Kresch, E., Isenberg, H.D., Singer, C., Cooke, O., and Greenwald, R.A., 1985, Disseminated *Sporothrix schenckii* infection with arthritis in a patient with acquired immunodeficiency syndrome, J. Rheumatol., 12:805.

Micozzi, M.S., and Wetli, C.V., 1985, Intravenous amphetamine abuse, primary cerebral mucormycosis and acquired immunodeficiency, .J. Forensic. Sci., 30:504.

Mostaza, J.M., Barbado, F.J., Fernandez-Martin, J., Pena-Yanez, J., and Vazquez-Rodriguez, J.J., 1989, Cutaneoarticular mucormycosis due to *Cunninghamella bertholletiae* in a patient with AIDS, Rev. Infect. Dis., 11:316.

Peeters, P., Depré, G., Rickaert, F., Coremans-Pelseneer, J., and Serruys, E., 1987, Disseminated african histoplasmosis in a white heterosexual male patient with the acquired immune deficiency syndrome, Mycosen, 30:449.

Peto, T.E.A., Bull, R., Millard, P.R., Mackenzie, D.W.R., Campbell, C.K., Haines, M.E., and Mitchell, R.G., 1988, Systemic mycosis due to Penicillium marneffei in a patient with antibody to human immunodeficiency virus, J. Infect., 16:285.

Piehl, M.R., Kaplan, R.L., and Haber, M.H., 1988, Disseminated penicilliosis in a patient with acquired immunodeficiency syndrome, Arch. Pathol. Lab. Med., 112:1262.

Romana, CA, Stern, M, Chovin, S, Drouhet, E, and Pays, JF, 1989, Penicilliose pulmonaire a Penicillium marneffei chez un patient atteint d'un syndrome immunodéficitaire acquis, Bull. Soc. Fr. Mycol. Med., 18:311.

Sathapatayavongs, B., Damrongkitchaiporn, S., Saengditha, P., Kiatboonsri, S., and Jayanetra, P., 1989, Disseminated penicilliosis associated with HIV infection, J. Infect., 19:84.

Sethi, N., and Mandell, W., 1988, Saccharomyces fungemia in a patient with AIDS, N. Y. State J. Med., 88:278.

Smith, A.G., Bustamante, C.I., and Gilmor, G.D., 1989, Zygomycosis (Absidiomycosis) in an AIDS patient, Mycopathol., 105:7.

So, S.Y., Chau, P.Y., Jones, B.M., Wu, P.C., Pun, K.K., Lam, W.K., and Lawton, J.W.M., 1985, A case of invasive penicilliosis in Hong Kong with immunologic evaluation, Am. Rev. Resp. Dis., 131:662.

Stenderup, A., Schonheyder, H., Ebbesen, P., and Melbye, M., 1986, White piedra and Trichosporon beigelii carriage in homosexual men, J. Med. Veter. Mycol., 24:401.

Tawfik, O.W., Papasian, C.J., Dixon, A.Y., and Potter, L.M., 1989, Saccharomyces cerevisiae pneumonia in a patient with acquired immune deficiency syndrome, J. Clin. Microbiol., 27:1689.

Torssander, J., Carlsson, B., and Von Krogh, G., 1985, Trichosporon beigelii: increased occurrence in homosexual men, Mycosen, 28:355.

Wetli, C.V., Weiss, S.D., Cleary, T.J., and Gyori, E., 1984, Fungal cerebritis from intravenous drug abuse, J. Forensic. Sci., 29:260.

Wiest, P.M., Wiese, K., Jacobs, M.R., Morrissey, A.B., Abelson, T.I., Witt, W., and Lederman, M.M., 1987, Alternaria infection in a patient with acquired immunodeficiency syndrome: case report and review of invasive Alternaria infections, Rev. Infect. Dis., 9:799.

MECHANISM OF PATHOGENESIS - WHY ARE CERTAIN MYCOSES RARE IN AIDS PATIENTS?

H.P.R. Seeliger and K. Tintelnot

Institute of Hygiene and Microbiology, University of
Würzburg, D-8700 Würzburg, Josef-Schneider-Str. 2

INTRODUCTION

The response of the human body to fungi varies with the fungal species
involved, with their virulence and pathogenicity, with infection promoting
factors, but also with the host involved, i.e. the age, health condition,
the immunity status, susceptibility to infection or allergy etc.

Fungal infection may be almost obligatory with certain species
prevalent in the human environment, such as the agents causing the
geographically wide spread systemic mycoses - coccidioidomycosis,
histoplasmosis and others although the response of the human body may vary
a great deal from almost inapparent infection to fatal outcome. T-cell
dependant immunity with positive reactions in respective skin tests usually
indicates past or present infection as well as various degrees of
protection. In such cases AIDS may cause serious concern as reason for
severe relapse of apparently cured infections.

MAIN PART

The high incidence of cryptococcosis in AIDS patients and the leading
role of T-cell dependant immunity reactions of the human body in the
pathogenesis of this infection is generally accepted and also supported by
experimental evidence (Chandler, 1985).

With regard to the occurrence of hyphomycetic infection and certain
systemic mycoses in HIV infected individuals the opinions are not uniform.

As the aforementioned AIDS-associated infections due to *Histoplasma
capsulatum* and *Coccidioides immitis* are dealt with separately the following
report will be devoted to infections caused by molds, particularly by
Aspergillus fumigatus.

The number of respective cases as reported in the literature is
relatively small compared to infections by *Pneumocystis carinii*,
Cryptococcus neoformans, *Candida albicans* or *Histoplasma capsulatum* in
endemic areas.

The report of Chandler (1985) from the Centers for Disease Control
(CDC) in Atlanta covering the years 1983 and 1984 revealed only 5 cases
(0.16%) of invasive aspergillosis among 3170 AIDS patients. Lesbordes et

Mycoses in AIDS Patients, Edited by
H. Vanden Bossche *et al.,* Plenum Press, New York, 1990

al. (1986) mentioned 2 cases of invasive aspergillosis out of 49 AIDS victims of which one had suffered previously from an aspergilloma in a tuberculous cavity.

Still rarer are reports on other disseminated hyphomycoses in AIDS patients: Apparently neither Mucorales – occasionally complicating diabetes – nor *Fusarium* species seem to play a role in AIDS.

Only one case of *Alternaria* infection in a 7 years old girl has been reported by Levy-Klotz et al. (1985) as a consequence of a blood transfusion in acute lymphoblastic leukemia and subsequent AIDS.

Four single cases of sporotrichosis in AIDS patients have become known (Lipstein-Kresch et al., 1985; Bibler et al., 1986; Fitzpatrick and Eubanks, 1988; Kurosowa et al., 1988). Systemic sporotrichosis is rare and was occasionally reported in patients with hematologic and neoplastic diseases, alcoholism, diabetes mellitus and after prolonged corticosteroid therapy. The latter may have induced sporotrichosis in a 71 years old lady (Bibler et al., 1986) who was infected with AIDS by blood transfusion therapy for an acquired factor VIII deficiency.

Infections due to dematiaceous fungi are rare, such as the case reported by Rothe et al. (1989).

Although the CDC has eliminated aspergillosis from the list of AIDS associated opportunistic infections (Jaffe and Selik, 1984) some reports on the occasional occurrence of aspergillosis in this group (cf. Henochowicz et al., 1985; Woelki et al., 1989) point to a possible disposition of AIDS victims for this ubiquitous fungal species.

In the following the pathogenesis of some typical clinical syndromes caused by aspergilli will be briefly summarized, before an attempt will be made to evaluate the risk of aspergillosis in AIDS patients under particular consideration of their immunity status.

A superficial infection by aspergilli (and other fungi) may result in colonization of mucous membranes with resulting otomycosis which is saprophytic in nature and without a serologically demonstrable immune reaction.

Another form of saprophytic hyphomycosis may lead to aspergilloma (Boeckh et al., 1989) or mycetoma due to other fungi such as *Pseudallescheria boydii*. After inhalation the spores of the respective fungal species settle in preformed cavities, cysts or bronchiectases and start to grow on the surfaces. After contact of the resulting fungus ball with immunocytes without any involvement of T-cells, precipitins and other specific antibodies are produced and may become valuable in diagnosis. The wall of the cavity is formed by fibrous tissue and is vascularized. Resulting hemorrhage and hemoptysis may lead to a fatal outcome without previous dissemination.

Spores of aspergilli, of other hyphomycetes and thermoactinomycetes may cause IgE-induced type I allergy, but also type III allergy and thus exogenic allergic alveolitis (EAA) (cf. Reese et al., 1989). The frequent inhalation of *Aspergillus* spores from the environment will provoke in predisposed individuals the formation of antibodies of class IgG and subsequently of immune complexes which cause inflammatory reactions of the lung.

The pathogenesis of allergic bronchopulmonary aspergillosis (ABPA) is not yet completely clear. In patients with bronchial asthma the inhalation of *Aspergillus* spores may result in saprophytic growth and pulmonary infiltration. According to Rosenberg et al. (1977) five clinical stages can be distinguished, finally leading to pulmonary fibrosis. Specific IgE is observed in ABPA as well as IgG of all four subclasses (cf. Reese et al., 1989).

What these aspergillus-associated allergic syndromes have in common is that they result from humoral immune reactions. Their therapy includes, besides prevention of inhalation by various means (cf. Tintelnot & Seeliger, 1988), symptomatic corticosteroid therapy.

Entirely different pathological mechanisms, however, prevail in invasive aspergillosis (cf. Rinaldi, 1983). After uptake of *Aspergillus* spores by the body and their germination, hyphal growth invades the surrounding tissue. Invasive pulmonary aspergillosis is almost exclusively observed in immunosuppressed individuals and after severe disturbance of the leukocytic defence by other reasons (cf. Round table, Grenoble 1988, Proceedings in press).

Invasive aspergillosis leads to infiltration with central necrosis and epitheloid bodies which may resemble tuberculosis lesions. Other primary localizations of such lesions are possible. When dissemination occurs a lethal course is often inevitable in spite of antimycotic therapy.

Experimental investigations on the cellular and humoral immune reactions in mice have elucidated the pathogenesis of this syndrome. Inhalation of germinable *Aspergillus* spores results in immediate phagocytosis by alveolar macrophages (Merkow et al., 1971). Macrophages activated thereby proved to be more efficient than resident macrophages (Schaffner et al., 1982). If the macrophages are, however, unable to eliminate the spores, these can germinate and produce hyphae. In this system, an oxidative granulocytic killing system is required to prevent invasion of hyphae (Schaffner, 1989). Schaffner demonstrated in a mouse model that pretreatment with cortisone alters the function of macrophages to the extent that they will no longer be able to prevent the germination of spores. In addition the function of neutrophils is suppressed: High fatality rates among cortisone treated mice after intravasal inoculation of *Aspergillus* spores have been reported by Dixon, Polak and Walsh (1989). Moreover the amount of inhaled spores may have influence on and even exhaust the ability of macrophages, as revealed by reports of severe, acute necrotizing aspergillosis in primary healthy individuals after exposure to massive doses of mold spores (Karam & Griffin 1986, Schweisfurth 1972, Van de Wyngaert et al., 1986). Fungus dose-dependant primary pulmonary aspergillosis in immunsuppressed mice has been described by Dixon et al. (1986).

That polymorphonuclear phagocytes (PMs) in the blood and monocytes may damage hyphae of molds has been shown by *in vitro*-experiments of Diamond et al. (1978, 1983).

Neutrophils are able to kill the yeast cells of *Candida albicans*. But in contrast it has been suggested by Schaffner et al. (1982) that neutrophilic granulocytes will be involved in the defence against molds when the macrophage defence fails and hyphal growth infiltrates into the tissues. Levitz and Diamond (1985) found that neutrophils could kill *A. fumigatus* conidia that had been preincubated in broth so that they swelled and became metabolically active before germination. Of decisive influence for the susceptibility of AIDS patients for mold infections are the results obtained by Schaffner (1982), with athymic mice: no protective role was found for T-lymphocytes because there was no difference between healthy mice and athymic mice in resistance for infection with *Aspergillus* spores.

Thus the primary response to *A. fumigatus* spores leading to invasive aspergillosis is invariably bound to poorly functioning granulocytes and macrophages. It is obvious that the granulocytopenia or agranulocytosis resulting from natural disease or connected with chemotherapy or treatment with corticosteroids (or sometimes other remedies leading to agranulocytosis) as well as bone marrow irradiation etc., create conditions which make the host highly susceptible to infections by aspergilli.

When the spores germinate, microabscesses will usually develop with great numbers of leukocytes and macrophages or giant cells. The rapid progression of such infection, particularly in brain abscesses, in such uncontrolled infections almost excludes *per se* an intervention of T-cells and helper cells to build up an effective local defence and immunity at a later stage. Consequently, the suppression or elimination of helper cells in AIDS cannot be expected to exert any influence on the outcome or course of acute invasive aspergillosis – although there are reports of a few cases like the cerebral abscess mentioned in the beginning of this report of Woelki et al. (1989). The victim had suffered from cerebral toxoplasmosis and the aspergillosis was a subsequent infection. Similarly the case of Henochowicz et al. (1985) was an intravenous drug abuser who had also acquired pulmonary pneumocystosis complicated by *Aspergillus* pneumonia and endocarditis.

We support in this respect wholeheartedly the conclusions of Staib et al. (1989) who observed various species of *Aspergillus* in the sputum of 10 out of 32 patients suffering from AIDS-associated cryptococcosis, but who never developed any bronchial or pulmonary sequelae. They state that in these cases the cultivated aspergilli were of no pathogenic significance and lacked primary pathogenicity for the AIDS patient.

This is in line with similar observations in the University Hospital of Würzburg. There 12 out of 29 AIDS patients have died, but only in 2 of them aspergilli were cultivated in routine sputum examinations, without any indication of pathogenetic importance.

The seeming discrepancies of these observations may find their explanation in the use of high-doses of corticosteroids in combination with antibiotics for treatment in typical AIDS-associated infections such as pneumocystosis etc.

CONCLUSIONS

Most infections caused by hyphomycetes are circumstantial or conditioned by a variety of factors including diabetes which eventually promote invasion and propagation of fungal elements in various parts of the body. Those mycoses which lead to a primary leukocytic response, abscess formation, fistulation and/or granulomas will thus not be typical sequelae of AIDS, for instance, infections due to Mucorales, Dematiaceae or *Fusarium* and the group of fungal agents of maduromycosis, and possibly chromomycosis. *Sporothrix schenckii* may hold a similar position as it causes lymphangitis and lymphadenitis, but a number of reports deal with the finding of this mycosis in AIDS patients. – Likewise *Malassezia* which usually results in neither suppuration nor tissue reactions may belong to this category, although the authors are not certain in this respect.

This is a rather brief and possibly incomplete account of certain mycoses which are usually not directly associated with the consequences of HIV infection. The somewhat still limited number of individuals suffering from AIDS almost excludes any major occurrence of infections caused by hyphomycetic environmental fungi. On the other hand it is clear that the therapy of AIDS may in some cases induce infections such as aspergillosis after or in connection with the use of corticosteroids.

In some areas of mycotic research more experimental work may be required in order to evaluate the role of AIDS and more clinical data may in the future help in assessing the importance of T-helper cells in the response of the human being to fungal invasion.

REFERENCES

Bibler, M.R., Luber, H.J., Glueck, H.I., and Estes, S.A., 1986,
Disseminated sporotrichosis in a patient with HIV infection after
treatment for acquired factor VIII inhibitor, J.A.M.A., 256:3125.

Boeckh, M., Hoffken, G., and Lodl, H., 1989, Spektrum pulmonaler
Manifestation von Aspergillosen, Dtsch. med. Wschr., 113:1706.

Chandler, F.W., 1985, Pathology of the mycoses in patients with the
acquired immunodeficiency syndrome (AIDS), in: "Berlin Current
Topics in Medical Mycology", M.R. McFiunis, ed., Springer Verlag,
New York.

Diamond, R.D., Huber, E., and Haudenschild, C.C., 1983, Mechanisms of
destruction of Aspergillus fumigatus hyphae mediated by human
monocytes, J. Infect. Dis., 147:474.

Diamond, R.D., Krzesicki, R., Epstein, B., and Wellington, J., 1978, Damage
to hyphal forms of fungi by human leucocytes in vitro. A possible
host defense mechanism in Aspergillosis and Mucor mycosis, Am. J.
Pathol., 91:313.

Dixon, D.M., Polak, A., and Walsh, T.J., 1989, Fungus dose-dependent
primary pulmonary aspergillosis in immunosuppressed mice, Infect.
Immun., 57:1452.

Fitzpatrick, J.E., and Eubanks, S., 1988, Acquired immunodeficiency
syndrome presenting as disseminated cutaneous sporotrichosis, Int.
J. Dermatol., 27:406.

Henochowicz, S., Mustafa, M., Lawrinson, W.E., Pistole, M., and Lindsay, J.
Jr., 1985, Cardiac aspergillosis in acquired immune deficiency
syndrome, Am. J. Cardiol., 55:1239.

Jaffe, H.W., and Seik, R.M., 1984, Acquired immune deficiency syndrome: is
disseminated aspergillosis predictive of underlying cellular immune
deficiency? J. Infect. Dis., 149:828.

Karam, G.H. and Griffin, F.M. Jr., 1986, Invasive pulmonary Aspergillosis
in nonimmunocompromised nonneutropenic hosts, Rev. Infect. Dis.
8:357.

Kurosawa, A., Pollock, S.C., Collins, M.P., Kraff, C.R., Tso, M.O.M., 1988,
Sporothrix schenckii endophthalmitis in a patient with human
immunodeficiency virus infection, Arch. Ophthalmol., 106:376.

Lesbordes, J.L., Meunier, D.M.Y., and Georges-Courbot, M.C., 1986, Les
mycoses au cours du S.I.D.A. en Republique Centrafricaine, Med.
Trop., 46:257.

Levitz, S.T., and Diamond, R.D., 1985, Mechanisms of resistance of
Aspergillus fumigatus conidia to killing by neutrophils in vitro, J.
Infect. Dis., 152:33.

Levy-Klotz, B., Badillet, G., Cavalier-Balloy, B., Chemaly, P., Leverger,
G., and Civatte, J., 1985, Alternariose cutanee au cours d'un SIDA,
Ann. Dermatol. Venerol., 112:739.

Lipstein-Kresch, E., Isenburg, H.D., Singer, C., Cooke, O., and Greenwald,
R.A., 1985, Disseminated Sporothrix schenckii infection with
arthritis in a patient with acquired immunodeficiency syndrome, J.
Rheumatol., 12:805.

Merkow, L.P., Epstein, S.M., Sidransky, H., Verney, E., and Pardo, M.,
1971, The pathogenesis of experimental pulmonary aspergillosis. An
ultra structural study of alveolar macrophages after phagocytosis of
A. flavus spores in vivo, Am. J. Pathol., 52:57.

Reese, G., Tromplet, J., Becker, W.-M., and Schlaak, M., 1989, Exogen-
allergische Alveolitis: IgG-Subklassenreaktivitaten und monoklonale
Antikorper gegen Micropolyspora faeni und Aspergillus fumigatus,
Immun. Infekt., 17:165.

Rinaldi, M.G., 1983, Invasive aspergillosis, Rev. Infect. Dis., 5:1061.

Rosenberg, M., Patterson, R., Mintzer, R., Cooper, B.J., Roberts, M., and
Harris, K.E., 1977, Clinical and immunological criteria for the
diagnosis of allergic bronchopulmonary aspergillosis, Am. Intern.
Med., 86:405.

Rothe, M.J., et al., 1989, Cited after Mycology Observer, Dematiaceous
     fungi associated with AIDS, S. Kemper, ed., New York, 9, 1-2
Round Table: "Agranulocytosis and deep mycoses". Proceedings of meeting 30.
     Sept. and 1. Oct. 1988 in Grenoble, organized by Becton & Dickinson
     Comp. France (in press).
Schaffner, A., 1989, Experimental basis for the clinical epidemiology of
     fungal infections. A review, Mycoses, 32:499.
Schaffner, A., Douglas, H., and Braude, A., 1982, Selective protection
     against conidia by mononuclear and against mycelia by
     polymorphonuclear phagocytes in resistance to *Aspergillus*
     observations on these two lines of defense *in vivo* and *in vitro* with
     human and mouse phagocytes, J. Clin. Invest., 69:617.
Schimpff, S.C., and Bennett, J.E., 1975, Abnormalities in cell-mediated
     immunity in patients with *Cryptococcus neoformans* infection, J.
     Allergy Clin. Immunol., 55:430.
Schweisfurth, R., 1972, Unpublished personal communication.
Staib, F., Seibold, M., and Grosse, G., 1989, *Aspergillus* findings in AIDS
     patients suffering from cryptococcosis, Mycoses, 32:516.
Tintelnot, K., and Seeliger, H.P.R., 1988, Aktuelle Gesichtspunkte zur
     Epidemiologie der Aspergillose, Mycoses, 31:245.
Woelki, U., Schlote, W., and Stille, W., 1989, AIDS - von Fall zu Fall, Die
     Neue Arztliche, 39.
van de Wyngaert, F.A., Sindic, C.J.M., Rousseau, J.J., Fernandes Xavier,
     F.G., Brucher, J.M., and Laterre, E.C., 1986, Spinal arachnoiditis
     due to *Aspergillus meningitis* in a previously healing patient, J.
     Neurol., 233:41.

Dimorphic Fungi

HISTOPLASMA IN AIDS PATIENTS

Michael G. Rinaldi

Department of Pathology, Fungus Testing Laboratory,
University of Texas Health Science Center, San Antonio,
Texas, U.S.A.

INTRODUCTION

It is clear that histoplasmosis is a significant infectious
complication in AIDS patients and has been recognized as one of the
defining opportunistic infections for this syndrome. An increasing body of
data has accumulated and been published regarding histoplasmosis in AIDS
victims. One particularly fascinating aspect of this mycosis in AIDS
patients concerns its epidemiology.

GENERAL CONSIDERATIONS

We are concerned here with the epidemiology of histoplasmosis
capsulati caused by *Histoplasma capsulatum* variety *capsulatum* (as
contrasted to histoplasmosis duboisii incited by *H. capsulatum* var.
duboisii or histoplasmosis farciminosi, an equine disease caused by
H. capsulatum var. *farciminosum*), see Tables 1 and 2. Histoplasmosis is a
protean disease of cosmopolitan distribution occurring worldwide, but with
particularly noteworthy endemic areas located in the U.S.A. The AIDS
epidemic has allowed and necessitated additional epidemiologic studies of
H. capsulatum var. *capsulatum*.

Table 1. Anamorphic classification of the genus *Histoplasma*

Superkingdom:	Eukaryota
Kingdom:	Fungi
Form-Division:	Fungi Imperfecti
Form-Class:	Hyphomycetes
Form-Order:	Hyphomycetales
Form-Family:	Moniliaceae
Form-Genus:	*Histoplasma* Darling, 1906
Form Species/Varieties:	*H. capsulatum* variety *capsulatum* Darling, 1906
	H. capsulatum variety *duboisii* (Vanbreuseghem) Ciferri 1960
	H. capsulatum variety *farciminosum* (Rivolta) Weeks, Padhye, et Ajello, 1985

Mycoses in AIDS Patients, Edited by
H. Vanden Bossche *et al.*, Plenum Press, New York, 1990

Table 2. Diseases caused by *Histoplasma*

Histoplasmosis capsulati

  Synonyms = classical histoplasmosis, small form histoplasmosis, American
              histoplasmosis, Darling's disease, Tingo Maria fever,
              reticuloendotheliosis, reticuloendotheliosis cytomycosis

Histoplasmosis duboisii

  Synonyms = African histoplasmosis, large form histoplasmosis

Histoplasmosis farciminosi

  Synonyms = epizootic lymphangitis, equine epizootic lymphangitis

---

EPIDEMIOLOGY

It is firmly established that *H. capsulatum* var. *capsulatum* grows in soil with high nitrogen content usually associated with the fecal material of birds and bats (Rippon, 1988). It has been isolated many times from bat-inhabited caves, bird roosts, chicken coops and like habitats. The fungus appears to grow in decaying fecal materials mixed with soil rather than in fresh droppings and may also grow on bird feathers. Not all guano serves equally as a substrate and often the fungus is restricted to small areas where birds/bats congregate in large numbers (DiSalvo and Johnson, 1979).

The most highly endemic areas in the U.S.A. (perhaps in the world) are located in the Ohio-Mississippi River Valley (Missouri, Kentucky, Tennessee, southern Illinois, Indiana and Ohio) which are also the areas with the greatest concentration of starling birds. These avians congregate in huge numbers with resultant formation of large excreta deposits. Interestingly, in South America, where starlings have not been introduced, most *H. capsulatum* var. *capsulatum* is associated with chicken and bat habitats (Negroni, 1940).

Despite abundant growth only in relatively restricted endemic areas, the fungus may be found throughout the world. Factors favoring growth include a mean temperature of 68 to 90°F (22 to 29°C), annual precipitation of 35 to 50 inches (ca. 1,000 mm), and relative humidity of 67 to 87% or more during the growing season (Carmona, 1971; Furcolow, 1958). These conditions occur in many areas of tropical, subtropical, and temperate locales of the globe where there is adequate moisture. It is intriguing that most of Europe seems to have escaped heavy infestation.

There also exist scattered areas of high endemicity within areas lacking the aforementioned environmental conditions. These "closed" areas are inevitably associated with caves inhabited by birds and/or bats. The protected and stable conditions inside the caves allow fungal proliferation. While birds and bats must certainly disseminate *H. capsulatum* var. *capsulatum* in caves/chicken coops, it is the wind which serves as the chief vector of fungal dispersion in most areas. Of note, *H. capsulatum* var. *capsulatum* actually lives in the gastrointestinal tract of bats while birds do not appear to become infected, due to their high body temperature which precludes fungal tissue invasion (Carmona, 1971; Emmons, 1949; Klite and Dierchel, 1965).

Other epidemiologic features influencing histoplasmosis include wind direction, limestone in the soil, flooding of riverbanks, etc. (Furcolow, 1958). Using histoplasmin skin testing as an epidemiologic tool, it appears that the largest area with high prevalence (80 to 90%) is the mid-section of the North American continent (Manos et al., 1956). However, there are

other areas where the disease occurs in high prevalence including southern
Mexico, northern Panama, Honduras, Guatemala, Nicaragua, Venezuela,
Colombia, Peru, Brasil, Surinam, Jamaica, Puerto Rico, Belize, Alaska,
Burma, Indonesia, The Philippines, Turkey, Israel, Italy, Switzerland,
Australia, parts of Asia, and others (Rippon, 1988). Skin tests in endemic
U.S. areas have demonstrated that by age 20, between 80 to 90% of a
population has a positive skin test, e.g., Kansas City, Missouri,
Cincinnati-southern Ohio, southern Illinois, central Missouri, certain
areas of Kentucky, Tennessee, and Arkansas, as well as focal areas of high
endemicity in Michigan, Minnesota, Georgia, Louisiana, and Texas. Epidemics
of acute pulmonary histoplasmosis capsulati are frequently noted. Thus far,
about 200 "epidemics" have been recorded. Often such epidemics involve
exposure of a small number of people to a large presence of fungus in a
small area and usually at the same time. Some epidemics, however, have
involved large numbers of individuals in a defined location, i.e., the
Indianapolis, Indiana outbreaks (Sathapatayavongs et al., 1983; Schlech et
al., 1983). Table 3 indicates some salient epidemiologic features of
histoplasmosis.

While the vast majority of individuals infected with *H. capsulatum*
var. *capsulatum* experience no or subclinical disease, disseminated
infection may occur following exposure to heavy inocula. However, prior to
the AIDS epidemic, most disseminated histoplasmosis occurred as
opportunistic disease in the immunocompromised host. Disseminated
histoplasmosis is manifested as mildly to moderately chronic disease in
adults and as fulminant disease in infants/children and in adults with
associated immunologic dysfunction. The case rate of disseminated disease
in the above groups is 1:100,000 to 1:500,000 of persons infected per year
(Rippon, 1988; Table 4).

Table 3. Epidemiology of histoplasmosis capsulati prior to the AIDS epidemic

---

- a cosmopolitan disease in humans and lower animals

- diagnosed in all continents but with greatest concentration of cases
  located in the eastern half of the U.S.A. and most of Latin America

- occurs in tropical as well as temperate regions, e.g., cases in Africa:
  Cameroon, Gabon, Ivory Coast, Kenya, Rwanda, Sudan, Tanzania, Uganda,
  South Africa, Zaire, and Zimbabwe among others

- Austral-Asian-Oceania cases in Australia, Cambodia, India, Indonesia,
  Japan, Malaya, Sri Lanka, Thailand, Vietnam, and New Caledonia

- the Americas aside from the U.S.A. and Canada: cases in Belize, Costa
  Rica, Guatemala, Honduras, Panama, El Salvador, Argentina, Brasil,
  Chile, Colombia, Ecuador, Guyana, Paraguay, Peru, Venezuela, Cuba,
  Jamaica, Martinique, Puerto Rico, and other countries

- *H. capsulatum* var. *capsulatum* is a saprobic soil fungus flourishing in
  habitats enriched by feces of birds: pigeons, grackles, oil birds,
  starlings, as well as chickens and a large variety of bats

- infections following activity disturbing soil or droppings harboring the
  fungus

- the disease is not contagiousor transmissible. Except for direct
  inoculation, all infections result from inhalation of airborne conidia
  of the saprobic (mould) form of the organism

- a large variety of domestic and wild animals are susceptible; cases
  reported in badgers, bats, cattle, cats, dogs, foxes, horses, bear,
  rodents, and skunks. Birds are not infected

Table 4. Clinical forms of histoplasmosis capsulati

I. Benign or Asymptomatic – represents 90–95% of all infections

II. Acute Pulmonary – usually self-limited; night sweats, cough, fever, weight loss, etc.

III. Chronic Pulmonary – adults are the prime victims; symptoms similar to cavitary tuberculosis

IV. Disseminated – serious and potentially fatal opportunistic infection in immunocompromised adults and as a rapidly-fulminant disease in infants/children with cell mediated immune dysfunction. Case rate of disseminated disease is 1:100,000 to 1:500,000 of persons infected per year

---

Table 5. Features of the epidemiology of histoplasmosis in AIDS patients

---

- a "new" type of disseminated disease accompanying the severe immunologic dysfunction of AIDS

- more than 150 reported cases of disseminated histoplasmosis to date: this figure continues to escalate

- histoplasmosis capsulati is now one of the defining opportunistic infections of AIDS (CDC, 1985)

- some areas have observed histoplasmosis as the major opportunistic infection encountered in their AIDS patients

- early in the epidemic, most AIDS was diagnosed in New York City and San Francisco; histoplasmosis capsulati was not often seen

- with time, the mycosis has become a major affliction of AIDS patients who resided in or visited endemic areas

- however, an increasing number of AIDS patients with histoplasmosis capsulati living in nonendemic areas and who have not visited such locales, e.g., New York City, N.Y., Denver, CO., and the state of Oregon

- explanation? "New" niches for the fungus (if so, why not observed early on in the same locales). Is the fungus actually living in areas not previously thought or known to be infested?

- because of disease severity and often fatal outcome, it is important to continue epidemiologic investigations of histoplasmosis capsulati in the AIDS population

- particularly important to study disease in those AIDS patients who live in non-endemic areas and have never visited endemic locales

---

With the advent of AIDS a "new" type of disseminated histoplasmosis capsulati has emerged (Hazelhurst and Vismer, 1985; Bartholomew et al., 1984; Small et al., 1983; Kaur and Myers, 1983; Wheat and Small, 1984; Taylor et al., 1984; Wheat et al., 1985; Bonner et al., 1984; Huang et al., 1987; Anaissie et al., 1988; Mandell et al., 1986; Johnson et al., 1986; Haggerty et al., 1985; Pasternak and Bolivar, 1983; Johnson and Sarosi, 1987; Johnson et al., 1988; Freeman et al., 1989; Kalter et al., 1985; Tomita and Chiga, 1988; Graybill, 1988; Macher, 1988). Of course, such individuals are severely immunocompromised as regards cellular immunity and it is not surprising that disseminated mycoses result following exposure to the etiologic agent. More than 150 cases of disseminated histoplasmosis have now been reported in AIDS patients and in some locations this disease is the major opportunistic infection encountered, e.g., 2.7% disseminated histoplasmosis in all AIDS patients in Houston, Texas; 21% in Alabama; 53% in Indianapolis, Indiana; 29% in Trinidad. Table 5 summarizes features of the epidemiology of histoplasmosis in AIDS patients

Early in the AIDS epidemic, with the bulk of patients residing in New York City, New York, and San Francisco, California, histoplasmosis was not frequently observed and was not included among opportunistic infections which helped to define AIDS (Graybill, 1988). As the epidemic raged on, it became clear that this fungal infection was escalating significantly and, in fact, soon was included among the defining infections for AIDS (CDC, 1985). Histoplasmosis capsulati is now well-recognized as a major opportunistic mycosis in AIDS patients; particularly those who reside in or visited an endemic area. There have been, however, an increasing number of reports of *H. capsulatum* var. *capsulatum*-caused infections in AIDS patients living in non-endemic areas and who have not visited such locales. For example, AIDS patients from New York City, Denver, Colorado, and the state of Oregon have had documented disseminated histoplasmosis and have not visited the classic endemic areas (Huang et al., 1987; Salzman et al., 1988; Baptist et al., 1985). It is possible that *H. capsulatum* var. *capsulatum* resides in locations not previously felt or known to be infested, e.g., inner city New York.

In the AIDS patients histoplasmosis is severe, often refractory to currently available antifungal therapy; it often relapses following (or even during) treatment, and is frequently fatal (Graybill, 1988). It is important to continue investigations of the epidemiology of histoplasmosis in the AIDS population; especially the disease occurring in AIDS patients who live in non-endemic areas and have never visited endemic locales. This etiologic fungus may live in ecologic niches previously not considered. It is of additional interest to ascertain if infections caused by *H. capsulatum* var. *duboisii* occur in AIDS patients from Africa as well as epidemiologic features thereof (Peeters et al., 1987).

REFERENCES

Anaissie, E., Fainstein, V., Samo, T., Body, G.P., and Sarosi, G.A., 1988, Central nervous system histoplasmosis. An unappreciated complication of the acquired immunodeficiency syndrome, Am. J. Med., 84:215.
Baptist, S.J., Montana, J.B., Arden, S.B., Leon, L., and Dutel, M.,1985, Disseminated histoplasmosis in a man with AIDS, N.Y. State J. Med., 85:664.
Bartholomew, C., Raju, C., Patrick, A., Penco, F., and Jankey, N., 1984, AIDS on Trinidad, Lancet, 1:103.
Bonner, J.R., Alexander, W.J., Dismukes, W.E., App, W., Griffin, F.M., Little, R., and Shin, M.S., 1984, Disseminated histoplasmosis in patients with the acquired immunodeficiency syndrome, Arch. Intern. Med., 144:2178.
Carmona, F.J., 1971, Analisis estadistico y ecologio-epidemiologico de la sensibilidad a la histoplasmina en Colombia, 1950-1968, Antioquia Med., 21:109.

Centers for Disease Control, 1985, Revision of the case definition of
    acquired immune deficiency syndrome for national reporting - United
    States, MMWR, 34:373.
DiSalvo, A.F., and Johnson, W.M., 1979, Histoplasmosis in South Carolina:
    support for the microfocus concept, Am. J. Epidemiol., 109:480.
Emmons, C.W., 1949, Isolation of *Histoplasma capsulatum* from soil, Public
    Health Rep., 64:892.
Freeman, W.E., O'Quinn, J.L., and Lesher, J.L. Jr., 1989, Fever and
    hyperpigmented papules in an intravenous drug abuser. Disseminated
    histoplasmosis in acquired immunodeficiency syndrome (AIDS), Arch.
    Dermatol., 125:689.
Furcolow, M.L., 1958, Recent studies on the epidemiology of histoplasmosis,
    Ann. N.Y. Acad. Sci., 72:127.
Graybill, J.R., 1988, Histoplasmosis and AIDS, J. Infect. Dis., 158:623.
Haggerty, C.M., Britton, M.C., Dorman, J.M., and Marzoni, F.A. Jr., 1985,
    Gastrointestinal histoplasmosis in suspected acquired
    immunodeficiency syndrome, West. J. Med., 143:244.
Hazelhurst, J.A., and Vismer, H.F., 1985, Histoplasmosis presenting with
    unusual skin lesions in acquired immunodeficiency syndrome (AIDS),
    Br. J. Dermatol., 113:345.
Huang, C.T., McGarry, T., Cooper, S., Saunders, R., and Andavolu, R., 1987,
    Disseminated histoplasmosis in the acquired immunodeficiency
    syndrome. Report of five cases from a nonendemic area, Arch. Intern.
    Med., 147:1181.
Johnson, P.C., Sarosi, G.A., Septimus, E.J., and Satterwhite, T.K., 1986,
    Progressive disseminated histoplasmosis in patients with the
    acquired immune deficiency syndrome: A report of 12 cases and a
    literature review, Semin. in Resp. Infect., 1:1.
Johnson, P.C., and Sarosi, G.A., 1987, AIDS and progressive disseminated
    histoplasmosis, J.A.M.A., 258:202.
Johnson, P.C., Khardori, N., Najjar, A.F., Butt, F., Mansell, P.W., and
    Sarosi, G.A., 1988, Progressive disseminated histoplasmosis in
    patients with acquired immunodeficiency syndrome, Am. J. Med.,
    85:152.
Kalter, D.C., Tschen, J.A., and Klima, M., 1985, Maculopapular rash in a
    patient with acquired immunodeficiency syndrome. Disseminated
    histoplasmosis in acquired immunodeficiency syndrome (AIDS), Arch.
    Dermatol., 121:1455.
Kaur, J., and Myers, A.M., 1983, Sexuality, steroid therapy, and
    histoplasmosis, Ann. Intern. Med., 99:567.
Klite, P.D., and Dierchel, F.H., 1965, *Histoplasmosis capsulatum* in fecal
    contents and organs of bats in the Canal Zone, Am. J. Trop. Med.
    Hyg., 14:433.
Macher, A.M., 1988, The pathology of AIDS, Public Health Rep., 103:246.
Mandell, W., Goldberg, D.M., and Neu, H.C., 1986, Histoplasmosis in
    patients with the acquired immune deficiency syndrome, Am. J. Med.,
    81:974.
Manos, N.E., Ferebee, S.H., and Kerschbaum, W.F., 1956, Geographic
    variation in the prevalence of histoplasmin sensitivity, Dis. Chest,
    29:649.
Negroni, P., 1940, Estudio micologico del primer caso Argentino de
    histoplasmosis, Rev. Inst. Bacteriol. Malbran., 9:239.
Pasternak, J., and Bolivar, R., 1983, Histoplasmosis in acquired
    immunodeficiency syndrome (AIDS): diagnosis by bone marrow
    examination, Arch. Intern. Med., 143:2024
Peeters, P., Deprie, G., Richaert; F., Coremans-Pelseneer, J., and Serruys,
    E., 1987, Disseminated African histoplasmosis in a white
    heterosexual male patient with the acquired immune deficiency
    syndrome, Mykosen, 30:449.
Rippon, J.W., 1988, "Medical Mycology", 3rd edition, W.B. Saunders,
    Philadelphia.

Salzman, S.H., Smith, R.L., and Aranda, C.P., 1988, Histoplasmosis in patients at risk for the acquired immunodeficiency syndrome in a nonendemic setting, Chest, 93:916.

Sathapatayavongs, B., Batteiger, B.E., Wheat, J., Slama, T.G., and Wass, J.L., 1983, Clinical and laboratory features of disseminated histoplasmosis during two large urban outbreaks, Medicine, 62:263.

Schlech, W.F., Wheat, L.J., Ho, J.L., French, M.L.V., Weeks, R.J., Kohler, R.B., Deane, C.E., Eitzen, H.E., and Band, J.D., 1983, Recurrent urban histoplasmosis, Indianapolis, Indiana, 1980-1981, Am. J. Epidemiol., 118:301.

Small, C.B., Klein, R.S., Friedland, G.H., Moll, B., Emeson, E.E., and Spiglandi, I., 1983, Community-acquired opportunistic infections and defective cellular immunity in heterosexual drug abusers and homosexual men, Am. J. Med., 74:433.

Taylor, M.N., Baddour, L.M., and Alexander, J.R., 1984, Disseminated histoplasmosis associated with the acquired immune deficiency syndrome, Am. J. Med., 77:579.

Tomita, T., and Chiga, M., 1988, Disseminated histoplasmosis in acquired immunodeficiency syndrome: light and electron microscopic observations, Hum. Pathol., 19:438.

Wheat, L.J., and Small, C.B., 1984, Disseminated histoplasmosis in the acquired immune deficiency syndrome, Arch. Intern. Med., 144:2147

Wheat, L.J., Slama, T.G., and Zeckel, M.L., 1985, Histoplasmosis in the acquired immune deficiency syndrome, Am. J. Med., 78:203.

COCCIDIOIDES IMMITIS IN AIDS PATIENTS

John N. Galgiani, Neil M. Ampel, Cinthia L. Dols, and Douglas
G. Fish

Veterans Administration Medical Center and the University of
Arizona College of Medicine, Tucson, Arizona 85723

INTRODUCTION

Coccidioidomycosis is a systemic fungal infection endemic to parts of
the southwestern United States and elsewhere in the Western Hemisphere
(Bronnimann and Galgiani, 1989). In most patients, *Coccidioides immitis*
produces relatively minor and transient symptoms typical of a subacute
respiratory illness that may not even need medical attention. Occasionally,
such early infections are identified by seeing characteristic spherules in
respiratory secretions, recovering *C. immitis* in culture, or detecting
highly specific anticoccidioidal antibodies in a patient's serum. For this
stage of infection, treatment is not required to resolve the illness if the
patient is otherwise healthy.

In contrast to this usual benign course, coccidioidal infection can
lead to chronic, morbid, even lethal complications (Drutz and Catanzaro,
1978; Galgiani and Wack, 1988). Pulmonary complications such as cavity
formation or persistent pneumonia are rare in children but occur in
approximately 5% of adults, especially those with pre-existing lung disease
or diabetes mellitus. Other complications ensue when fungal invasion
spreads beyond the chest, typically producing combinations of skin,
skeletal, or meningeal lesions. Therapy for these problems often includes
antifungal drugs such as amphotericin B, ketoconazole, or newer triazoles
now in clinical trials such as itraconazole, fluconazole, and SCH 39304
(Knoper and Galgiani, 1988; Sagg and Dismukes, 1988). Surgical procedures
are also combined with drug therapy. However, limited comparative data and
the protean manifestations of these more complicated patients has produced
diverse opinions regarding optimal therapy for the various forms of
treatment.

Although extrathoracic dissemination can occur without underlying
disease, severe manifestations of infection are common when *C. immitis*
infects immunosuppressed patients such as recipients of organ transplants
(Cohen et al., 1982), patients with lymphoma and other malignancies
(Deresinski and Stevens, 1975), or patients treated with high doses of
corticosteroids (Rutula and Smith, 1976). Experimental evidence supports
the concept that immune control is critically dependent upon T lymphocyte
recognition of and response to coccidioidal antigens (Beaman, et al.,
1976). Moreover, nonspecific defenses such as those mediated by human
neutrophils appear to be insufficient in limiting fungal proliferation
(Galgiani, 1986). Given the profound T lymphocyte deficiency that develops

Mycoses in AIDS Patients, Edited by
H. Vanden Bossche *et al.,* Plenum Press, New York, 1990

171

in HIV-infected patients, it might be anticipated that patients with AIDS would be particularly susceptible to coccidioidomycosis.

This review will survey first in chronological order the currently available published reports of coccidioidomycosis that have complicated the course of HIV-infected patients. Discussed next will be the important questions that these papers raise regarding the virulence of coccidioidal infection and its likelihood of recrudescence in immunodeficiency states many years after initial exposure. Finally, systematic and prospective studies will be outlined that have recently been initiated to find answers to these important questions.

LITERATURE REVIEW

The first three cases of coccidioidal infection in AIDS were published as separate reports in 1984. The first report (Kovacs et al., 1984) was from Los Angeles and described a 30 year-old hispanic homosexual male farm worker from the San Joaquin Valley who sought medical attention because of a three-month history of fever, night sweats, and weight loss. Three years earlier the patient had had a calcified pulmonary nodule, a reactive coccidioidal skin test, and coccidioidal antibodies detectable in his serum. During the current evaluation, a coccidioidal skin test produced no induration, the pulmonary nodule was found to have cavitated, and *C. immitis* was recovered from cultures of sputum, a scalene node biopsy, and urine specimens. Amphotericin B was given and the patient improved. However, three months later the patient returned with *Pneumocystis carinii* pneumonia and *C. immitis* was found in tissue from an open lung biopsy. The patient died two weeks later and an autopsy revealed *C. immitis* in cervical, mediastinal and parapancreatic lymph nodes.

A second patient was treated in Phoenix, Arizona, an area endemic for coccidioidal infection (Roberts, 1984). The patient was a 39-year-old male intravenous-drug user who had developed fever, cough, dyspnea, and pleuritic chest pain over the prior five weeks. Additional and more protracted symptoms included weakness and anorexia. Chest roentgenography showed bilateral diffuse infiltrates and *P. carinii* was seen on stains of respiratory specimen obtained by bronchoalveolar lavage. Despite treatment with trimethoprim-sulfamethoxazole and pentamidine, the patient's respiratory status deteriorated. Specimens from a repeat bronchoscopic lavage procedure yielded spherules and subsequently *C. immitis* was recovered in culture. Despite instituting amphotericin B therapy, the patient died and no autopsy was obtained. Serologic evidence of coccidioidal infection was present on the first specimen tested.

A third patient was described by Abrams et al. (1984). A 30-year-old white male with AIDS developed fever and lassitude. A chest x-ray demonstrated bilateral interstitial infiltrates which were treated initially with trimethoprim-sulfamethoxazole for presumed *P. carinii* infection. However, a bone marrow biopsy demonstrated endospores of *C. immitis*. Despite treatment with amphotericin B, the patient died two days later. At an autopsy, coccidioidal infection was evident in the lungs, liver, spleen, and periaortic lymph nodes above and below the diaphragm. A premortem blood culture also grew *C. immitis*. Although the patient had apparently visited the San Joaquin Valley his stays had not been extensive and it was not clear when infection had first been acquired.

During the next two and one half years, two additional patients were reported. One was a 23-year-old white male intravenous-drug user who developed fever and sputum production and sought care in a hospital in Michigan (Salberg and Venkatachalam, 1986). Although a chest x-ray was unremarkable, computerized tomography of the chest demonstrated hilar nodes and bilateral pulmonary nodules (one of which was calcified). Spherules were seen on an axillary lymph node specimen, *C. immitis* was grown from cultures of sputum, and serum contained anticoccidioidal antibodies.

Amphotericin B was given with some improvement in his fever and cough. However, the patient became dyspneic six months later and despite aggressive support died of respiratory failure. Recurrence of coccidioidal infection was not documented premortem, and an autopsy was not performed. Although the exact time of the original infection was not known, two years previously the patient had had a coccidioidal skin test which was reactive.

A second patient was a 36-year-old homosexual male who was hospitalized in St. Louis with a five day history of cough, pleuritic chest pain, dyspnea, and fever (Wolf et al., 1986). *P. carinii* pneumonia was diagnosed from examination of specimens obtained by bronchoscopy and treated. Although radiographic improvement ensued, the patient's fever persisted. One month later, a supraclavicular node was discovered, and a biopsy specimen yielded *C. immitis*. Treatment with amphotericin B was begun and the fever resolved. However, two weeks after completing 2.6 g of amphotericin B therapy, fluid was noted in the right trochanteric bursa adjacent to a gluteal mass. Aspiration of this fluid again yielded *C. immitis*. No further follow-up was provided.

Seven patients were reported from Tucson, Arizona, a city near the center of the endemic area (Bronnimann et al., 1987). The total number of patients with AIDS in that region was 27, suggesting a high frequency of coccidioidal infections. The symptoms were generally nonspecific and included fever, weight loss, cough, and fatigue. Even though the symptoms were relatively mild, roentgenographic evidence of pulmonary involvement was very extensive with five of the patients exhibiting diffuse reticulonodular infiltrates on their initial chest films. Such a diffuse pulmonary pattern is indicative of hematogenous spread but is not specific for coccidioidal infection. Indeed, some of the patients in this series and in the earlier case reports had *P. carinii* identified simultaneously with *C. immitis*. In a related investigation of the histopathology of coccidioidal lung involvement, the density of spherules was found to be much higher than that seen in non-AIDS patients with diffuse pulmonary lesions (Graham et al., 1988). In addition, autopsies of many of these patients demonstrated widespread dissemination to liver, spleen, kidneys, bone marrow, thyroid, and brain. Patient were treated with either amphotericin B or ketoconazole. Improvement with either agent was not dramatic and there were instances of drug failures with both treatments. However, the interval between diagnosis of coccidioidal infection to death and the interval between AIDS diagnosis and death was not appreciably different, suggesting that mortality was more dependent upon the underlying immunodeficiency state than upon uncontrolled coccidioidal infection.

Reports of three additional patients have been published. One concerned a 28-year-old hispanic male living in Ventura, California, with fever, anemia, right upper lobe pneumonia, and several 3-5 mm papulopustules, mostly on the extremities (Prichard, et al., 1987). Biopsy of one of the skin lesions as well as a culture of bone marrow yielded *C. immitis*. The patient had been raised in the San Joaquin Valley and ten years prior to this illness had been told he had a "lung cyst" on a routine chest roentgenogram. The patient received ketoconazole and amphotericin B. Within ten days the patient's fever had dissipated and the cutaneous lesions had undergone considerable involution. However, the patient refused further therapy and died three weeks later.

Another report described a 26-year-old gay male from San Diego who developed morning headaches, fever, and weight loss (Jarvik et al., 1988). A supraclavicular node was noted but a biopsy was declined by the patient. Computerized tomography of the head revealed a 4 mm midbrain lesion. Coccidioidal serology was positive and ketoconazole therapy was begun. Although fever improved, headaches worsened and the concentration of coccidioidal antibodies increased. Sixteen months after the onset of his symptoms, the patient developed anisocoria, coarse right-sided tremor, and confusion. Soon thereafter he became unresponsive and died. An autopsy demonstrated an abscess in the right midbrain, 5 mm in diameter, containing spherules of *C. immitis*, with focal necrosis of parenchyma and chronic inflammatory reaction.

A third article reported a 57-year-old male with a colostomy for 12 years and multiple prior blood transfusions who developed intestinal bleeding, fever, headaches, and bilateral interstitial and alveolar infiltrates (Byrne and Dietrich, 1989). Sputum and cerebrospinal fluid grew *C. immitis* and amphotericin B was administered intravenously and intrathecally for cumulative doses of 2.7 g and 34.5 mg, respectively. Two years after amphotericin B had been discontinued, fever recurred with abdominal pain and distention. Repeated culture of ascites yielded *C. immitis*, amphotericin B was reinstituted, and subsequent cultures of ascites were sterile. The patient subsequently died of hepatic encephalopathy and aspiration pneumonia and an autopsy was not performed.

ANALYSIS OF CURRENT REPORTS

To summarize the experience from the 15 patients included in the above reports, several patterns emerge. First, the manifestations of coccidioidal infections in AIDS patients are strikingly worse than generally found in immunocompetent patients. Whereas otherwise healthy patients infrequently develop progressive forms of coccidioidal infection, diffuse pulmonary infiltrates and other evidence of hematogenous dissemination seems to be the rule in AIDS patients.

A second point is that a significant discrepancy exists between the frequencies of coccidioidal infections in HIV-infected patients residing within and beyond the regions endemic for *C. immitis*. In Tucson, for example, over 20% of patients with AIDS developed clinically apparent coccidioidal infection (Bronnimann et al., 1987), whereas from other published reports, it seems that coccidioidal infections beyond the endemic area are much less common. This discrepancy has been surprising to some experts since some of the most intense areas of AIDS prevalence within the United States have been Los Angeles and San Francisco, areas adjacent to regions endemic of *C. immitis*.

These differences between endemic and nonendemic rates could be accounted for in two ways. One possibility is that immunodeficient patients are extraordinarily sensitive to exogenous infection. Although possible, this explanation would contradict air sampling observations in Southern Arizona which indicate extremely low density of arthroconidia of *C. immitis* in the ambient air. Moreover, most sporadic coccidioidal infections in immunocompetent persons are thought to be caused by a very small inocula because experimental infections in mice can be caused by very small numbers of arthroconidia (Kong et al., 1964).

An alternative possibility is that the high rate of endemic coccidioidal infections is predominantly the result of endogenous reactivation. For several of the patients reported from beyond the endemic area, prior exposure was known to have occurred several years earlier. Thus, endogenous reactivation clearly may account for some of the infections. Moreover, using life time exposure as the basis for calculating rates of infection, Bronnimann et al. (1987) estimated that annual infection frequency would be only 2.7% per year of endemic residence, a result very similar to estimates in other patient groups in Tucson and much less then the alternative calculation of 27% per year of immunodeficiency. However, such estimates are indirect at best and corroboration with new data would be very desirable.

The implications of these two explanations differ considerably. If hypersensitivity to new exogenous infection is an especially important contributor to the high rate of infection within the endemic area, then counselling HIV-infected patients about this risk would be valuable. It may be that these patients would choose to not move to or relocate from coccidioidal endemic regions. On the other hand, if reactivation of prior infection is the predominant factor responsible for the rate of coccidioidal disease, then the current place of residence would have much less influence on developing overt coccidioidomycosis. Moreover, testing HIV-infected patients for markers of prior coccidioidal infection before a severe immunodeficiency state develops may be helpful in future counselling and management decisions.

RECENTLY INITIATED STUDIES

Prospective Surveillance

To better define the factors responsible for immunodeficient patients developing coccidioidomycosis, we have embarked upon a prospective longitudinal surveillance of HIV-infected persons for evidence of coccidioidal infection. Enrollment was begun in 1988 and is open to any HIV-infected patient residing within the coccidioidal endemic area. After providing written consent to participate, a blood specimen is obtained and skin tests are applied. Each patient is stratified with respect to his immunologic competence as estimated by T4 lymphocyte counts. Patients with peripheral blood T4 counts ≥200 cells/mm^3 are categorized as immunocompetent and those with counts <200 cells/mm^3 are categorized as not immunocompetent. Patients in the former category are further classified by the presence or absence of i) conventional anticoccidioidal antibodies, ii) dermal hypersensitivity to spherulin, or iii) *in vitro* response of peripheral-blood mononuclear leukocytes to spherule-derived coccidioidal antigens. Those with skin test or lymphocyte transformation evidence of prior coccidioidal infection but without active infection currently are categorized as "Group I" subjects. Those without cellular immune markers of prior infection are categorized as "Group II" subjects. "Group III" subjects are those with low T4 cell numbers and in whom immunologic markers could not be relied upon. Follow-up evaluations are repeated every 4 months, and changes in immunologic status will be correlated with other clinical events.

Of primary interest to this study is the frequency of coccidioidal infection in the subgroup of subjects that develop an immunodeficient state. Since skin test conversion rates within Southern Arizona are approximately 3% per year in a normal population (Dodge et al., 1985), one might find relatively few (perhaps 5% or less per year) active coccidioidal infections in Group II subjects and significantly greater numbers of coccidioidal infections in Group I subjects. This would be interpreted to indicate that endogenous reactivation of prior disease was a major source of disease in our subjects. Alternatively, if primary inoculation was predominantly responsible for the frequency of coccidioidal infections, relatively frequent (perhaps 20 to 30% per year) active coccidioidal infections would be detected in both Group I and Group II subjects. A related correlate is that the frequency of coccidioidal infections due to exogenous exposure would be distributed more evenly over time for both Group I and Group II subjects whereas endogenous reactivation would be expected to become more frequent with more profound immunodeficiency in group I but not in Group II.

From April, 1988, to August, 1989, 106 patients have been enrolled to this study. The median age of the subjects was 37 years, 98% were male, 84% were caucasian, and virtually all had recognized risk factors for acquiring HIV infection. Of the first 58 immunocompetent subjects enrolled, 21 have immunologic markers of prior coccidioidal infection and 37 do not. The median endemic residence for Groups I, II, and III was 10.0, 3.4, and 8.1 years, respectively. To date, peripheral blood T4 counts have decreased below 200 cells/mm^3 in one subject from group I and three subjects from group II. There have been no clinically active coccidioidal infections in Group I (18.5 subjects-years of surveillance), two active infections in

Group II (37.5 subject-years; both patients had T4 counts <200 cells/mm^3), and one only one coccidioidal infection in Group III (31.3 subject-years). Clearly, these observations are premature to draw reliable conclusions. However, maintenance of this prospective study is likely improve our understanding of the importance of endogenous and exogenous infection as factors contributing to the high incidence of coccidioidal morbidity in AIDS patients.

Retrospective Survey

In order to better understand the spectrum of coccidioidal infection in HIV-infected patients, we initiated a survey to identify as yet unreported patients with these two infections. Within Pima county much of the AIDS treatment is provided by the University of Arizona Medical Center and the Veterans Administration Medical Center. All patients with HIV and coccidioidal infections at these institutions were identified. Members of the Arizona Infectious Disease Society were invited to contribute information regarding patients under their direct care. Additional information was obtained from colleagues in California and elsewhere in the United States. Data collection onto a standardized form was accomplished by chart audits by physicians familiar with each specific patient. The abstracted information included the patient's age, sex, race, risk group for contracting HIV infection, HIV clinical status classification, peripheral blood T4 counts, demographic data concerning residence in the endemic areas for *C. immitis*, clinical characteristics of the initial manifestations of coccidioidal infection, anticoccidioidal antibody test results, treatment of the coccidioidal infection, and subsequent course of the coccidioidal infection. Forms were then entered into a microcomputer database program (Rbase, Microrim, Belview, WA) for initial analysis.

As a result of this effort, we have been able to identify 77 patients with coccidioidal infections complicating their HIV disease. The vast majority of these infections have been diagnosed in the past two years. Although most of these patients were identified within the southern deserts of Arizona, increasing numbers of patients were diagnosed in Los Angeles. The full details of these patients will be published elsewhere (Ampel, et al., 1989). However, as a preliminary finding it is apparent that the clinical manifestations of coccidioidal infection are much more diverse than noted previously. Many of the patients had positive coccidioidal antibodies indicating that standard serologic tests are often helpful in detecting coccidioidomycosis in HIV-infected patients.

Additionally, although most of the patients were treated at least initially with amphotericin B, a few have benefited from newer oral agents such as itraconazole and fluconazole. In particular, one patient with meningitis, who was included in a previous report (Tucker, et al., 1989), was treated successfully with fluconazole as sole therapy. At this time, 19 months after beginning fluconazole, he remains on therapy without symptoms of his coccidioidal infection. After 17 months of treatment, his cerebral spinal fluid showed only 5 leukocytes per mm^3, a glucose of 35 mg/dL, and a protein of 44 mg/dL.

SUMMARY

Coccidioidomycosis poses frequent, significant, and diverse problems in patients already infected with HIV. Most coccidioidal infections are presently concentrated in regions highly endemic for the fungus. However, if reactivation of latent coccidioidal infections turns out to be commonplace as some suppose, then this distribution could change. Studies now underway should help in guiding future recommendations concerning minimizing the impact of coccidioidal complications.

REFERENCES

Abrams, D.I., Robia, M., Blumenfeld, W., Simonson, J., Cohen, M.B., and
    Hadley, W.K., 1984, Disseminated coccidioidomycosis in AIDS, N.
    Engl. J. Med., 310:986.
Ampel, N.M., Fish, D.G., Dols, C.L., and Galgiani, J.N., 1989,
    Coccidioidomycosis (coccy) in HIV-infected subjects. A prospective
    and retrospective evaluation, in: "Program and Abstracts of the
    Twenty-Ninth Interscience Conference on Antimicrobial Agents and
    Chemotherapy", American Society for Microbiology, Washington.
Beaman, L., Benjamini, E., and Pappagianis, D., 1979, Mechanisms of
    resistance to infection with Coccidioides immitis in mice, Infect.
    Immun., 23:681.
Bronnimann, D.A., Adam, R.D., Galgiani, J.N., Habib, M.P., Petersen, E.A.,
    Porter, B., and Bloom, J.W., 1987, Coccidioidomycosis in the
    acquired immunodeficiency syndrome, Ann. Intern. Med., 106:372.
Bronnimann, D.A., and Galgiani, J.N., 1989, Coccidioidomycosis, Eur. J.
    Clin. Microbiol. Infect. Dis., 8:466.
Byrne, W.R., and Dietrich, R.A., 1989, Disseminated coccidioidomycosis with
    peritonitis in a patient with acquired immunodeficiency syndrome,
    Arch. Intern. Med., 149:947.
Cohen, I.M., Galgiani, J.N., Potter, D., and Ogden, D.A., 1982,
    Coccidioidomycosis in renal replacement therapy, Arch. Intern. Med.,
    142:489.
Deresinski, S.C., and Stevens, D.A., 1975, Coccidioidomycosis in
    compromised hosts: experience at Stanford University Hospital,
    Medicine (Baltimore), 54:377.
Dodge, R.R., Lebowitz, M.D., Barbee, R., and Burrows, B., 1985, Estimates
    of C. immitis infection by skin test reactivity in an endemic
    community, Am. J. Public Health, 75:863.
Drutz, D.J., and Catanzaro, A., 1978, Coccidioidomycosis: Part 2, Am. Rev.
    Respir. Dis., 117:727.
Galgiani, J.N., 1986, Inhibition of different phases of Coccidioides
    immitis by human neutrophils or hydrogen peroxide, J. Infect. Dis.,
    153:217.
Galgiani, J.N., and Wack, J.N., 1988, Coccidioidomycosis, in: "Pulmonary
    Diseases and Disorders", A.P. Fishman, ed., McGraw-Hill, New York
Graham, A.R., Sobonya, R.E., Bronnimann, D.A., and Galgiani, J.N., 1988,
    Quantitative pathology of coccidioidomycosis in acquired
    immunodeficiency syndrome, Hum. Pathol., 19:800.
Jarvik, J.G., Hesselink, J.R., Wiley, C., Mercer, S., Robbins, B., and
    Higginbottom, P., 1988, Coccidioidomycotic brain abscess in an HIV-
    infected man, West. J. Med., 149:83.
Knoper, S.R., and Galgiani, J.N., 1988, Coccidioidomycosis, Infect. Dis.
    Clin. N.A., 2:861.
Kong, Y.M., Levine, H.B., Madin, S.H., and Smith, C.E., 1964, Fungal
    multiplication and histopathologic changes in vaccinated mice
    infected with Coccidioides immitis, J. Immunol., 92:779
Kovacs, A., Forthal, D.N., Kovacs, J.A., and Overturf, G.D., 1984,
    Disseminated coccidioidomycosis in a patient with acquired immune
    deficiency syndrome, West. J. Med., 140:447.
Prichard, J.G., Sorotzkin, R.A., and James, R.E., III, 1987, Cutaneous
    manifestations of disseminated coccidioidomycosis in the acquired
    immunodeficiency syndrome, Cutis, 39:203.
Roberts, C.J., 1984, Coccidioidomycosis in acquired immune deficiency
    syndrome. Depressed humoral as well as cellular immunity, Am. J.
    Med., 76:734.
Rutala, P.J., and Smith, J.W., 1978, Coccidioidomycosis in potentially
    compromised hosts: the effect of immunosuppressive therapy in
    dissemination, Am. J. Med. Sci., 275:283.
Salberg, D.J., and Venkatachalam, H., 1986, Disseminated coccidioidomycosis
    presenting in AIDS, V.A. Pract., (April):89.

Tucker, R.M., Galgiani, J.N., Denning, D.W., Hanson, L.H., Graybill, J.R., Sharkey, K., Eckman, M.R., Salemi, C., Libke, R., Klein, R.A., and Stevens, D.A., Treatment of coccidioidal meningitis with fluconazole, *Rev. Infect. Dis.*, in press.

Wolf, J.E., Little, J.R., Pappagianis, D., and Kobayashi, G.S., 1986, Disseminated coccidioidomycosis in a patient with the acquired immune deficiency syndrome, *Diagn. Microbiol. Infect. Dis.*, 5:331.

THE MAJOR ENDEMIC MYCOSES IN THE SETTING OF AIDS:

CLINICAL MANIFESTATIONS

John R. Graybill, Patricia Kay Sharkey, Philip Johnson,
Steven Nightingale

University of Texas Health Science Center in San Antonio and
Audie L. Murphy Memorial Veterans Hospital, San Antonio, TX,
Univ. of Texas Health Science Center, Houston, TX, Univ. of
Texas Southwestern, Dallas, TX

INTRODUCTION

There are four major endemic systemic mycoses, including
coccidioidomycosis, paracoccidioidomycosis, blastomycosis, and
histoplasmosis. All of these are commonly found in geographically
circumscribed areas of the western hemisphere, but histoplasmosis is
uncommonly scattered about the eastern hemisphere as well. None of these
infections were initially associated with AIDS, and as recently as 1985
histoplasmosis, the most commonly seen, was reported in only 0.5% of AIDS
patients. The situation has changed considerably since 1985. At present
there are increasing numbers of patients who have developed
coccidioidomycosis in the setting of HIV exposure (Ampel et al., 1989).
These are discussed elsewhere in this book (see Galgiani et al.). Two
others, paracoccidioidomycosis and blastomycosis, remain rarities in the
setting of AIDS. Of 2 patients with North American blastomycosis, neither
had other concurrent infections, neither had diagnostic serologic titers,
and both had disseminated disease (Chiu et al., 1988). One reported patient
with paracoccidioidomycosis also had disseminated disease (Bakos et al.,
1988). Remarkably, this developed while he was receiving ketoconazole for
esophageal candidiasis. The disease did respond to intravenously
administered amphotericin B. There are other scattered unreported cases of
these infections in AIDS patients. However, there are no obvious
explanations for their remarkable infrequency. One reason might be the
strong association with both infections with outdoor activity, which may
not be strongly associated with AIDS risk behavior. Another reason might be
the relatively slow spread of AIDS into the more rural areas where many of
those exposed reside. Even so, it is surprising that paracoccidioidomycosis
is not abundant in Brazilian AIDS patients, because this country is
unhappily distinguished as the leader for AIDS in South America. Another
factor may be the relative rarity of late relapses after the initial
infection and immune response has occurred.

The major endemic mycosis associated with AIDS is histoplasmosis.
Initially, histoplasmosis was considered quite uncommon. Most patients were
either imported cases from the Caribbean or late reactivation in people who
developed AIDS long after they had left the endemic areas (Tiley et al.,
1988; Haggerty et al., 1985; Bartholomew et al., 1985; Mandell et al.,

Mycoses in AIDS Patients, Edited by
H. Vanden Bossche *et al.*, Plenum Press, New York, 1990

1986). However, AIDS has also sharply increased in the midwestern United States, the endemic region for *Histoplasma capsulatum*. Not surprisingly, in this area, including much of the state of Texas, there has been a sharp increase in histoplasmosis in general, and AIDS patients in particular (Gustafson and Henson, 1985; Wheat and Small, 1984; Bonner et al., 1984; Johnson et al., 1988; Graybill, 1988; Jones et al., 1983; Taylor et al., 1984). At present in San Antonio, Dallas, and Houston, histoplasmosis is seen almost as frequently as cryptococcosis, and the vast majority of our patients with histoplasmosis also have AIDS (Johnson et al., 1988; Graybill, 1988).

H. capsulatum infects man by inhalation of microaleuriospores. These are ingested by alveolar macrophages and convert to yeast cells within several days. The intracellular survival of *H. capsulatum* is critically dependent on ingestion mediated by C3 receptors, which do not stimulate the macrophage or monocyte to release superoxide, to which the yeasts are susceptible (Bullock and Wright, 1987). Intracellular organisms are then able to "hitch a ride" with monocytes to other organs of the body. Because monocytes densely populate the reticuloendothelial organs, *H. capsulatum* is preferentially transported to these sites via the hematogenous route. *H. capsulatum* replicates in the lungs and in these peripheral sites until controlled by cell mediated immunity. Excessive immune response produces dense fibrotic or even caseating lesions such as histoplasmoma, fibrosing mediastinitis, or hepatosplenic calcified granulomas (Goodwin et al., 1972; Goodwin et al., 1976; Goodwin et al., 1969). A deficient cell mediated immune response allows *H. capsulatum* to proliferate within the reticuloendothelial organs, peripheral circulating monocytes and tissue macrophages (Goodwin et al., 1980). For reasons as yet unclear, the adrenal glands and oral mucosal surfaces are also attractive targets for growth of *H. capsulatum*.

CLINICAL MANIFESTATIONS OF HISTOPLASMOSIS IN AIDS

Given this albeit simplified understanding of the pathogenesis of histoplasmosis, the form taken by the disease in AIDS patients is quite predictably the aggressive disseminated disease pattern, reflected by dense parasitization of the reticuloendothelial system, and no caseation necrosis (Goodwin et al., 1980).

The manifestations of histoplasmosis at presentation are given in Table 1.

Fever and weight loss were the only symptoms found in over half of our patients. These findings are nonspecific, and are commonly a manifestation of AIDS-related complex, or ARC. They may also be seen with tuberculosis, atypical mycobacterial infection, or lymphoma. However, on physical and laboratory examination there is often evidence of diffuse reticuloendothelial system involvement. This takes the form of hepatomegaly, splenomegaly, and abnormal liver function studies. These, and mucocutaneous lesions are suggestive but not pathognomonic for histoplasmosis. Also, we have seen 3 patients present with diarrhea or constipation caused by lesions of the colon.

The adrenal gland may be a target of histoplasmosis (Goodwin et al., 1980), tuberculosis, or cytomegalovirus in patients with AIDS, and patient symptoms of weakness, anorexia, weight loss, and orthostatic dizziness may be caused by adrenal involvement (Greene et al., 1984; Tapper et al., 1984). All patients suspected for disease caused by these pathogens should undergo minimally a 60 or 90 minute ACTH stimulation test. The serum cortisol should either double or increase to above 20. Of our 14 patients in San Antonio, 2 had adrenal insufficiency associated with (but not necessarily caused by) histoplasmosis.

Table 1. Clinical findings of histoplasmosis in AIDS as % of 78 Texas
patients (Johnson et al., 1989; Sharkey et al., 1989)

Symptoms		Signs	
Fever	69	Splenomegaly	33
Weight loss	50	Lymphadenopathy	24
Abdominal pain	4	Hepatomegaly	15
Malaise	5	Skin lesions	15
Headache	1	Colitis	3
Arthralgia	1	Abdominal mass	1

Laboratory abnormalities in 50 patients

Thrombocytopenia (platelets $<10^5$)	16/33
Anemia (Hct<40)	45/47
Leukopenia (WBC$<5x10^3$)	28/46
Abnormal liver function	9/?
Depressed cortisol response to ACTH	2/?

? = No data on how many patients were tested

Table 2. Chest roentenographic findings in patients with histoplasmosis in
the setting of AIDS (Tapper et al., 1984; Wheat, 1989)

Finding	San Antonio (Tapper et al., 1984)	Houston (Wheat, 1989)	Dallas*	Total
Total	14	64	35	113
Normal	5	16	12	33
Interstitial infiltrates	6	34	11	51
Nodular densities	3	1	6	10
Apical thickening	1	0	5	6
Pleural effusions	0	4	1	5
Granulomas	0	2		2
Pneumothorax	0	1	0	

* = unpublished data, S. Nightingale – Data not available on all patients

Table 3. Recovery of *H. capsulatum* in 115 Texas AIDS patients with histoplasmosis

Site	No. tested	% positive
Skin	14	100
Intestine	7	100
Lymph node	14	93
Liver	5	80
Bone marrow	66	74
Respiratory*	48	65
Blood culture	79	63
Blood smear	33	58
CSF	30	17

* = Includes both sputum and bronchoscopically obtained specimens

Sources: Johnson et al., 1989; Sharkey et al., 1989, and 36 cases from Dr. S. Nightingale, Univ. Texas Southwestern, Dallas, TX, unpublished observations.

Although *H. capsulatum* is frequently recovered from sputum or bronchoscopically obtained washings and biopsies, and although pulmonary infiltrates may be seen, respiratory manifestations of histoplasmosis do not predominate in these patients. Only 19 of 78 Texas patients had cough and 9 had dyspnea at the time of presentation. Radiographic findings more commonly indicated pulmonary involvement than symptoms. Chest radiographic findings are indicated in Table 2.

According to 3 studies of Texas patients, one third of patients have normal radiographs and 2/3 show diffuse interstitial infiltrates, with other findings being less frequently seen. In addition to the above, Wheat (1989) found that 43% of his patients had normal chest radiographs, while 44% had diffuse infiltrates. Cavitary and fibrotic disease was distinctly uncommon in these patients. The findings on chest radiographs were in general nondiagnostic.

DIAGNOSIS OF HISTOPLASMOSIS IN AIDS

Microbiologic diagnosis of histoplasmosis in most patients is relatively easy (Table 3).

The highest yields, close to 100%, came from biopsies of organs suspected to be infected, or lesions. However, the numbers of patients with such findings were relatively low.

For a more general sampling in patients without specific lesions, the fungal burden is frequently so dense that simple Wright-Giemsa smear of the blood will reveal yeasts in peripheral blood monocytes. Half of our patients tested were positive. Blood cultures were frequently positive. The isolator technique is superior for fungi, and should be used routinely for suspected fungal or mycobacterial infection in these patients. The most common site for diagnosis was the bone marrow, which was positive in 50 of 79 patients. The histopathology is virtually as sensitive as the culture. With examinations of the blood and bone marrow most of the 64 patients of Dr. Wheat in Indianapolis (Wheat, 1989) and our 115 Texas patients were diagnosed. Additional sites which were positive included sputum and bronchoalveolar secretions, lung biopsy, and biopsies of mucocutaneous lesions, lymph nodes, urine, and cerebrospinal fluid. Patients who presented with gastrointestinal symptoms and signs required colonoscopic biopsy to make the diagnosis. Because of the difficulty of culturing stool for *H. capsulatum*, histopathology was the basis of diagnosis.

We are unaware of AIDS patients in whom *H. capsulatum* was vigorously sought and not promptly found using culture and histopathology as noted above. However, many of our patients underwent invasive diagnostic procedures relatively late in the course of their illness, and it is possible that noninvasive screening tests could have led to a diagnosis much earlier, and before major clinical deterioration had taken place. In our experience, *H. capsulatum* antibodies have frequently if not usually been negative in patients with AIDS, and the nonspecificity of low titers further limits interpretation. A radioimmunoassay has recently been developed for *H. capsulatum* polysaccharide antigens (Wheat et al., 1986). Wheat et al. have found this to be sensitive and specific in the diagnosis of histoplasmosis in non-AIDS patients. More recently they have enlarged their experience to include 61 patients with AIDS and histoplasmosis (Wheat et al., 1989). They have found that 37 of 47 patients with histoplasmosis and AIDS had antigen detectable in their blood: antigen was found in urine specimens in 59 of 61. In pretreatment urine specimens obtained from 35 patients, only 1 was negative. Some of the patients were treated with amphotericin B. All of 10 patients responding to amphotericin B reduced their antigenemia, and 19 of 21 had reduced antigenuria. Further, among 8 patients who relapsed, all increased their urine antigen, and all of 5 with serum tested had increased antigen. Thus we have at last available a test which can be used not only for diagnosis but also for prognosis in these patients with widely disseminated disease. The antigen test is at present available only through Dr. Wheat at the University of Indianapolis (Department of Medicine, 1100 West Michigan Street Indianapolis, IND, 46223). However, Dr Wheat is able and willing to test specimens from outside physicians.

TREATMENT OF HISTOPLASMOSIS IN AIDS

Once histoplasmosis is diagnosed, what is the best approach to therapy? Until recently, amphotericin B has been the mainstay of treatment for histoplasmosis. The mortality for untreated disseminated histoplasmosis in the pre-AIDS era was above 80%, and this was lowered to less than 10% by a course of 25-35 mg/kg of amphotericin B. In immunocompetent patients there are few relapses after a successful course of amphotericin B (Goodwin et al., 1980). However, patients with defective cell-mediated immunity, such as those with lymphoma or acute lymphocyte leukemia, tended to have recurrences (Dismukes et al., 1978; Wheat et al., 1982). This problem was in essence a predictor of the situation that would arise with AIDS, wherein the defective immune response is severe and permanent. Initial reports of AIDS patients were few and scattered, but suggested that relapses and failures were significant problems (Chiu et al., 1988; Bakos et al., 1988; Tiley et al., 1988; Haggerty et al., 1985; Bartholomew et al., 1985; Mandell et al., 1986; Gustafson and Henson, 1985; Wheat and Small, 1984; Bonner et al., 1984). The first major report was that of Johnson et al., who described 48 patients with histoplasmosis occurring in the setting of AIDS (Johnson et al., 1988). Their updated experience, combined with the experience of Nightingale et al. from the University of Texas Southwestern, B was clearly discouraging for amphotericin B, and is summarized in Table 4.

Of the total 84 outcomes noted above, only 20 patients were alive and had improved at the time of these reports. Forty eight patients were not considered evaluable. Of these, 25 had less than 100 mg of amphotericin B, and were not considered to have received enough treatment to represent a fair trial for amphotericin B by the time they died. Postmortem examinations were not done on 19 patients who succumbed after longer treatment periods. The majority were considered to have succumbed to histoplasmosis, but because AIDS patients can have a multiplicity of infections, we could not confirm histoplasmosis as the cause of death. Four patients were lost to followup.

Table 4. Results of antifungal treatment in 84 AIDS patients with
disseminated histoplasmosis (Modified from Johnson et al., 1988
and Nightingale)

| | Amphotericin B | | Amphotericin B followed by ketoconazole | | |
	A	B	A	B	Total
Source					
Inevaluable					
Death before effective therapy (<100 mg)	12	13	0	0	25
Death without autopsy	4	6	6	3	19
Lost to followup	0	0	4	0	4
Improved	7	11	1	1	20
Failed					
Positive autopsy	4	0	2	0	6
Relapsed	6	1	2	1	10
Total	33	31	15	5	84

Legend: A = Johnson et al., 1988; B = S. Nightingale, unpublished
observations.

Therefore, of 84 episodes of treatment for histoplasmosis, 48 were not
evaluable for results of chemotherapy. Of the remaining 36 patients, 20
were improved on continuing amphotericin B or ketoconazole suppressive
treatment. Sixteen failures includes 10 with relapses, and 6 with positive
autopsies for *H. capsulatum*. Follow up ketoconazole suppressive therapy did
not shift the balance toward clinical improvement. We suspect this is
because ketoconazole is clearly malabsorbed in relatively achlorhydric AIDS
patients (Lake-Bakaar et al., 1988).

Another problem was the propensity to relapse. This occurred in 10
(12%) of the 84 outcomes, but was a higher 29% of 36 evaluable outcomes.
Ketoconazole suppression was used after amphotericin B in 20 of the 84
episodes presented in Table 4, but only 2 of those patients were continuing
as improved at the time of reporting the data. Another alternative was a
primary course of amphotericin B followed by indefinitely continued
suppression with 1 mg/kg dosing once or twice weekly. This regimen was
reported by McKinsey et al. in 16 patients, 13 of whom had successful
suppression for a median course of 14 months at the time of the report
(McKinsey et al., 1989). Wheat's observations on outcome include 81
episodes in AIDS patients treated similarly for histoplasmosis (Wheat,
1989). Of those patients, 4 died with disease other than histoplasmosis, 7
died having received less than 0.5 grams of amphotericin B, 4 died having
received more than 0.5 grams of amphotericin B, and the remaining 66
outcomes were improved, without relapses.

The as yet unpublished results of Wheat et al. are so strikingly
different from others reported that they bear a further comment. While most
physicians diagnosed histoplasmosis using microbiological means, Wheat was
able to utilize a serum or urine antigen test for histoplasmal cell
products. It is possible that this test, if sufficiently sensitive, when
employed earlier in the course of disease, might provide a diagnosis at a
time when the patient would be more responsive to chemotherapy. Another
alternative is the use of amphotericin B suppressive therapy similar to
that employed by McKinsey et al. This is speculative, and it is still
unclear why some patients have had consistently good responses to induction
therapy with amphotericin B, and others have clearly failed. One
possibility might be earlier diagnosis caused by heightened clinical

suspicion of physicians in the endemic area, and another would be early
screening for histoplasmal antigenuria or antigenemia.

Even with control of histoplasmosis by amphotericin B, there are still
formidable problems in patient management. One is the cumulative toxicity
of amphotericin B is to the kidneys. A second problem is the potential of
amphotericin B induced anemia aggravating anemia caused by concurrent
azidothymidine (AZT, Zidurudine). A third problem is caused by the presence
of Hickman catheters for central administration of amphotericin B. These
are very commonly used for venous access, and may be in place longer than a
year. These catheters are a particularly high risk as foci for sepsis in
AIDS patients. In one recent report 16 of 44 Hickman catheters placed in
AIDS patients were complicated by bacterial sepsis, versus only 2 of 25 in
non-AIDS patients (Raviglione et al., 1989). This was recently confirmed by
Schlamm et al. (1989) in another study. One predisposition to catheter
sepsis may be the marginal or clearly depressed neutrophil counts that are
a common concomitant of AZT or ganiciclovir treatment. Therefore, even when
successful amphotericin B therapy is not benign.

Because of some failures in patients on amphotericin B, and because of
the inconvenience and complications of amphotericin B therapy, alternative
new approaches to treatment were sought. The first of these was
ketoconazole. The experience with ketoconazole is small and scattered
through a number of small reports. There is especially sparse experience
with ketoconazole as primary treatment, and it is not encouraging. In their
updated series, of 7 patients that Johnson et al. (1989) initially treated
with ketoconazole, there were 6 relapses. This is significantly worse than
amphotericin B initial treatment, in which there were 14 relapses among 40
patients treated (p<0.025 by x^2). Ketoconazole has also been used after
induction treatment with amphotericin B, in an effort to find a more benign
alternative for prolonged suppression of histoplasmosis. In the same update
by Johnson et al., 21 patients treated initially with amphotericin B
followed by ketoconazole suppression had a mean followup of 269 days, which
was not significantly different from the follow up of 12 patients treated
with amphotericin B alone. The reason for this is probably the same as the
cause of failures for thrush treatment with ketoconazole. Ketoconazole
requires an acid intragastric environment, and AIDS has been appreciably
associated with a state of hypochlorhydria and malabsorption of the drug
(Lake-Bakaar et al., 1988). Poor absorption could be corrected by
administering ketoconazole dissolved in 0.1 normal HCl, but the efficacy of
such a regimen has not been tested in patients with histoplasmosis. At
present, if ketoconazole is to be used, patients should be treated with
high doses, probably 600 mg per day. Antacids and H2 blockers and rifampin
should be rigorously avoided. Ketoconazole should be given with some form
of acid. Finally, blood concentrations should be checked within a few days
to be sure the drug is being absorbed.

An additional consideration with ketoconazole is its propensity to
suppress endocrine function, including predominantly testosterone
production, but also at higher concentrations adrenal synthesis of cortisol
(Pont et al., 1984). In normal subjects this is probably not a major
consideration, but there is a potential for aggravating adrenal
insufficiency in patients who may have adrenal function compromised by
invasive histoplasmosis or concurrent mycobacterial or cytomegalovirus
infection. Certainly every patient with AIDS and histoplasmosis should have
adrenal function evaluated by an ACTH stimulation test, and if even
minimally abnormal one should consider replacement corticosteroids if
ketoconazole is to be used.

A final concern regarding ketoconazole is that the most common and
most frequently dose-limiting adverse reactions are nausea and vomiting
(Sugar et al., 1987). Because these are dose related in ketoconazole, and
because ketoconazole is suggested at rather high doses in these patients,
and because patients may already have some element of this from other AIDS

related problems or their treatments, the addition of ketoconazole could significantly worsen a tendency for "failure to thrive."

At present there are two triazole alternatives to ketoconazole, and there may soon be a third. The first alternative is fluconazole, a drug that is relatively water soluble, is well absorbed from the gastrointestinal tract, and penetrates most tissues very well (Graybill, 1989; Dismukes, 1988; Humphrey et al., 1985). Unfortunately, the experience with histoplasmosis, both in non-AIDS and in AIDS patients is very small. Also, all of this experience has been acquired at doses which we now know are suboptimal. In an unpublished Mycoses Study Group experience with non-AIDS patients treated with 50-100 mg per day, fewer than half responded to fluconazole. The only experience with AIDS patients has been 6 patients whom we treated in San Antonio, using similar doses. Five of these patients ultimately failed fluconazole, and a sixth has been maintained in remission for more than 2 years (Sharkey et al., 1989). While these numbers are discouraging at first consideration, 3 of the patients had prolonged remissions, more than 4 months, before recurrence of the histoplasmosis. Therefore, it is possible that higher doses of fluconazole may be effective in AIDS patients. A Mycoses Study Group multicenter trial with doses up to 400 mg per day is underway, but there are no data available. The status of fluconazole thus is indeterminate.

Another azole antifungal is itraconazole. Itraconazole requires some acid for absorption, is much less water soluble than fluconazole, and has prolonged clearance (up to 40+ hours biological half life) by hepatic degradation (Hardin et al., 1988; Fromtling, 1988). Itraconazole is extremely potent *in vitro* and in animal models has been found very effective against *Histoplasma capsulatum*. The Mycoses Study Group has found in non-AIDS patients that itraconazole is highly effective for treatment of histoplasmosis (Saag and Dismukes, 1988). They treated 28 patients with histoplasmosis. Of 28 patients treated, 27 were successes, and 16 patients continue on therapy. At doses of 400 mg per day itraconazole is not associated with the cortisol suppression that has marked ketoconazole, and thus may be safer to use in a patient with potential adrenal insufficiency. Also, gastrointestinal intolerance is much less common with itraconazole than with ketoconazole.

Despite this marked evidence of potency, there is very little experience with itraconazole in treatment of histoplasmosis in AIDS. Virtually the only reported experience is our treatment of 9 patients with 400 mg per day (Graybill, 1988; Sharkey et al., 1989). Our patients were generally as ill with histoplasmosis as those reported elsewhere. They were not selected for severity of illness. All patients were offered itraconazole, and no patient refused to take it. Of our 9 evaluable patients, 2 had previously failed significant courses of 500 and 700 mg of amphotericin B, and 3 had failed ketoconazole. We considered clinical remission to be greater than 2/3 of all signs improved significantly, all signs attributable to histoplasmosis improving or resolved, and negative follow up cultures were done. Improvement was defined as fewer than 2/3 of symptoms improved, the majority but not all of signs resolving, and cultures negative or not available. Of our 14 patients treated for 16 evaluable episodes of histoplasmosis, there were 10 courses of itraconazole. Of these, 7 entered remission, and 3 improved on itraconazole. Of the 5 itraconazole recipients who have succumbed, causes were *Mycobacterium avium-intracellulare* in 2 (1 of whom also had evidence of active histoplasmosis at autopsy), lymphoma in 1, suicide in 1, and Kaposi's sarcoma in 1. At present, 7 patients remain in remission, and our median follow up time is >9 months. Of the 2 failures, both occurred in patients who had sustained prior improvement, and then became protocol inadherent, presumably because of AIDS dementia. Although cerebrospinal fluid abnormalities were seen in 4 patients, they could not be attributable to histoplasmosis, and cerebrospinal fluid cultures were negative for fungi. No other diagnoses accounted for the progressive dementia.

In addition, one patient with colonic histoplasmosis entered remission on itraconazole therapy but later relapsed with diarrhea and positive colonic biopsies for *H. capsulatum*. This was associated with his developing dyspepsia and self-prescribing antacids for this. When the antacids were discontinued, the histoplasmosis again entered remission. This patient had itraconazole serum concentrations consistently less than 1 mcg/ml, and these were undetectable when he was taking antacids. Parenthetically, we have seen one other similar event. This occurred in a bone marrow recipient who developed pulmonary and cutaneous aspergillosis which persisted unchanged after 2 weeks of amphotericin B at doses reaching 60 mg per day. Amphotericin B was discontinued and itraconazole was begun at 400 mg per day. The patient's primary physicians were unwilling to discontinue antacids and ranitidine, and a few days after commencing itraconazole the aspergillosis assumed a fulminating ultimately lethal course. Serum concentrations of itraconazole were undetectable. These 2 events have prompted us to measure itraconazole concentrations in all patients with conditions associated with hypochlorhydria, including AIDS.

Itraconazole was extremely well tolerated. None of our patients has had itraconazole discontinued because of adverse reactions. Of interest, antifungal azoles, including itraconazole, have the potential for causing abnormal liver functions. All but one of our 8 patients who presented with abnormal liver function tests had improvement of these tests on itraconazole. We considered that these laboratory abnormalities were attributable to histoplasmosis, and did not consider them a contraindication for use of this potentially hepatotoxic drug. The patient with persistently abnormal liver function tests had concurrent hepatitis B antigenemia and was thought to have chronic active hepatitis.

This is a limited experience at one institution, and cannot be generalized as a recommendation for itraconazole treatment of histoplasmosis. However, our experiences are encouraging, and form the basis of several ongoing investigations. In one, the AIDS Clinical Treatment Group centers located in the endemic area for histoplasmosis have begun a trial which involves initial treatment with amphotericin B, and then after remission of disease is induced, randomization to either continuing weekly amphotericin B, after the method of McKinsey et al. (1989), or oral itraconazole at 400 mg per day. A second study, by the same groups, is about to commence. This is an open label primary therapy trial with itraconazole at 400 mg per day.

These 2 studies should acquire sufficient additional data to enable us to determine whether itraconazole is effective as suppressive treatment or as primary therapy. One present limitation of itraconazole is that it must be given orally. This may limit the number of patients enrolled. However, an intravenous form prepared in cyclodextrins is under development, and should be available by the time of this publication.

There is one additional antifungal triazole that may be useful in histoplasmosis occurring in the setting of AIDS. This is SCH39304. This drug has some characteristics intermediate between those of itraconazole and fluconazole. It has great *in vitro* potency and a broad antifungal spectrum that includes such species as *Aspergillus*, like itraconazole. SCH39304 is well absorbed orally, does not require acid for absorption, penetrates the cerebrospinal fluid well, and is primarily excreted in the urine, like fluconazole. Like itraconazole, SCH39304 must be given orally at present, but an intravenous form is also under development.

At present the only experience with histoplasmosis is in animal models (Kobayashi et al., 1988). However, the great efficacy in murine histoplasmosis suggests that this drug may be highly effective in histoplasmosis. One of the unique characteristics of SCH39304 is remarkably prolonged half life, more than 60 hours. This may permit dosing SCH39304 as infrequently as once or twice weekly. The reverse side of this coin is the potential for adverse reactions to be more sustained than for the other azoles. At present there are too few data to comment on whether SCH39304 will be comparable in toxicity or other characteristics to other antifungal azoles.

SUMMARY

In summary, of the 4 endemic mycoses, coccidioidomycosis and histoplasmosis are increasingly associated with HIV infection, while paracoccidioidomycosis and blastomycosis are rare in this patient population. This is probably not a simple matter of geography. The endemic zone of blastomycosis overlaps most of that for histoplasmosis; as AIDS has moved into the midwest endemic area for *H. capsulatum*, these regions of the United States are reporting increased frequency of histoplasmosis, but not blastomycosis. In Texas the great majority of our cases of histoplasmosis are in patients with AIDS. The disease characteristically takes its most severe widely disseminated form. Diagnosis is relatively easy, and can be done with peripheral blood smears or cultures, bone marrow examination, and examination of respiratory secretions in most patients. New antigen detection tests may further accelerate the speed of diagnosis, hopefully confirming this at a much earlier stage of illness. With absent cell mediated immunity, the goals of treatment have shifted from cure to suppression. This can be accomplished very effectively with amphotericin B in many, but not all patients. Treatment with amphotericin B has significant limitations in the forms of cumulative nephrotoxicity and catheter associated infections. Less noxious alternatives have been tried, including ketoconazole, fluconazole, and itraconazole. Based on very limited and noncomparative studies, itraconazole appears the most promising among the azole drugs studied to date. An acid intragastric environment appears desirable for itraconazole. Fulminating disseminated histoplasmosis, an increasingly common opportunistic infection in AIDS, no longer needs to be regarded as an invariably relapsing and ultimately fatal illness.

REFERENCES

Ampel, N.M., Fish, D.G., Dols, C.L., and Galgiani, J.N., 1989, Coccidioidomycosis (coccy) in HIV-infected subjects. A prospective and retrospective evaluation. Abstract 376, Twenty-ninth Interscience Conference on Antimicrobial Agents and Chemotherapy, Houston TX.

Bakos, L., Kronfeld, M., Hampe, S.V., Castro, I., Zampese, M.S., and Coiro, J.R.R., 1988, Paracoccidioidomycosis with skin lesions in a patient with AIDS. Abstract 5553, International AIDS Meeting, Stockholm.

Bartholomew, C., Raju, C., Patrick, A., Penco, F., and Jankey, N., 1985, AIDS on Trinidad, Lancet, 1:103.

Bonner, J.R., Alexander, J.W., Dismukes, W.E., App, W., Griffin, F.M., Little, R., and Shin, M.S., 1984, Disseminated histoplasmosis in patients with the acquired immunodeficiency syndrome, Arch. Intern. Med., 144:2178.

Bullock, W.E., and Wright, S.D., 1987, Role of the adherence promoting receptors, CR3, LFA-1, and p150,95 in binding of *Histoplasma capsulatum* by human macrophages, J. Exp. Med., 165:195.

Chiu, J., Berman, S., Tan, G., and Tillis, J., 1988, Disseminated blastomycosis in HIV infected patients. Abstract 7209, International AIDS Meeting, Stockholm.

Dismukes, W.E., 1988, Azole antifungal drugs: old and new, Ann. Intern. Med.,109:177.

Fromtling, R.A., 1988, Overview of medically important antifungal azole derivatives, Clin. Microbiol. Rev., 1:187.

Dismukes, W.E., Royal, S.A., and Tynes, B.S., 1978, Disseminated histoplasmosis in corticosteroid-treated patients, J.A.M.A., 240:1495.

Greene, L.W., Cole, W., Greene, J.B., Levy, B., Louie, E., Raphael, B., Waitkevicz, H.J., and Blum, M., 1984, Adrenal insufficiency as a complication of the acquired immunodeficiency syndrome, Ann. Intern. Med., 101:497-8 (1984).

Goodwin, R.A., Jr., and Snell, J.D., Jr., 1969, The enlarging histoplasmoma, Am. Rev. Resp. Dis., 100:1.

Goodwin, R.A., Jr., Nickell, J.A., and Des Prez,R.M., 1972, Mediastinal fibrosis complicating healed primary histoplasmosis and tuberculosis, Medicine, 51:227.

Goodwin, R.A., Jr., Owens, F.T., Snell, J.D., Hubbard, W.W., Buchanan, R.D., Terry, R.T. and Des Prez, R.M., 1976, Chronic pulmonary histoplasmosis, Medicine, 55:413.

Goodwin, R.A., Jr., Shapiro, J.L., Thurman, G.H., Thurman, S.S., and Des Prez, R.M., 1980, Disseminated histoplasmosis: Clinical and pathologic correlations, Medicine, 59:1.

Graybill, J.R., 1988, Histoplasmosis and AIDS, J. Infect. Dis., 158:623.

Graybill, J.R., 1989, New antifungal agents, Eur. J. Clin. Microbiol. Infect. Dis., 8:402.

Gustafson, P.R., and Henson, A., 1985, Ketoconazole therapy in AIDS patients with disseminated histoplasmosis, Arch. Intern. Med., 145:2272.

Haggerty, C.M., Britton, M.C., Dorman, J.M., and Marzoni, F.A., 1985, Gastrointestinal histoplasmosis in suspected acquired immunodeficiency syndrome, Western J. Med., 143:244.

Hardin, T.C., Graybill, J.R., Fetchick, R., Woestenborghs, R., Rinaldi, M.G., and Kuhn, J.G., 1988, Pharmacokinetics of itraconazole following oral administration to normal volunteers, Antimicrob. Agents Chemother., 32:1310.

Humphrey, M.J., Jevons, S., and Tarbit, M.H., 1985, Pharmacokinetic evaluation of UK49,858, a metabolically stable triazole antifungal drug, in animals and humans, Antimicrob. Agents Chemother., 28:648.

Johnson, P.C., Hamil, R.J., and Sarosi, G.A., 1989, Clinical Review: progressive disseminated histoplasmosis in the AIDS patient, Semin. Respir. Infect., 4:139.

Johnson, P.C., Khardori, N., Najjar, A.F., Butt, F., Mansell, P.W.A., and Sarosi, G.A., 1988, Progressive disseminated histoplasmosis in patients with acquired immunodeficiency syndrome, Am. J. Med., 85:152.

Jones, P.G., Cohen, R.L., Batts, D.H. and Silva, J., Jr., 1983, Disseminated histoplasmosis, invasive aspergillosis, and other opportunistic infections in a homosexual patient with acquired immune deficiency syndrome, Sex. Transm. Dis., 10:202.

Kobayashi, G.S., Travis, S.J., and Medoff, G., 1988, In vitro evaluation of SCH39304, fluconazole (FLU) and amphotericin B (Amb) against 12 clinical isolates of Histoplasma capsulatum and comparison of therapeutic efficacies in treating histoplasmosis in immunocompetent and immunosuppressed mice. Abstract 172. Twenty eighth Interscience Conference on Antimicrobial Agents and Chemotherapy, Los Angeles.

Lake-Bakaar, G., Tom, W., Lake-Bakaar, D., Gupta, N., Beidas, S., Elsakr, M., and Straus, E., 1988, Gastropathy and ketoconazole malabsorption in the acquired immunodeficiency syndrome (AIDS), Ann. Intern. Med., 109:471.

Mandell, W., Goldberg, D.M. and Neu, H.C., 1986, Histoplasmosis in patients with the acquired immunodeficiency syndrome, Am. J. Med., 81:974.

McKinsey, D.S., Gupta, M.R., Riddler, S.A., and Driks, M.R., 1989, Long-term amphotericin B therapy for disseminated histoplasmosis in patients with the acquired immunodeficiency syndrome (AIDS), Ann. Intern. Med. 111:655.

Pont, A., Graybill, J.R., Craven, P.C., Galgiani, J.N., Dismukes, W.E., Reitz, R.E., and Stevens, D.A., 1984, High-dose ketoconazole and adrenal and testicular function in humans, Arch. Intern. Med., 144:2150.

Raviglione, M.C., Battan, R., Pablos-Mendez, A., AvecesCasillas, P., and Mullen, M.P., 1989, Infections associated with Hickman catheters in patients with acquired immunodeficiency syndrome, Am. J. Med., 86:780.

Saag, M.S., and Dismukes, W.E., 1988, Treatment of histoplasmosis and blastomycosis, Chest, 93:848.

Saag, M., Bradsher, R., Chapman, S., Kauffman, C., Girard, W., Stevens, D., Bowles-Patton, C., and Dismukes, W.E., 1988, Itraconazole (I) therapy (Rx) for blastomycosis (B), Histoplasmosis (H), and sporotrichosis (S). Abstract 574, Twenty Eighth Interscience Conference on Antimicrobial Agents and Chemotherapy. Los Angeles, CA.

Schlamm, H.T., Rebdel, M., Roth, J., and Seidlin, M., 1989, Infectious complications of Hickman catheters in HIV infected patients, Abstract 378, Twenty-Ninth Interscience conference on antimicrobial Agents and Chemotherapy, Houston, TX.

Sharkey, P.K., Fetchick, R., Smith, T., Dietrich, J.F. and Graybill, J.R., 1989, Histoplasmosis in the acquired immunodeficiency syndrome (AIDS): treatment with new azoles, Journal of Aids, in Press.

Sugar, A.M., Alsip, S., Galgiani, J.N., Graybill, J.R., Dismukes, W.E., Cloud, G.A., Craven, P.C., and Stevens, D.A., 1987, Pharmacology and toxicity of high dose ketoconazole, Antimicrob. Agents Chemother., 31:1874.

Tapper, M.L., Rotterdam, H.Z., Lerner, C.W., Al'Khafaji, K., and Seitzman, P.A., 1984, Adrenal necrosis in the acquired immunodeficiency syndrome, Ann. Intern. Med., 100:239.

Taylor, M.N., Baddour, L.M., and Alexander, J.R., 1984, Disseminated histoplasmosis associated with the acquired immune deficiency syndrome, Am. J. Med., 77:579.

Tiley, K., Antoniskis, D., Evans, S., and Larsen, R.A., 1988, AIDS-related histoplasmosis in a nonendemic area (Abstract F-97) Annual Meeting for the American Society of Microbiology, Washington, DC.

Wheat, L.J., 1989, Fungal infections in AIDS patients, Symposium "Fungal Infections in the Immunocompromised host". Session 24. Twenty-Ninth Interscience Conference on Antimicrobial Agents and Chemotherapy. Houston, TX.

Wheat, L.J., and Small, C.B., 1984, Disseminated histoplasmosis in the acquired immunodeficiency syndrome, Arch. Intern. Med., 144:2147.

Wheat, L.J., Connely-Stringfield, P., Kohler, R.B., and Gupta, M.R., 1989, Histoplasma capsulatum polysaccharide antigen detection in diagnosis and management of disseminated histoplasmosis in patients with acquired immunodeficiency syndrome, Am. J. Med., 396.

Wheat, L.J., Kohler, R.V.B., and Tewari, R.P., 1986, Diagnosis of disseminated histoplasmosis by detection of Histoplasma capsulatum antigen in serum and urine specimens, N. Engl. J. Med., 314:83.

Wheat, L.J., Slama, T.G., Norton, J.A., Kohler, R.B., Eitzen, H.E., French, L.M.V., and Sathapatayavongs, B., 1982, Risk factors for disseminated or fatal histoplasmosis, Ann. Intern. Med., 96:159.

IMMUNOLOGICAL ASPECTS OF DIMORPHIC FUNGI IN AIDS

Juneann W. Murphy

Dept. of Microbiology and Immunology, University of Oklahoma
Health Sciences Center, Oklahoma City, Oklahoma 73190 USA

The hallmark of an infection with human immunodeficiency virus (HIV)
is a selective depletion of CD4$^+$ lymphocytes and an associated progressive
decline of immunological function to the point the host becomes overwhelmed
by opportunistic infections or malignancies (McChesney and Oldstone, 1989;
Schupbach, 1989). Since CD4$^+$ cells are essential components of effective
host defense against most of the systemic mycotic agents, patients with
acquired immunodeficiency syndrome (AIDS) frequently suffer from
fulminating fungal infections (Armstrong, 1988; Spencer and Jackson, 1989).
The two dimorphic fungi which are most commonly isolated from individuals
infected with HIV are *Histoplasma capsulatum* and *Coccidioides immitis*,
(Armstrong, 1988; Spencer and Jackson, 1989; Cairns, 1988) but less
frequently other dimorphic fungi such as *Sporothrix schenckii*, (Bibler et
al., 1986; Kurosawa et al., 1988) and *Paracoccidioides brasiliensis* (Bakos
et al., 1989) have been reported to be secondary invaders in AIDS patients.
To understand the AIDS patients' immunological responses to the
opportunistic fungi, it is essential to first recognize the immunological
status of the host at the onset of the fungal infection. So, before
discussing the specific immune responses during systemic dimorphic fungal
infections in AIDS patients, I want to define the state of immune function
as an HIV infection progresses from the asymptomatic, seroconversion stage
to AIDS and identify the stage(s) at which the host is most vulnerable to
opportunistic infections. For this purpose, I will use the classification
scheme proposed by the Walter Reed group which is based upon the classical
immunological and clinical status of HIV infected persons.

The onset of infection with HIV which would be the acute phase or the
Walter Reed Stage 1 (WR1) is characterized by viremia, seroconversion
(detection of anti-HIV antibodies), and an acute inversion of the CD4$^+$/CD8$^+$
lymphocyte ratio (less than 1.0) usually due to an increase in the numbers
of CD8$^+$ cells (McChesney and Oldstone, 1989; Schupbach, 1989). The acute
phase of an HIV infection may last for months and can range from
subclinical to a syndrome similar to infectious mononucleosis (McChesney
and Oldstone, 1989). As the HIV infection progresses from the acute phase
through the asymptomatic infection stage, the viremia subsides, the CD4$^+$
lymphocyte numbers and functions begin a persistent decline (numbers are
15-40% below seronegative controls, the rate of decline varies from patient
to patient and plateaux may occur at intervals), and the CD8$^+$ cells remain

Mycoses in AIDS Patients, Edited by
H. Vanden Bossche *et al.,* Plenum Press, New York, 1990

high or increase (35-55% above controls) (McChesney and Oldstone, 1989). In the Walter Reed staging classification, chronic lymphadenopathy occurs at stage WR2 when the $CD4^+$ cell numbers are still greater than $400/mm^3$ and the patient typically displays positive delayed-type hypersensitivity (DTH) to a battery of 4 recall antigens such as *Candida*, mumps, trichophytin, and tetanus toxoid (Schupbach, 1989). As the $CD4^+$ cell numbers drop below $400/mm^3$, the HIV infected individuals continue to react with DTH reactivity to the 4 recall antigens for a time (WR3 stage), but later they become partially anergic in that they react to only 1 of the 4 skin test antigens (WR4 stage) (Schupbach, 1989). As the disease progresses to the WR5 stage, the $CD4^+$ lymphocyte numbers remain below $400/mm^3$ and the DTH responses to the recall antigens may be absent or oral candidiasis may be a complicating feature (Schupbach, 1989). The WR6 stage, which is referred to as AIDS, is defined by $CD4^+$ numbers being less than $400/mm^3$, partial to complete anergy to recall antigens, and the presence of an AIDS-related opportunistic disease. With progression of the HIV infection to AIDS, the features that impact on immune surveillance mechanisms are: (i) the critically low level of $CD4^+$ cell numbers which results in dramatically reduced lymphokine production and in a generalized immunological unresponsiveness, (ii) the increase in serum levels of viral antigen which in some cases leads to immune complex formation in antigen excess with a concomitant loss of titer for anti-HIV antibody, (iii) a drop in $CD8^+$ cell numbers, (iv) elevated concentrations of serum IgG and IgA due to polyclonal activation of B cells which is usually associated with impaired specific antibody responses, and (v) varying levels of dysfunction of macrophages/monocytes, dendritic cells, and Langerhans cells which range from reduced monokine production to a diminished ability to serve in antigen presentation. Thus, a consequence of prolonged HIV infection is a serious generalized immunodeficiency which can be explained only partially by the loss of $CD4^+$ lymphocytes.

When the CD4 counts reach $200-250/mm^3$ in the HIV infected individuals or the percent of CD4 cells is 20-25, the patient is at high risk for developing one or more of the HIV-related opportunistic infections (Masur et al., 1989) In such patients, the immunosurveillance mechanisms effective in preventing secondary infections are absent or drastically reduced in most of their organ systems. For examples, not only is the protective immune capacity abrogated in the lymphoid organs but also in the brain, skin, lungs, and gastrointestinal tract. It is at this stage of immunodeficiency that the host is likely to succumb to an acute secondary infection which frequently affects organs not usually involved. The systemic fungal pathogens typically invade as the CD4 cell levels fall below $200/mm^3$, and they may or may not be the first or only secondary invaders in AIDS patients. Concurrent infections with *Pneumocystis*, *Mycobacterium*, *Candida*, or *Cryptococcus* are common, and these additional insults to the host may further alter the immune system (Macher, 1988).

Immunosurveillance in the HIV-infected host with or without opportunistic infections is so minimal that infections with *H. capsulatum* or *C. immitis* are typically progressive and disseminated. The immunosuppressed patient can acquire the dimorphic fungal infection through primary exposure to the organism from an exogenous source or can develop clinical disease by reactivation of a quiescent infection with the mycotic agent (Spencer and Jackson, 1989). In the absence of anti-fungal therapy, the fungal organisms generally multiply rapidly and unchecked in multiple tissues of the AIDS patients. This is evident by the fact that approximately 85% of the AIDS patients with histoplasmosis have progressive disseminated disease; whereas, only 5% of the individuals from the normal population who have histoplasmosis have the disseminated form (Wheat et al., 1985). In fact, due to the severe immunoincompetency of HIV infected individuals, the dimorphic fungi are often isolated in high numbers from organ systems that are not routinely infected by the fungal agent (Graham et al., 1988; Graybill, 1988; Anaissie et al., 1988; Alterman and Cho, 1988). Frequently, *H. capsulatum* can be observed in and isolated from peripheral blood monocytes and neutrophils from AIDS patients with histoplasmosis (Graybill, 1988; Tomita and Chiga, 1988; Johnson et al., 1988). Moreover, the large numbers of organisms in the tissues translate, in immunological terms, to an overwhelming antigen overload for the weakened immune system.

Host tissue responses to an invading dimorphic fungus are different in AIDS and non-AIDS patients (Graham et al., 1988). Generally, granuloma development in response to the fungus is minimal in AIDS patients, a situation that might be expected in individuals whose CD4$^+$ T lymphocyte numbers are significantly reduced (Graham et al., 1988). Without CD4$^+$ cells, certain lymphokines are not produced at the antigen deposition site to signal migration of other mononuclear cells into the region or to activate the macrophages, which is a sequence of events that eventually results in a histological pattern of granuloma. In fact, it is entirely possible that some of the CD8$^+$ T cells may be suppressor cells that suppress any remaining CD4$^+$ T cells which are reactive to the fungal antigens and which have the potential to trigger granuloma formation. Further investigations are required to gain a complete understanding of the cellular interactions which occur in the HIV-infected host with an opportunistic dimorphic fungal infection.

The humoral or cell-mediated immune responses that the HIV infected individual mounts against *H. capsulatum*, *C. immitis*, or other dimorphic fungi are dependent on the level of immunocompetence of the patient. As one might expect in the profoundly immunosuppressed AIDS patient, the immunological response to the fungal organism is minimal. The fungal-specific immune reactivity is inversely related to the severity of the immunosuppression of the patient.

Although there are only a few reports on skin testing of AIDS patients with fungal antigens, AIDS patients with dimorphic fungal infections would not be expected to demonstrate DTH reactivity to the fungal antigens since they have so few CD4$^+$ lymphocytes which are responsible for the DTH reactivity. The fact that most AIDS patients display a generalized anergy leads one to believe that AIDS patients with fungal disease would also be unable to react to the specific fungal antigens (Schupbach, 1989; Wheat et al., 1985; Bonner et al., 1984). However, there is a report by Byrne and Dietrich (1989) in which a patient with disseminated coccidioidomycosis and AIDS was reactive to coccidioidin two months prior to showing antibodies for HIV. Fourteen months after diagnosis of AIDS, the patient did not manifest a DTH reaction to coccidioidin, but was reactive 6 months after the anergic response was noted. Since this patient survived for 29 months after initial diagnosis of AIDS, in retrospect, the positive skin test reaction was considered to have been a favorable sign. Although one might predict, based on our current understanding of immunology, that patients with positive skin tests to the relevant fungal antigen would have some level of protective response and thus survive longer than those that are anergic, considerably more experience in this area will be required to validate this concept.

In reviewing the literature concerning the serological reactions of histoplasmosis and coccidioidomycosis patients with AIDS, it is apparent that the trends indicate immunodiffusion reactions and complement fixing

(CF) antibody titers of such patients to the appropriate fungal antigen(s) are not significantly different from the reactions and titers that might be expected with serum samples from patients who do not have AIDS but have the disseminated mycotic disease (Wheat et al., 1985; Graham et al., 1988; Graybill, 1988; Johnson et al., 1988; Bonner et al., 1984; Salzman et al., 1988). It is not unusual to expect 30% of the patients who have disseminated histoplasmosis without AIDS to have negative CF antibody titers to *H. capsulatum* antigens, and this is about the same percentage of negative reactors reported for AIDS patients with disseminated histoplasmosis (Graybill, 1988). The antifungal humoral immune responses of AIDS patients infected with a dimorphic mycotic agent would not be expected to be significantly reduced since B lymphocytes in AIDS patients are polyclonally activated. Considering that many of the AIDS patients may have had a quiescent fungal infection reactivated, they may have memory cells directed against the fungal antigens, and the polyclonal activation of B cells would be expected to result in some antibody production to the relevant fungal antigen(s).

In summary, immunological responses and the course of disease in HIV-infected individuals who acquire a dimorphic fungal infection are dependent on the level of immunocompetence of the patient. Generally, HIV infected individuals do not display active dimorphic fungal disease until there is evidence of severe leukopenia and the numbers of $CD4^+$ T lymphocytes are below $200/mm^3$ (Masur et al., 1989). Under these conditions, the fungal organisms multiply rapidly, overwhelming any protective fungal-specific cell-mediated immunity that may be present in the host. The heavy antigen burden has the potential to induce suppressor cells which in turn could abrogate any remaining cell-mediated reactivity that the patient might have to the fungal agent. In the early stages of AIDS with a superimposed mycotic infection, antifungal chemotherapy will usually reduce the fungal burden to the point that the host can overcome the clinical episode (Wheat et al., 1985; Graham et al., 1988; Graybill, 1988; Johnson et al., 1988; Bonner et al., 1984; Salzman et al., 1988). But as the patients' immune responses deteriorate, relapses of the fungal infection occur and antifungal drugs are less effective. After one or more relapses, the patients frequently succumb to the fungal infection. Thus, it seems that measures to augment the hosts immune responses must be developed to prevent the fatal outcome. Development of such immunotherapeutic protocols is certainly a challenge for the future.

REFERENCES

Alterman, D.D., and Cho, K.C., 1988, Histoplasmosis involving the omentum in an AIDS patient: CT demonstration, _J. Comput. Assist. Tomogr._, 12:664.
Anaissie, E., Fainstein, V., Samo, T., Bodey, G.P., and Sarosi, G.A., 1988, Central nervous system histoplasmosis, _Am. J. Med._, 84:215.
Armstrong, D., 1988, Life-threatening opportunistic fungal infection in patients with acquired immunodeficiency syndrome, _Ann. N.Y. Acad. Sci._, 544:443.
Bakos, L., Kronfeld, M., Hampe, S., Castro, I., and Zampese, M., 1989, Disseminated paracoccidioidomycosis with skin lesions in a patient with acquired immunodeficiency syndrome, _J. Am. Acad. Dermatol._, 20:854.
Bibler, M.R., Luber, H.J., Glueck, H.I., and Estes, S.A., 1986, Disseminated sporotrichosis in a patient with HIV infection after treatment for acquired factor VIII inhibitor, _J.A.M.A._, 256:3125.

Bonner, J.R., Alexander, W.J., Dismukes, W.E., App, W., Griffin, F.M., Little, R., and Shin, M.S., 1984, Disseminated histoplasmosis in patients with the acquired immune deficiency syndrome, *Arch. Intern. Med.*, 144:2178.

Byrne, W.R., and Dietrich, R.A., 1989, Disseminated coccidioidomycosis with peritonitis in a patient with acquired immunodeficiency syndrome, *Arch. Intern. Med.*, 149:947.

Cairns, M.R., 1988, Fungal infections in the acquired immunodeficiency syndrome, *J. Electron. Microsc. Tech.*, 8:115.

Graham, A.R., Sobony, R.E., Bronnimann, D.A., and Galgiani, J.N., 1988, Quantitative pathology of coccidioidomycosis in acquired immunodeficiency syndrome, *Hum. Pathol.*, 19:800.

Graybill, J.R., 1988, Histoplasmosis and AIDS, *J. Infect. Dis.*, 158:623.

Johnson, P.C., Khardori, N., Najjar, A.F., Butt, F., Mansell, P.W.A., and Sarosi, G.A., 1988, Progressive disseminated histoplasmosis in patients with acquired immunodeficiency syndrome, *Am. J. Med.*, 85:152.

Kurosawa, A., Pollock, S.C., Collins, M.P., Kraff, C.R., and Tso, M.O.M., 1988, *Sporothrix schenckii* endophthalmitis in a patient with human immunodeficiency virus infection, *Arch. Ophthalmol.*, 106:376.

Macher, A.M., 1988, The pathology of AIDS, *Public Health Rep.*, 103:246.

Masur, H., Ognibene, F.P., Yarchoan, R., Shelhamer, J.H., Baird, B.F., Travis, W., Suffredini, A.F., Deyton, L., Kovacs, J.A., Falloon, J., Davey, R., Polis, M., Metcalf, J., Baseler, M., Wesley, R., Gill, V.J., Fauci, A.S., and Lane, H.C., 1989, CD4 counts as predictors of opportunistic pneumonias in human immunodeficiency virus (HIV) infection, *Ann. Intern. Med.*, 111:223.

McChesney, M.B., and Oldstone, M.B.A., 1989, Virus-induced immunosuppression: infection with measles virus and human immunodeficiency virus, *Adv. Immunol.*, 45:335.

Salzman, S.H., Smith, R.L., and Aranda, C.P., 1988, Histoplasmosis in patients at risk for the acquired immunodeficiency syndrome in a nonendemic setting, *Chest*, 93:916.

Schupbach, J., 1989, Human retrovirology. Facts and concepts, *Curr. Top. Microbiol. Immunol.*, 142:36.

Spencer, P.M., and Jackson, G.G., 1989, Fungal and mycobacterial infections in patients infected with the human immunodeficiency virus, *Antimicrob. Chemother.*, 23 Suppl. A:107.

Tomita, T., and Chiga, M., 1988, Disseminated histoplasmosis in acquired immunodeficiency syndrome: light and electron microscopic observations, *Hum. Pathol.*, 19:438.

Wheat, L.J., Slama, T.G., and Zeckel, M.L., 1985, Histoplasmosis in the acquired immune deficiency syndrome, *Am. J. Med.*, 78:203.

Chemotherapy

CURRENT STATUS AND PERSPECTIVES OF ANTIFUNGAL THERAPY

John E. Bennett

Mycology Section, Laboratory of Clinical Investigation,
National Institute of Allergy and Infectious Disease,
Bethesda, Md. 20892

The last decade has seen a quickening pace of antifungal drug
development, driven in part by the success of imidazole compounds and in
part by the increased number of patients with mycoses. There seems to be no
signs of slackening in the rising numbers of immunosuppressed patients,
driven by the AIDS epidemic and the broadening indications for
immunosuppressive therapy. Successful therapy of bacterial, parasitic and
viral infections has allowed survival not only of immunosuppressed patients
but also other fragile patients, such as very low birth weight neonates,
extensively burned patients and those postoperative from complicated
abdominal surgery.

A large proportion of the drugs under study are N-substituted azoles.
These drugs have attempted to circumvent the limitations of ketoconazole,
the first widely used systemic antifungal azole (Table 1).

Although the list of advantages looks small, ketoconazole was truly a
remarkable advance. There is no need to list the indications for this drug
because they have become well known. The limitations are somewhat more
subtle but are the driving force for new drug development. Failure to
achieve a suitable intravenous formulation meant that ketoconazole was
never suitable for a sizable proportion of hospitalized patients. There are
too many uncertainties about oral agents in severely ill patients to trust

Table 1. Perspective on ketoconazole

Advantages	Disadvantages
Oral adsorption	No intravenous preparation
Broad spectrum	Blood levels impaired by
Low toxicity	several other drugs
	Hepatotoxicity
	Hormonal suppression
	Nausea
	Poor efficacy in fungal meningitis,
	aspergillosis, mucormycosis
	and sporotrichosis

Mycoses in AIDS Patients, Edited by
H. Vanden Bossche *et al.*, Plenum Press, New York, 1990

oral administration. Use of H2 receptor antagonists is particularly common in intensive care units, postoperative recovery units and medical wards. These agents block oral adsorption of ketoconazole. A major subgroup of postoperative patients are at high risk of deep candidiasis but cannot take oral medications. These are the patients recovering from complicated abdominal surgery. Even for the postoperative patient who is taking oral medications, emesis may make for irregular dosing. These limitations hindered trials of candidemia. I am still uncertain whether ketoconazole is useful for candidemia or deep candidiasis.

Hepatotoxicity is a rare complication that only looms large when the condition being treated causes only modest morbidity. Recurrent *Candida* vaginitis, griseofulvin-resistant ringworm and extensive tinea versicolor are all valid indications for ketoconazole but hepatotoxicity has prevented this usage from being routine. Hormonal suppression, like nausea, has put the cap on the upper limits of dosing. I am not convinced that ketoconazole has caused symptomatic adrenal insufficiency and know of no good evidence of permanent damage to adrenal or reproductive organs.

The failure of ketoconazole in cryptococcal meningitis has been explained by the very low cerebrospinal fluid drug levels. Yet, amphotericin is effective in that disease and provides unmeasurable CSF concentrations. The efficacy of itraconazole in experimental cryptococcosis of rabbits also cannot be explained by good penetration into the CSF (Perfect, et al, 1986). With coccidioidal meningitis, CSF drug appears to be necessary, making the story clearer as to why conventional doses of ketoconazole failed to be useful. Limitations in the spectrum of ketoconazole, not pharmacology, appear to explain failure in aspergillosis, mucormycosis and extracutaneous sporotrichosis.

Solutions to some of the problems posed by ketoconazole are being addressed by two triazoles now being introduced to the market in several countries (Table 2).

Itraconazole and fluconazole are both available as oral formulations. Blood concentrations of itraconazole are more than doubled when the drug taken after ingestion of a substantial meal compared to the fasting state (Heykants et al., 1987). Levels also are depressed transiently during bone marrow transplantation (Heykants et al, 1987). Considering the mucositis

Table 2. Comparison of itraconazole and fluconazole

Properties	Itraconazole	Fluconazole
Formulations	oral*	oral, I.V.
Decreased azole blood conc.		
H2 antagonists	0?	0
rifampin	++	±
phenytoin	++	±
Raises blood level		
warfarin	0	±
sulfonylureas	0	++
phenytoin	0	++
cyclosporine	±	±
Dose adjustment for		
azotemia	0	++
hemodialysis	0	++

*See also Cauwenbergh and Heykants, this book

that these patients can have, I wonder if the patients ingested and retained the drug reliably. There has been a suggestion that low blood levels might account for some therapeutic failures, but this issue remains unclear (van't Wout et al., submitted). As shown in Table 2, several drugs lower itraconazole levels. I have not seen the data on histamine H2 receptor antagonists, but am told that itraconazole bioavailability is not decreased, as is the case with ketoconazole. Data on phenytoin were supplied to me by Dr. Richard Tucker in David Stevens' laboratory. Dr. Tucker also confirmed the published data on rifampin, which causes a profound reduction in itraconazole levels (Heykants et al, 1987). The import of lower blood levels depends upon the gravity of the patient's illness and the possibility of increasing the dose of compensate. Assuming that itraconazole blood assays were not readily available, I would not like to treat a rapidly fatal disease with an oral compound that might not be well absorbed. In such a situation, I prefer a parenteral drug.

I would not like my concern about variable blood levels to be interpreted to mean that I know what a therapeutic level should be. Direct interpretation of itraconazole blood levels is obfuscated by the substantial differences between assay results by HPLC and bioassay (Warnock et al., 1988) as well as the disparity between serum levels and tissue levels (Heykants et al, 1987) . Further, no comparison between serum levels and *in vitro* susceptibility is possible because the latter result depends a great deal upon the method.

Fluconazole is available both orally and as a intravenous infusion of 2 mg/ml. In obtunded or confused patients I assume one could give the drug intravenously or administer the drug through a nasogastric tube. Itraconazole tends to clog nasogastric tubes because it is poorly soluble. Absorption of orally ingested fluconazole appears to be essentially complete (Bennett et al., 1989) . The effect of cimetidine and phenytoin on fluconazole blood levels is a measurable but trivial reduction (Lazar et al., 1989). Surprisingly, rifampin has a definite, thought modest effect of accelerating the metabolism of fluconazole (Lazar et al., 1989) . I say the effect is surprising because about 70% of the dose is excreted in the urine, at least in patients not taking rifampin.

Azole antifungals have the capacity to activate certain hepatic microsomal mixed oxidase functions. When itraconazole (Damanhouri et al., 1988) and fluconazole (Bennett, J., et al, 1989) were tested in rats for these effects, none was found. Patient data are not so simple. Results have varied between reports, depending at least in part on dosage and duration of triazole therapy. Cyclosporine is a case in point. Ketoconazole was found to elevate the levels of cyclosporine enough to cause nephrotoxicity. Although both itraconazole (Trenk et al., 1987) and fluconazole (Lazar et al., 1989), (Collignon et al., 1989) have been said to raise cyclosporine blood levels, nephrotoxicity is usually not mentioned. Many things influence cyclosporine blood levels, leaving unclear whether the triazoles have important interactions with cyclosporine. Similarly, fluconazole can elevate prothrombin times of some patients receiving warfarin or the blood levels of some patients receiving tolbutamide or glipizide, but the effect is not of great clinical magnitude (Lazar et al., 1989). Phenytoin blood levels are raised 75% by fluconazole 200 mg per day, indicating a more important interaction (Blum et al., 1989). I am not aware that itraconazole interacts with warfarin, sulfonylureas or phenytoin. Nor does the dose of itraconazole have to be adjusted for renal failure or hemodialysis (Boelaert et al., 1988) . Although fluconazole dosage should be adjusted in both situations, no dose-dependent toxicity has been discovered yet.

The hormonal effects of high dosage ketoconazole were to decrease the synthesis of cortisol and testosterone, as well as to increase the quantity of precursors with mineralocorticoid activity, such as deoxycorticosterone (De Coster et al., 1986) . Although decreased serum cholesterol might have been anticipated, this has not been seen in patients treated for mycoses. *In vitro* tests of fungal and mammalian enzyme systems has found the new

triazoles, such as itraconazole and fluconazole, to be far more active
against fungal 14-alphademethylase than against mammalian cytochromes P450
(Vanden Bossche, 1987) . Patient studies have confirmed these results, in
that no effect on testosterone or cortisol levels have been seen with
itraconazole up to 200 mg per day or fluconazole up to 400 mg per day.
There has been an occasional report of gynecomastia, such as in one of 60
patients given 50 -200 mg daily for coccidioidomycosis (Tucker et al.,
1988) . When 8 patients were given 600 mg of itraconazole for a mean of
five months' duration, hypokalemia was a significant and occasionally
severe complication of this high dosage (Sharkey et al., 1989). Other side
effects in those patients included hypertension in 5 patients,
rhabdomyolysis in one patient and reversible adrenal insufficiency in one
patient. Hypokalemia was seen in only one of 79 patients treated with 200 -
400 mg itraconazole (Saag et al., 1988) . It is conceivable that these
aldosterone-like effects could have resulted from increased
mineralocorticoid precursors.

Another limitation of ketoconazole was hepatotoxicity. Because serious
liver damage was so rare, it will take a long time to be sure that the
newer agents have less hepatotoxicity. At present, neither fluconazole nor
itraconazole have been clearly linked to serious liver disease, to my
knowledge. Both appear to be safe drugs. The overall incidence of side
effects with both drugs at doses up to 200 mg daily is less than 10%. The
most common side effect with either drug is gastrointestinal distress, a
dose-dependent effect. Less than 1% of the non-AIDS patient's have had the
drug discontinued because of adverse reactions. AIDS patients have had more
adverse reactions, including rash. Although the incidence of rash has been
a little under 1% overall with either drug, in AIDS patients fluconazole
has been associated with two cases of Stevens Johnson syndrome and one of
toxic epidermal necrolysis. Stevens Johnson syndrome contributed to one of
these patient's death (Bennett et al., 1989).

These two new azole antifungals are offering options to ketoconazole
in terms of both safety and expanded spectrum. Both drugs are effective in
superficial mycoses, such as candidiasis of the mouth and vagina.
Itraconazole, like ketoconazole, is contraindicated during pregnancy
because of teratogenesis and embryotoxicity when high doses are given to
pregnant rats. Fluconazole has not been teratogenic but has been
embryotoxic at doses toxic to the mother. Whether this result will
translate into differences in pregnant humans is not clear. At present, it
would seem preferable to treat *Candida* vaginitis with topical drugs during
pregnancy and during breast feeding. Itraconazole has been well studied in
dermatophytosis and tinea versicolor and found to be effective. In deeper
mycoses, itraconazole has been effective for the same infections as
ketoconazole, such as blastomycosis, histoplasmosis and coccidioidomycosis
(Tucker et al., 1988; Saag et al., 1988). The numbers of patients with
these mycoses has been relatively small, compared to the experience with
ketoconazole. Fluconazole was poor at doses of 50-100 mg in small numbers
of patients with blastomycosis, histoplasmosis and coccidioidomycosis.
Using larger doses, fluconazole has looked more promising in small numbers
of patients with disseminated coccidioidomycosis, including a few with
meningitis.

The expanded spectrum of the new triazoles may be in aspergillosis,
where itraconazole has occasionally been of value (Viviani et al,. 1989),
and in cryptococcosis, where both triazoles may have a role in AIDS
patients. In the rabbit model of cryptococcal meningitis, both itraconazole
and fluconazole have comparable activity (Perfect et al., 1986). There are
reports that both drugs are useful in preventing relapse of AIDS patients
treated with amphotericin. Data are best for fluconazole, which has
reported successful in at least 83% of cases (Table 3). This is quite
comparable to the 85% success reported with amphotericin B. The exact
figure depends on many variables, including how intensive is the original
course of amphotericin B and how well the patient's other diseases can be
controlled. Larsen has reported that prostatic massage specimens are often

Table 3. Fluconazole in cryptococcal meningitis of AIDS patients

1st author	# cases	% success
Prevention of relapse after culture conversion		
Dupont*	18	83
Sugar	18	89
Stern	15	87
Initial therapy of acute infection		
Dismukes	64	63
Larsen	14	43
Kalambayi	17	29

* See also Dupont et al. (this book)

positive in patients receiving amphotericin B prophylaxis. If those
cultures are counted as failures, then the success rate of amphotericin B
prophylaxis falls. It is not clear to me whether this site tends to relapse
during azole prophylaxis. Most of the failures with fluconazole have
discontinued the drug or were taking it erratically at time of failure.
Experience with itraconazole as maintenance therapy is small but
encouraging (Table 4). Problems interfering with drug intake are inherent
in the patients, such as AIDS dementia, allergic reactions in patients
taking multiple medications, and abnormal liver function which could have
been due to a multiplicity of causes. Ability to prevent recurrence of
cryptococcosis by an oral drug is a very important advance for AIDS
patients, avoiding the catheter-acquired bacteremias, the toxicity and
expense of weekly intravenous amphotericin B.

The data for initial treatment with fluconazole and itraconazole are
also provided in Tables 3 and 4. Because these patients were kept on
therapy until toxicity or death from other causes, I included as successes
those patients who not only survived the initial therapy but also did not
relapse. These figures are less promising. I am not convinced that either
fluconazole or itraconazole should be given to patients with previously
untreated cryptococcal meningitis, with or without AIDS. At least one
fourth of these patients die of the infection during the first month,
making imperative that the patients have uninterrupted drug in their blood.
Oral drugs are not satisfactory in patients who are becoming obtunded or
vomiting. Variable bioavailability of itraconazole is an additional worry.

Table 4. Itraconazole in Cryptococcal meningitis of AIDS patients

1st author	# cases	% success
Prevention of relapse in culture negative patients		
Viviani	4	100
de Gans	5	100
Initial therapy of acute infection		
Denning*	10	40

* See also Denning et al. (this book)

I anticipate that experience will eventually sort out the acutely ill cryptococcosis patients who are good candidates for oral itraconazole or fluconazole initially and those who might benefit from receiving this drug later on in the course of their disease.

As an area for the future, it may be possible to identify which patients are at such risk of acquiring cryptococcosis that one could consider evaluating the newer triazoles as prophylactic agents. The goal would not be to prevent recurrent disease but to prevent even the first episode. To the extent this was successful, one would no longer lose the 25 or more percent that now die despite therapy. Considering that at least one third of the patients have cryptococcosis as their first presenting manifestation of AIDS, one would like to identify high risk patients before they have clinical AIDS. The CD4 cell count is almost uniformly 200 or less at time cryptococcosis is diagnosed. But I believe we would need further identifiers to know who was at sufficient risk to warrant prophylaxis. Perhaps time will show us the way.

The newer triazoles deserve mention because several appear promising, such as Sch 39304, a compound now entering clinical trial. This agent has a very long half life, which may be useful in evening out drug levels in patients taking drug at home, such as for prophylaxis. There is also cilofungin, the lipopeptide with activity against many species of *Candida* (Hanson et al, 1989). The narrow spectrum, necessity for intravenous administration and relatively poor activity in some of the animal models, such as *Candida endocarditis* and disseminated disease, have made me wonder about the promise of this compound. The liposomal preparations of amphotericin B are in clinical trial. These preparations are difficult to manufacture in a uniform, stable formulation. Both the Vestar and Squibb formulations appear to give less acute febrile reactions and less nephrotoxicity than the same dose of the deoxycholate colloidal solution (Fungizone). Whether equivalent efficacy can be achieved is an unanswered problem. A safe guess is that both formulations will be used at higher doses than Fungizone. As far as has been studied, animals have developed nephrotoxicity with liposomal drugs if the dose is raised high enough. Only time will tell if the safety profile is better for these formulations or merely different. The impact of liposomal amphotericin B is likely to be less upon AIDS patients than upon intensive care patients. The latter are more fragile and less able to tolerate the azotemia and febrile reactions of amphotericin B.

This brief perspective should show why I am optimistic about the current developments in antifungal therapy. We are indeed fortunate to have promising developments in this area because the need has never before been so great.

REFERENCES

Bennett, J., and Grant, S., 1989, "Fluconazole. An overview," ADIS Press International Inc., Langhorne, Pa.
Blum, R.A., Wilton, J.H., Hilligoss, D.M., Foulds, G. and Schentag, J.J., 1989, The effect of fluconazole on the disposition of phenytoin, Abst. p. 30, in: "Major Development in the Management of Bacterial and Fungal Infections", Vancouver, B.C.
Boelaert, J., Schurgers, M., Matthys, E., Daneels, R., Van Peer, A., de Beule, K., Woestenborghs, R., and Heykants, J., 1988, Itraconazole pharmacokinetics in patients with renal dysfunction, Antimicrob. Ag. Chemother., 32:1595.
Collignon, P., Hurley, B., and Mitchell, D., 1989, Interaction of fluconazole with cyclosporin, Lancet i:1262.
Damanhouri, Z., Gumbleton, M., Nicholls, P.J., and Shaw, M.A., 1988, In-vivo effects of itraconazole on hepatic mixed-function oxidase, J. Antimicrob. Chemother., 21:187.

De Coster, R., Caers, I., Coene, M.-C., Amery, W., Beerens, D., and Haelterman, C., 1986, Effects of high dose ketoconazole therapy on the main plasma testicular and adrenal steroids in previously untreated prostatic cancer patients, Clin. Endocrinol., 24:675.

Dismukes, W., Cloud, G., Thompson, S., Sugar, S., Tuazon, C., et al, 1989, Fluconazole versus amphotericin B therapy of acute cryptococcal meningitis, Abst. #1065, Abstracts of the 29th Interscience Conference on Antimicrobial Agents and Chemotherapy, Washington, D.C.

Denning, D.W., Tucker, R.M., Hanson, L.H., Hamilton, J.R., and Stevens, D.A., 1989, Itraconazole therapy for cryptococcal meningitis and cryptococcosis, Arch. Intern. Med., 149:2301.

Dupont, B., Datry, A., Hilmarsdottir, I., Dellamonica, P., Bernard, E., Lefort, S., Frottier, F., Bleriot, J.P., Calamy, G., Coulaud, J.P., Decaze, J.M., Patey, O., Poirot, J.L., Roux, P., Rosenbaum, W., Testa, M., Vachon, F., and Vilde, J.L., Cryptococcal meningitis in AIDS patients., A pilot study of fluconazole therapy (submitted for publication.)

de Gans, J., Eeftinck Schattenkerk, J.K.M., and van Ketel, R.J., 1988, Itraconazole as maintenance treatment for cryptococcal meningitis in the acquired immune deficiency syndrome, Br. Med. J., 296:339.

Hanson, L.H., and Stevens, D.A., 1989, Evaluation of cilofungin, a lipopeptide antifungal agent, in vitro against fungi isolated from clinical specimens., Antimicrob. Ag. Chemother., 33:1391.

Heykants, J., Michiels, M., Meuldermans, M., Monbaliu, J., Lavrijsen, K., van Peer, A., Levron, J.C., Woestenborghs, R., and Cauwenbergh, G., 1987, The pharmacokinetics of itraconazole in animals and man: an overview, in: " Recent Trends in the Discovery, Development and Evaluation of Antifungal Drugs", R.E. Fromtling, ed., J.R. Prous Science Publishers.

Larsen, R.A., and Leal, M.E., 1989, Fluconazole compared with amphotericin B as treatment of cryptococcal meningitis, Abstr. #1062, Abstracts of the 29th Interscience Conference on Antimicrobial Agents and Chemotherapy, Washington, D.C.

Lazar, J.D., and Wilner, K.D., 1989, Key drug interaction studies with fluconazole (Abstr.), in: "Major Developments in the Management of Bacterial and Fungal Infections", Vancouver, B.C.

Perfect, J.R., Savani, D.V., and Durack, D.T., 1986, Comparison of itraconazole and fluconazole in treatment of cryptococcal meningitis and Candida pyelonephritis in rabbits, Antimicrob. Ag. Chemother., 29:579.

Saag, M., Bradsher, R., Chapman, S., Kauffman, C., Girard, W., Stevens, D., Bowles-Patton, C., Dismukes, W., and the NIAID Mycoses Study Group., 1988, Abstr. #574, Abstracts of the Twenty-eighth Interscience Conference on Antimicrobial Agents and Chemotherapy", Washington,D.C.

Sharkey, P.J., Rinaldi, M.G., Lerner, C.J., Fetchick, R.J., Dunn, J.F. and Graybill, J.R., 1989, High dose itraconazole in the treatment of severe mycoses, Abstr. #575, Abstracts of the 29th Interscience Conference on Antimicrobial Agents and Chemotherapy, Washington, D.C.

Stern, J.J., Hartman, B.J., Sharkey, P., Rowland, V., Squires, K.E., Murray, H.W., and Graybill, R., 1988, Oral fluconazole therapy for patients with the acquired immunodeficiency syndrome and cryptococcosis : experience with 22 patients, Am. J. Med., 85:477.

Sugar, A.M., and Saunders, C., 1988, Oral fluconazole as suppressive therapy of disseminated cryptococcosis in patients with acquired immunodeficiency syndrome, Am. J. Med., 85:481.

Trenk, D., Brett, W., Jahnchen, E., and Birnbaum, D., 1987, Time course of cyclosporin/itraconazole interactions, Lancet, ii:1335.

Tucker, R.M., Denning, D.W., Rinaldi, M.R., Hanson, H.L., and Stevens, D.A., 1988, Itraconaozle therapy of progressive coccidioidomycosis. Abstr. #573, Abstracts of the 29th Interscience Conference on Antimicrobial Agents and Chemotherapy, Washington, D.C..

Vanden Bossche, H., 1987, Itraconazole: a selective inhibitor of the cytochrome P-450 dependent ergosterol biosynthesis, in: "Recent Trends in the Discovery, Development and Evaluation of Antifungal Agents, R.A. Fromtling, ed., J.R. Prous Science Publishers.

Viviani, M.A., Tortorano, A.M., Langer, M., Almaviva, M., Negri, C., Cristina, C., Scoccia, S., De Maria, R., Fiocchi, R., Ferrazzi, P., Goglio, A., Gavazzeni, G., Faggian, G., Rinaldi, R., and Cadrobbi, P., 1989, Experience with itraconazole in cryptococcosis and aspergillosis, J. Infect., 18:151.

van't Wout, J.W., Novakova, I., Verhagen, S., Fibbe, W.E., de Pauw, B.E., and van der Meer, J.W.M., The efficacy of itraconazole against systemic fungal infections in neutropenic patients: a randomized comparative study with amphotericin B, Eur. J. Clin. Microbiol. Infect. Dis., (submitted).

Warnock, D.W., Turner, A., and Burke, J., 1988, Comparison of high performance liquid chromatographic and microbiological methods for determinations of itraconazole, J. Antimicrob. Chemother., 21:93.

FUNGAL MODELS IN IMMUNOCOMPROMISED ANIMALS

Jan Van Cutsem

Department of Bacteriology and Mycology, Janssen Research
Foundation, B-2340 Beerse, Belgium

INTRODUCTION

Fungi may be responsible for various superficial, systemic and
disseminated infections in man and animals. The frequency of mycoses, the
extension and the tendency to invasion by fungi of large body areas and
deeper tissues are increasingly due to the presence of predisposing,
favouring or stimulating factors (Warnock and Richardson, 1982; Viviani et
al., 1989; Neijens et al., 1989). A new group of patients, AIDS patients,
have recently joined the great number of people who are highly sensitive to
fungal infections and are anxiously demanding active and safe therapeutics.

Against this background some animal models have been developed for the
evaluation of antifungal drugs. Drug activity was compared in
immunocompetent and immunocompromised animals. The immunodeficiency in AIDS
patients is different from that induced in animals. Nevertheless
immunodepression in laboratory animals may be considered to be a standard
for the preclinical evaluation of the activity of drugs in
immunocompromised patients.

Various animal models have been used, such as: mice undergoing
sublethal X-irradiation before infection with *Cryptococcus*, *Candida*,
Histoplasma and *Blastomyces*, or athymic mice or rats having deficient cell-
mediated immunity, infected with *Cryptococcus*, *Candida*, *Histoplasma*, other
dimorphic fungi or phaeohyphomycetes. Corticosteroids and various other
substances and interventions have been used to depress defense mechanisms
in several mouse and rat strains, in rabbits, hamsters, guinea-pigs and
also in larger animals for a wide range of superficial and systemic
mycoses. Moreover, immunomodulation has often been useful to obtain more
homogeneous infections with pathogenic fungi and to establish infections
with saprophytes or poorly invasive organisms (Graybill, 1986; Miyaji,
1987; Odds, 1988; Perfect et al., 1986; Polak, 1986; Rippon, 1986; Van
Cutsem, 1989b; Walzer, 1986; Williams, 1986).

In some of the models, the fungus became more invasive due to the
reduced resistance of the animal. However, other factors may influence the
vulnerability of the animal to the fungus. Levine (1986) reported large
differences in sensitivity of mouse strains to *Coccidioides*. He found the

Mycoses in AIDS Patients, Edited by
H. Vanden Bossche *et al.,* Plenum Press, New York, 1990

NANRU mouse strain the most receptive, although this was immunologically competent and without endocrinologic disorders. Also Green and Balish (1979) were unable to find differences in intensity or duration of dermatophytosis between unpretreated-normal, cortisone-treated and germ-free guinea-pigs. In this study the effects on the leucocyte counts of a number of immunocompromising agents will be studied. The activity of antifungals in various *in-vivo* models will be compared.

The effects of immunodepressing agents on the leukocyte counts of guinea-pigs

Albino guinea-pigs received cyclophosphamide at 10 or 20 mg.kg^{-1}, hydrocortisone acetate or prednisolone acetate at 10 mg.kg^{-1}, or mechlorethamine hydrochloride by the IP, IM and IP routes respectively. This resulted in serious effects on the leucocyte counts (Table 1) at the moment of infection which persisted in general for the duration of the experimental period. The most pronounced effects were obtained for mechlorethamine hydrochloride with neutropenia, lymphocytosis and monocytosis still present at the end of the experiment.

Streptozotocin was administered by the IP route at a dosage of 40 mg.kg^{-1} for three consecutive days to Wistar rats. The mean number of leucocytes increased rapidly from 7,500 per mm^3 for females to 10,400 after one week and remained stable for a five-week observation period at weekly control at 12,000 ($\pm$ 150). In male rats 7,700 leucocytes per mm^3 were counted before streptozotocin and this number increased to 9,500 after one week, remaining stable at subsequent weekly controls for a five-week observation period at 13,500 ($\pm$ 800). The mean serum glucose level increased from 142 (F) and 147 (M) mg.dl^{-1} before streptozotocin administration to 511 (F) and 439 (M) mg.dl^{-1} after one week. It varied between 486 and 566 mg.dl^{-1} during a thirteen-week observation period. After eight weeks the first fatalities occurred and after thirteen weeks the emaciated survivors were sacrificed.

In Swiss mice the onset of the effects of streptozotocin at 40 mg.kg^{-1} for three consecutive weeks was slower. The mean number of leucocytes (9,500 per mm^3 before) remained stable after one week, increased to 12,200 after two, reached 19,300 per mm^3 from the third week on and remained statistically highly significant ($p \leq 001$) for the whole period of observation. The mean serum glucose level of the mice was 175 mg.dl^{-1} before streptozotocin administration, 187 after one week, 235 after two, 411 after three, 509 after five, 673 after 10 and 714 after 13 weeks. By that time 37% of the mice had died and the survivors were cachectic.

Dermatophyte infections

Microsporum canis and *Trichophyton mentagrophytes* infections are easily induced on the scarified skin of guinea-pigs. A series of predisposing regimens was used: IM administration of prednisolone acetate at 10 mg.kg^{-1} from day minus 7 to plus 7; IP metronidazole at 20 mg.kg^{-1} from day minus 7 to plus 7; IM alloxan at 200 mg.kg^{-1} 24 hours before infection; oral chloramphenicol at 50 mg.kg^{-1} and streptomycin at 40 mg.kg^{-1} on alternate days, combined or not with 10 mg.kg^{-1} prednisolone acetate IM from day minus 7 to plus 7; four SC weekly injections of 2 mg.kg^{-1} oestrogens in female guinea-pigs starting one week before infection; two IP injections of 0.25 mg.kg^{-1} mechlorethamine hydrochloride

Table 1. Influence of predisposing factors on the leucocyte counts of the guinea-pig

Parameter	Control data before administration (N 208)	Cyclophosphamide (IP)		Hydrocortisone acetate (IM)	Prednisolone acetate (IM)	Mechlorethamine HCl (IP)	
		% of leucocytic parameters versus controls					
		10 mg.kg^{-1} 4x (N 18)	20 mg.kg^{-1} 4x (N 17)	10 mg.kg^{-1} 7x (N 11)	10 mg.kg^{-1} 7x (N 12)	0.25 mg.kg^{-1} 2x (N 161)	0.5 mg.kg^{-1} 2x (N 47)
Leucocytes	7.8 (x10^3 per mm^3)	70***[a]	60***	98	132*	80***	63***
Neutrophils	42	172***	176***	197***	252***	62***	74***
Eosinophils	0.89	106	152	100	52	76*	26***
Basophils	0.08	25	71	71	33	175	213
Lymphocytes	56	71***	68***	57***	37***	123***	113*
Monocytes	1.2	75	90	140	80	342***	433***

[a] (70) Mann-Whitney U test (two-tailed) probability: * p ≤ .05

      ** p ≤ .01

      *** p ≤ .001

N: number

209

Table 2. Oral therapeutic treatment with itraconazole of disseminated trichophytosis in normal and in immunodepressed guinea-pigs (groups of 6)

Treatment* itraconazole mg.kg⁻¹	Mechlor. HCl 2x 0.25 mg.kg⁻¹	Reduction (%) of skin lesions versus solvent-treated at stated week of treatment					Percentage of animals with negative cultures on day 42			
		1	2	3	4	5**	skin	lungs	liver	kidneys
solvent	-	-	-	-	-	-	0	0	0	33
5	-	20	70	93	100	100	83	67	100	83
solvent	+	-	-	-	-	-	0	0	0	0
5	+	20	50	97	99.5	100	83	83	83	83

*Treatment: starting 7 days after infection od lasting 42 days

**Week 5: one week after the end of the treatment (42 days after infection)

210

on day 5 and 4 before infection; four daily IM injections of 50 mg
deferoxamine from day minus 1 to plus 2. None of these regimens was able to
modify the course of the infection.

In disseminated trichophytosis induced by the IV route in guinea-pigs
(Van Cutsem et al., 1985) the administration of mechlorethamine
hydrochloride was able to potentiate the infection, but oral treatment with
itraconazole was as efficacious as in normal animals (Table 2).

Skin candidosis

Diabetes was induced in albino guinea-pigs with alloxan 200 mg.kg^{-1} IM
24 hours prior to infection of the intact skin with *Candida albicans*. Both
topical and oral ketoconazole and topical and oral itraconazole were
efficacious (Table 3) (Van Cutsem et al., 1987).

Table 3. Treatment of skin candidosis with azoles in alloxan-diabetic
guinea-pigs

| Treatment oral/topical mg.kg^{-1}/conc. % | % cured + % markedly improved | | | |
| | ketoconazole | | itraconazole | |
	oral	topical	oral	topical
excipient	0 + 8	0 + 6	−	−
0.31 − 0.031	ND*	0 + 25	0 + 33	8 + 50
0.63 − 0.063	0 + 0	8 + 75	50 + 27	44 + 50
1.25 − 0.125	33 + 11	67 + 25	75 + 16	67 + 25
2.5 − 0.25	61 + 9	72 + 11	84 + 16	92 + 8
5 − 0.5	92 + 8	89 + 11	100 + 0	100 + 0
10 − 1	97 + 3	100 + 0	ND	ND

*ND: no data

Vaginal *Candida*- and *Torulopsis* infection

Ovariectomy and hysterectomy of Wistar rats, followed by weekly
subcutaneous injection of oestrogens (0.1 mg of oestradiol undecylate in
1 ml of an oily solution) produced a permanent pseudo-oestrus. These
animals were highly susceptible to vaginal infection by *C. albicans*, when
infected intravaginally with 8 x 10^5 CFU in a 0.2 ml volume. Excellent
therapeutic efficacy was obtained after topical and oral treatment with
various azoles (Van Cutsem, 1989a). IP streptozotocin at 40 mg.kg^{-1} was
administered for three consecutive days to rats in permanent pseudo-
oestrus: irreversible diabetes was obtained (serum glucose levels raised
from 125-156 mg.dl^{-1} before, to 300-577 mg.dl^{-1}). Vaginal candidosis in
non-diabetic and in diabetic rats was treated orally and responded to
ketoconazole and to itraconazole (Table 4). Intravaginal inoculation with
Torulopsis glabrata in diabetic rats in permanent pseudo-oestrus resulted
in a prolonged presence of the yeasts in the vagina. Oral treatment with
itraconazole eliminated *T. glabrata* completely (Table 5).

Table 4. Oral therapeutic treatment with azoles of vaginal candidosis in diabetic castrated rats* in permanent pseudooestrus

Treatment mg.kg^{-1}	% of cured + % of improved rats			
	ketoconazole		itraconazole	
	non-diabetic	diabetic	non-diabetic	diabetic
solvent	0.2 + 0.2	0		
1.25	0 + 0	-	51 + 10	50 + 0
2.5	19 + 0	0	91 + 4	83 + 17
5	38 + 4	33 + 0	94 + 3	100 + 0
10	95 + 1	83 + 17	100 + 0	ND

*Rats: ovariectomy and hysterectomy; weekly SC 0.1 mg of oestradiol undecylate in 1 ml of oily solution; streptozotocin 40 mg.kg^{-1} IP for 3 consecutive days
**Treatment: starting 3 days after infection od for 3 days

Table 5. Oral therapeutic treatment with itraconazole of vaginal *T. glabrata* infection in diabetic castrated rats** in permanent pseudooestrus (groups of 6)

Treatment*** itraconazole mg.kg^{-1}	No. of positive animals on stated days after infection		
	7	11	14
untreated (historical data)	6	5	2
solvent	6	4	3
1.25	2	0	0
2.5	1	0	0

Torulopsis infection: was induced by intravaginal infection on three consecutive days (*T. glabrata* strain M 13911-26.1)
**Castrated rats in permanent pseudooestrus received on three consecutive days 40 mg.kg^{-1} IP of streptozotocin
***Treatment: starting 3 days after the first infection o.d. for 3 days

Gastrointestinal candidosis

Prednisolone acetate was given IM at 10 mg.kg^{-1} to guinea-pigs, five doses weekly from day minus 8 on with, in addition, a simultaneous oral administration on alternating days of 50 mg.kg^{-1} of chloramphenicol and 40 mg.kg^{-1} of streptomycin (6 doses weekly) from day minus 2 on, during the whole experimental period. *C. albicans* (2 x 10^7 CFU) was given by gavage (Van Cutsem et al., 1987). Ketoconazole or itraconazole given for two weeks once daily (o.d.), starting 5 days after infection was highly efficacious (Table 6). The evaluation of the treatment was based on cultures of faeces collected weekly and on cultures at necropsy of tongue, oesophagus, stomach, small and large intestines and rectum.

Table 6. Oral therapeutic treatment with azoles of intestinal candidosis
in guinea-pigs*

Treatment mg.kg^{-1}	% cured + % improved	
	ketoconazole	itraconazole
excipient	0 + 5	–
2.5	17 + 0	50 + 50
5	75 + 14	94 + 0
10	96 + 4	100 + 0

*Guinea-pigs were predisposed by chloramphenicol and streptomycin on
alternate days and by prednisolone acetate

Systemic candidosis

Normal and mechlorethamine hydrochloride-immunocompromised guinea-pigs
were infected IV with 8,000 CFU of *C. albicans* per g BW (Van Cutsem et al.,
1985). The animals were treated orally with itraconazole or fluconazole and
parenterally with itraconazole (in solution in hydroxypropyl-ß-
cyclodextrin) or amphotericin B (Fungizone® Squibb). All drugs were
administered for 14 days o.d., starting on the day of infection. The
animals were observed regularly and skin folliculitis was scored. At
necropsy, cultures were performed from 1 g of skin and from the whole left
kidney. The results are given in Table 7. Oral itraconazole was equally
active in immunodepressed and normal animals and slightly superior to
fluconazole. Parenteral itraconazole was as active as oral and it was
superior to amphotericin B.

Aspergillosis

Normal and immunodepressed (mechlorethamine hydrochloride) guinea-pigs
were infected IV with 25,000 CFU of *Aspergillus fumigatus* per g BW (Van
Cutsem and Janssen, 1988). Treatment started on the day of infection (Table
8) or 24 hours later (Table 9) and was given o.d. for 14 consecutive days.
Itraconazole was given by gavage in PEG 200 or parenterally dissolved in
hydroxypropyl-ß-cyclodextrin. Amphotericin B (Fungizone® Squibb) was also
given parenterally. In all regimens amphotericin B and itraconazole were
active. No differences were seen between the results obtained in
immunodepressed and normal animals. Itraconazole was free of side-effects
and more efficacious than amphotericin B in terms of survival times as well
as in eradication of fungi from the organs.

Cryptococcosis

Normal and immunodepressed (mechlorethamine hydrochloride) guinea-pigs
were infected IV with 200 yeasts of *Cryptococcus neoformans* per g BW (Van
Cutsem et al., 1986). Treatment o.d. started three days after infection for
35 consecutive days. Invasion of almost all organs, including brain and
meninges, and formation of large granulomas occurred.

Table 7. Treatment of systemic candidosis in normal and in immunodepressed guinea-pigs

| Treatment od for 14 days | | Normal | | | | Immunodepressed* | | | |
| mg.kg⁻¹ | route | No. of animals | % of animals | | | No. of animals | % of animals | | |
			skin folliculitis cured + improved	negative cultures skin	kidneys		skin folliculitis cured + improved	negative cultures skin	kidneys
solvent**	oral	324	0 + 0	0	0	12	0 + 0	0	0
itraconazole 0.63	oral	70	36 + 26	33	11	12	50 + 33	33	33
itraconazole 1.25	oral	74	95 + 4	96	43	12	92 + 8	100	58
itraconazole 2.5	oral	76	100 + 0	99	57	12	100 + 0	100	67
itraconazole 5	oral	52	100 + 0	100	98	6	100 + 0	100	100
fluconazole 0.63	oral	12	0 + 42	33	25				
fluconazole 1.25	oral	12	25 + 58	50	58				
fluconazole 2.5	oral	12	67 + 25	67	67				
fluconazole 5	oral	12	84 + 8	75	83				
solvent	IP	36	0 + 0	0	0				
itraconazole 0.63	IP	12	66 + 17	67	50				
itraconazole 1.25	IP	18	100 + 0	89	56				
itraconazole 2.5	IP	18	100 + 0	100	78				
itraconazole 5	IP	18	100 + 0	100	100				
ampho B 0.63	IP	6	17 + 33	17	17				
ampho B 1.25	IP	12	42 + 25	50	25				
ampho B 2.5	IP	12	42 + 33	50	33				
ampho B 5	IP	12	42 + 17	42	42				

*Immunodepressed: mechlorethamine hydrochloride
**Solvent: oral treatment: PEG 200
       IP: itraconazole: hydroxypropyl-ß-cyclodextrin
       amphotericin B: saline

Table 8. Treatment* of systemic aspergillosis in normal and in immunodepressed guinea-pigs

Treatment			Normal				Immunodepressed**			
	mg.kg⁻¹	route	No. of animals	% survivors	mean survival in days on day 28	% of negative organs	No. of animals	% survivors	mean survival in days on day 28	% of negative organs
solvent		oral	324	0	5.5	1	66	0	5.6	1
itraconazole	0.63	oral	22	0	6.3	7	-***	-	-	-
	1.25	oral	28	21	11.5	33	6	17	11.3	24
	2.5	oral	52	46	17.5	69	18	44	16.1	64
	5	oral	92	82.6	24.7	94	30	93	26.9	93
solvent		IP	42	0	5.7	1	18	0	5.6	1
itraconazole	0.63	IP	12	0	5.7	5	-	-	-	-
	1.25	IP	12	25	12.8	45	6	33	17.5	52
	2.5	IP	12	67	22.2	86	6	67	20.8	74
	5	IP	12	92	27.1	98	6	83	26.3	94
amphotericin B	0.63	IP	12	0	6.6	3	-	-	-	-
	1.25	IP	18	17	10.4	11	6	33	15.7	33
	2.5	IP	36	39	15.8	30	6	50	19.0	46
	5	IP	24	67	21.0	69	12	83	24.8	72

*Treatment starting on the day of infection od for 14 days
**Immunodepressed: mechlorethamine hydrochloride
***-: no data

Table 9. Therapeutic treatment* of systemic aspergillosis in normal and in immunodepressed guinea-pigs

Treatment			Normal				Immunodepressed**			
	mg.kg^{-1}	route	No. of animals	% survivors	mean survival in days on day 28	% of negative organs	No. of animals	% survivors	mean survival in days on day 28	% of negative organs
solvent		oral	38	0	5.2	1	12	0	5.8	2
itraconazole	2.5	oral	12	33	13.3	39	12	33	14.1	23
	5	oral	30	70	24.6	91	30	80	24.6	92
	10	oral	12	83	25	95	-***	-	-	-
solvent		IP	24	0	5.8	2	18	0	5.6	2
itraconazole	2.5	IP	6	50	19	84	6	67	22.8	76
	5	IP	6	100	28	97	6	83	26.7	96
	10	IP	6	100	28	95	-	-	-	-
amphotericin B	1.25	IP	12	33	12.9	29	-	-	-	-
	2.5	IP	24	33	13.9	25	6	33	16.3	43
	5	IP	18	50	18.5	61	6	50	17.3	69

*Treatment: starting one day after infection od for 14 days

**Immunodepressed: mechlorethamine hydrochloride

***-: no data

Table 10. Oral therapeutic treatment* of meningocerebral and generalized cryptococcosis in normal and immunodepressed** guinea-pigs

Treatment			No. of animals								Brain and meningi		% of negative organs	
			Skin cryptococcomas										12 others per animal	
			Normal				Immunodepressed				Normal	Immuno-depressed	Normal	Immuno-depressed
mg.kg⁻¹	route		Total No.	Neg	Some	Massive	Total No.	Neg	Some	Massive				
solvent		oral	28	0	0	28	18	0	0	18	0	0	12.5	10.2
itra 5		oral	28	24	4	0	18	16	2	0	36	44	84.8	77.3
itra 10		oral	28	27	1	0	20	20	0	0	64	55	96.1	94.6
fluco 5		oral	8	4	1	3	-***	-	-	-	13	-	83	-
fluco 10		oral	8	7	1	0	-	-	-	-	0	-	81	-
solvent		IP	8	0	0	8	6	0	0	6	0	0	12	16.7
itra 5		IP	8	7	0	1	-	-	-	-	38	-	84	-
itra 10		IP	8	8	0	0	-	-	-	-	63	-	94	-
fluco 5		IP	8	3	1	4	-	-	-	-	13	-	67	-
fluco 10		IP	8	7	1	0	-	-	-	-	38	-	84	-
ampho B 1.25		IP	8	1	2	5	6	2	2	2	0	0	39	44.4
ampho B 2.5		IP	8	7	1	0	6	6	0	0	38	50	62	80.6

*Therapeutic treatment: starting 3 days after infection o.d. for 35 days
**Immunodepressed: mechlorethamine hydrochloride
***: no data

Itraconazole was given orally and amphotericin B parenterally to normal and to immunodepressed guinea-pigs. Fluconazole was administered orally and parenterally and itraconazole also parenterally to normal animals (Table 10) (Van Cutsem, 1989a). In another experiment (Table 11) oral itraconazole, oral 5-fluorocytosine and parenteral amphotericin B were administered alone or in combination to immunodepressed guinea-pigs infected with *Cr. neoformans*. Immunodepressed animals responded to treatment as well as normal ones. Itraconazole and amphotericin B were more active than fluconazole.

Table 11. Oral and parenteral therapeutic treatment of meningocerebral and generalized cryptococcosis in immunodepressed guinea-pigs (groups of 6)

Treatment o.d. for 35 days mg.kg^{-1}		No. of animals with					
		Skin granulomas			Negative organs		
oral	IP	Negative	Some	Massive	Brain and meningi	12 others (total 72)	(%)
untreated		0	0	6	0	11	(15.3)
PEG 200		0	0	6	0	8	(11.1)
itra 5		4	2	0	2	50	(69.4)
5-FC 10		0	1	5	0	18	(25)
5-FC 40		1	4	1	0	31	(43.1)
itra 5+5-FC 10		6	0	0	2	69	(95.8)
itra 5+5-FC 40		6	0	0	3	72	(100)
PEG	+saline	0	0	6	0	12	(16.7)
	ampho B 0.63	0	1	5	0	30	(41.7)
	ampho B 1.25	2	2	2	0	32	(44.4)
	ampho B 2.5	6	0	0	3	58	(80.6)
itra 5	+ampho B 0.63	5	1	0	1	53	(73.6)
itra 5	+ampho B 1.25	6	0	0	2	63	(87.5)
itra 5	+ampho B 2.5	6	0	0	4	72	(100)

Combination therapy of itraconazole with 5-fluorocytosine or with amphotericin B was synergistic.

Zygomycosis

Two zygomycetes, *Rhizopus microsporus, var. rhizopodiformis* isolated from a dialysis patient reported to have died from zygomycosis while on treatment with deferoxamine and *Rhizopus oryzae* isolated from a leukemic patient, were used for IV infections in guinea-pigs. The inoculum consisted of 293 CFU of *Rh. microsporus* per g BW and of 3,550 CFU of *Rh. oryzae* per g BW (Van Cutsem and Boelaert, 1989). Untreated guinea-pigs died in 4.2 and 8.8 days for *Rh. microsporus* and *Rh. oryzae* respectively (Table 12). Only treatment with amphotericin B was efficacious (Van Cutsem and Boelaert, 1989). When 50 mg deferoxamine was administered IM 24h before infection the animals became sick earlier and died sooner. This phenomenon was more pronounced with 2 or 4 doses of deferoxamine. The activity of amphotericin B was abrogated by deferoxamine.

Table 12. Experimental zygomycosis in guinea-pigs (groups of 6)
          Effects of deferoxamine (DFO)

DFO – 50 mg IM No. of doses	Rh. microsporus mean survival in days amphotericin B[a]		Rh. oryzae mean survival in days amphotericin B	
	–	+	–	+
0	4.2	25.5**[b]	8.8	27.7**
1	3.3*	4.8**	7.3*	16.2*
2	3.0**	3.2**	6.2**	10.7*
4	3.0**	3.2**	4.8**	4.8**

[a]Amphotericin B: 2.5 mg.kg^{-1} IP during 14 days starting on the day
of infection
[b]25.5: Mann-Whitney U test (two-tailed probability) comparing
survival on day 28   *p$\leq$.05
                      **p$\leq$.01

Table 13. Oral treatment of systemic infection with *Penicillium marneffei*
          B 55527 in normal and immunodepressed guinea-pigs (groups of 6)

Treatment Itraconazole mg.kg^{-1}	Start on day	Inoculum CFU/g BW	No. of evaluated organ with macroscopic lesion (Total No.30)		No. of evaluated organs with positive cultures (Total No.42)	
			Normal	Immunodepressed	Normal	Immunodepressed
solvent	0	12,500	23	26	38	35
2.5	0	12,500	3	1	1	0
5	0	12,500	1	2	0	0
solvent	0	25,000	23	28	41	40
2.5	0	25,000	2	2	0	0
5	0	25,000	1	0	0	0

*Immunodepressed: mechlorethamine hydrochloride

Penicilliosis

Normal and immunodepressed (mechlorethamine hydrochloride) guinea-pigs
were infected with 12,500 or 25,000 CFU of *Penicillium marneffei* B 55527
per g BW. This strain was isolated by Dr. M.A. Viviani, Milano, Italy, from
a drug-addicted AIDS patient (Table 13).

Oral treatment with itraconazole was given to the animals o.d. for 14
days starting on the day of infection. Controls were highly positive and
presented enlarged spleen and lymph nodes and necrotic foci on the
epididymis and the testicles. The lesions were more pronounced in
immunodepressed animals, where also necrotic foci were present in the
liver. All control animals also developed skin eruptions, from day 16 to 18
on. Organs were histologically positive for the specific septate cells of
P. marneffei. Cultures from all these organs were positive including the

lungs and the kidneys. In the itraconazole-treated guinea-pigs only in one animal a few colonies could be isolated.

Discussion

Immunodepression has been used by various authors to enhance the infection by several microorganisms in experimental animals in an attempt to mimic the clinical situation. Some of these agents are really promoting fungal invasion. The cumulative effect of various agents may be useful for infections such as vaginal torulopsidosis, where diabetes may act synergistically with the permanent pseudo-oestrus of the rat. On the other hand the guinea-pig is generally more susceptible to fungal infections than the mouse. Mouse data have not been discussed here because the extrapolation to human pathology is poor. The treatment data in the models in guinea-pigs presented in this study are more in agreement with results obtained in clinical conditions. The activity of the antifungals tested was not affected by the immunodepressing agents used except deferoxamine in zygomycosis for amphotericin B. It is well known that deferoxamine may enhance some bacterial infections such as *Yersinia_enterolytica* and protect against others (*Listeria monocytogenes*). This disparate effect of deferoxamine upon infections is reported to be due to ability or inability of the organism to use the iron-chelate of deferoxamine, feroxamine (Robins-Browne and Prpic, 1985). It is suggested that microorganisms having a receptor for feroxamine would be stimulated in their growth by the iron that can be provided by feroxamine. In other experimental infections we have seen that an analogous phenomenon occurred with *A. fumigatus*, but that superficial dermatophytosis, vaginal candidosis and systemic candidosis are not enhanced (in preparation).

It is interesting to see that itraconazole dissolved in hydroxypropyl-ß-cyclodextrin and administered parenterally possesses, in the various mycoses evaluated, a comparable activity to oral itraconazole. This may offer an efficacious and safe way for treatment of patients with absorption problems. Superiority of both oral and parenteral itraconazole to other reference compounds is clearly proven in these models. The activity of itraconazole in an *in-vivo* model of penicilliosis is a new addition to its spectrum of proven activity.

The synergistic activity in cryptococcosis of itraconazole and 5-fluorocytosine may be helpful in some patients who are difficult to treat and is in agreement with earlier described synergy between these two drugs (Polak, 1980). Also the synergistic activity in cryptococcosis of itraconazole and amphotericin B may have the same indications as with 5-FC. These data largely confirm the results obtained in experimental aspergillosis in guinea-pigs treated with ketoconazole or itraconazole followed by amphotericin B or the opposite. This combination appeared not to be contraindicated and may be beneficial in some circumstances (Van Cutsem and Janssen, 1988). Schaffner and Frick (1985) had shown in immunodepressed mice infected with *A. fumigatus* that ketoconazole pretreatment for 48 hours was able to abolish the protective effect of subsequent therapy with amphotericin B, whether ketoconazole therapy was stopped or not. This was not confirmed in these various experimental models.

It can be concluded that an important step forward was made in the therapeutic possibilities with the use of the new triazoles (itraconazole, fluconazole and saperconazole) with low risk for side-effects (Van Cutsem et al., 1989), especially the two compounds with the broadest-spectrum: itraconazole and saperconazole. The laboratory data have been largely confirmed by clinical experiences showing excellent results with itraconazole (Cauwenbergh and De Doncker, 1987; Viviani et al., 1989) and low risks for side-effects (Van Cauteren et al., 1989).

REFERENCES

Cauwenbergh, G., and De Doncker, P., 1987, The clinical use of itraconazole in superficial and deep mycoses, in: "Recent Trends in the Discovery, Development and Evaluation of Antifungal Agents", R.A. Fromtling, ed., Prous, J.R., Science Publ., Barcelona.

Graybill, J.R., 1986, Animal models for treatment of cryptococcosis, in: "Experimental Models in Antimicrobial Chemotherapy, O. Zak and M.A. Sande, eds., Academic Press, Harcourt Brace Jovanovich, Publ., London.

Green, F. III, and Balish, E., 1979, Suppression of in-vitro lymphocyte transformation during an experimental dermatophyte infection, Infect. Immun., 26:554.

Levine, H.B., 1986, Evaluation of antifungal drugs in a mouse model of coccidioidomycosis, in: "Experimental Models in Antimicrobial Chemotherapy, O. Zak and M.A. Sande, eds., Academic Press, Harcourt Brace Jovanovich, Publ., London.

Miyaji, M., 1987, "Animal Models in Medical Mycology", C.R.C. Press, Inc., Boca Raton, U.S.A.

Neijens, H.J., Frenkel, J., de Muinck Keizer-Schrama, S.M.P.F., Dzoljic-Danilovic, G., Meradji, M., and van Dongen, J.J.M., Invasive Aspergillus infections in chronic granulomatous disease: treatment with itraconazole, Proc. 2nd Intern. Symp. Itraconazole, Antwerp, Belgium, June 22-23, 1989, Mycoses, in press.

Odds, F.C., 1988, "Candida and candidosis", 2nd ed., Baillière Tindall, London.

Perfect, J.R., Savani, D.V., and Durack, D.T.,1986, Comparison of itraconazole and fluconazole in treatment of cryptococcal meningitis and Candida pyelonephritis in rabbits, Antimicrob. Ag. Chemoth., 29:579.

Polak, A., 1980, Détermination de la synergic entre la 5-fluorocytosine et trois dérivés de l'imidazole au moyen de différents modèles in vitro et in vivo, Bull. Soc. Mycol. Méd., 9:263.

Polak, A., 1986, Experimental Candida vaginitis, in: "Experimental Models in Antimicrobial Chemotherapy, O. Zak and M.A. Sande, eds., Academic Press, Harcourt Brace Jovanovich, Publ., London.

Rippon, J.W., 1986, Animal models of experimental dermatophyte infections, in: "Experimental Models in Antimicrobial Chemotherapy, O. Zak and M.A. Sande, eds., Academic Press, Harcourt Brace Jovanovich, Publ., London.

Robins-Browne, R.M., and Prpic, J.K., 1985, Effects of iron and desferrioxamine on infections with Yersinia enterocolitica. Infect. Immun., 47:774.

Schaffner, A., and Frick, P.G., 1985, The effect of ketoconazole on amphotericin B in a model of disseminated aspergillosis, J. Infect. Dis., 151:902.

Van Cauteren, H., Lampo, A., Vandenberghe, J., Vanparys, P., Coussement, W., De Coster, R., and Marsboom, R., 1989, Toxicological profile and safety evaluation of antifungal azole derivatives, Proc. 2nd Intern. Symp. Itraconazole, Antwerp, Belgium, June 22-23, 1989, Mycoses, in press.

Van Cutsem, J., Fransen, J., and Janssen, P.A.J., 1985, Animal models for systemic dermatophyte and *Candida* infection with dissemination to the skin, in: "Models in Dermatology", H.I. Maibach and N.J. Lowe, eds., S. Karger AG, Basel.

Van Cutsem, J., Fransen, J., Van Gerven, F., and Janssen, P.A.J., 1986, Experimental cryptococcosis: dissemination of *Cryptococcus neoformans* and dermotropism in guinea-pigs, Mykosen, 29:561.

Van Cutsem, J., Van Gerven, F., and Janssen, P.A.J., 1987, Activity of orally, topically and parenterally administered itraconazole in the treatment of superficial and deep mycoses. Animal models, Rev. Infect. Dis., 9:S15.

Van Cutsem, J., and Janssen, P.A.J., 1988, *In-vitro* and *in-vivo* models to study the activity of antifungals against *Aspergillus*, in: "*Aspergillus* and aspergillosis", H. Vanden Bossche, D.W.R. MacKenzie and G. Cauwenbergh, eds., Plenum Publ. Corp., New York.

Van Cutsem, J., 1989a, Oral, topical and parenteral antifungal treatment with itraconazole in normal and in immunocompromised animals, Proc. 2nd Intern. Symp. Itraconazole, Antwerp, Belgium, June 22-23, 1989, Mycoses, in press.

Van Cutsem, J., 1989b, Animal models for dermatomycotic infections, in: "Current Topics in Medical Mycology", M.R. McGinnis and M. Borgers, eds., Springer-Verlag New York, Inc., New York.

Van Cutsem, J., and Boelaert, J.R., 1989, Effects of deferoxamine, feroxamine and iron on experimental mucormycosis (zygomycosis), Kidney Internat., 36: in press.

Van Cutsem J., Van Gerven, F., and Janssen, P.A.J., 1989, Saperconazole, a new potent antifungal triazole: *in-vitro* activity spectrum and therapeutic efficacy, Drugs of the future, in press.

Viviani, M.A., Tortorano, A.M., Pagano, A., Vigevani, G.M., Gubertini, G., Cristina, S., Assaisso, M.L., Suter, F., Farina, C., Minetti, B., Faggian, G., Caretta, M., Di Fabrizio, N., and Vaglia, A., European experience with itraconazole in systemic mycoses, Proc. 2nd Intern. Symp. Itraconazole, Antwerp, Belgium, June 22-23, 1989, Mycoses, in press.

Walzer, P.D., 1986, *Pneumocystis carinii* infection, in: "Experimental Models in Antimicrobial Chemotherapy", O. Zak, and M.A. Sande, eds., Academic Press, Harcourt Brace Jovanovich, Publ., London.

Williams, D.W., 1986, Drug therapy in animal models of histoplasmosis, in: "Experimental Models in Antimicrobial Chemotherapy", O. Zak, and M.A. Sande, eds., Academic Press, Harcourt Brace Jovanovich, Publ., London.

Warnock, D.W., and Richardson, M.D., 1982, "Fungal Infection in the Compromised Patient, John Wiley & Sons Ltd., Chichester.

MODE OF ACTION OF ANTIFUNGALS OF USE IN IMMUNOCOMPROMISED PATIENTS.   FOCUS

ON *CANDIDA GLABRATA* AND *HISTOPLASMA CAPSULATUM*

Hugo Vanden Bossche, Patrick Marichal, Jos Gorrens, Danny
Bellens, Marie-Claire Coene, Wim Lauwers, Ludo Le Jeune,
Henri Moereels and Paul A.J. Janssen

Janssen Research Foundation,
B2340 Beerse, Belgium

INTRODUCTION

During recent years considerable advances have been made in the
identification of potential targets for antifungal agents. The most
important antifungals for use in immunocompromised patients interfere with
targets in the plasma membrane (polyenes), nucleus (5-fluorocytosine) or
smooth endoplasmic reticulum (allylamines, morpholines, azole derivatives).

The polyene macrolide antibiotics such as amphotericin B (Ampho B) and
nystatin interact with sterol-containing plasma membranes of sensitive
organisms and change the physical state of the membrane (for reviews see
Kerridge and Whelan, 1984; Kerridge, 1988). The original hypothesis that
the fungistatic and fungicidal effects originate from the association of
Ampho B with membrane sterols is no longer satisfactory in the light of
studies on intact cells, in which the membrane composition has been
modified, and on artificial membrane systems . Both the nature of the
sterol and the physical state of the phospholipids play an important role
in the disruptive interaction of polyenes with membranes. It is now
accepted that the primary growth inhibitory effect of Ampho B is the loss
of the proton barrier. However, the mechanism involved is not understood.
Some studies suggest that the involvement of amphotericin-induced oxidative
damage to the membrane is an important possibility to consider as a cause
of cell death (Sokol-Anderson et al., 1986).

5-Fluorocytosine (for reviews see Kerridge and Whelan, 1984; Polak,
1988a) is taken up by sensitive fungi by a cytosine permease and deaminated
to 5-fluorouracil (5-FU) by a cytosine deaminase. 5-FU is metabolised by
the pyrimidine salvage pathway to 5-fluorouridine triphosphate (5-FUTP) and
5-fluorodeoxyuridylate (5-FdUMP). The former is a precursor of aberrant
RNA and 5-FdUMP is a potent inhibitor of thymidylate synthase, a key enzyme
in DNA synthesis. The selectivity of 5-fluorocytosine results from the
failure of the host to metabolise the drug to an active derivative.
Indeed, cytosine deaminase is absent or present only in small amounts in
mammalian cells (Kerridge and Whelan, 1984). This well-tolerated drug
suffers from an important drawback i.e. the high frequency with which
resistant variants appear during therapy.

In this chapter the interaction of antifungals with ergosterol
biosynthesis will be discussed in greater detail. In the first part the
interaction with the squalene epoxidase will be reviewed; the 2nd part
discusses the interaction of azole antifungals with the cytochrome P-450-
dependent ergosterol synthesis and in the 3rd part data are presented on

Mycoses in AIDS Patients, Edited by
H. Vanden Bossche *et al.*, Plenum Press, New York, 1990

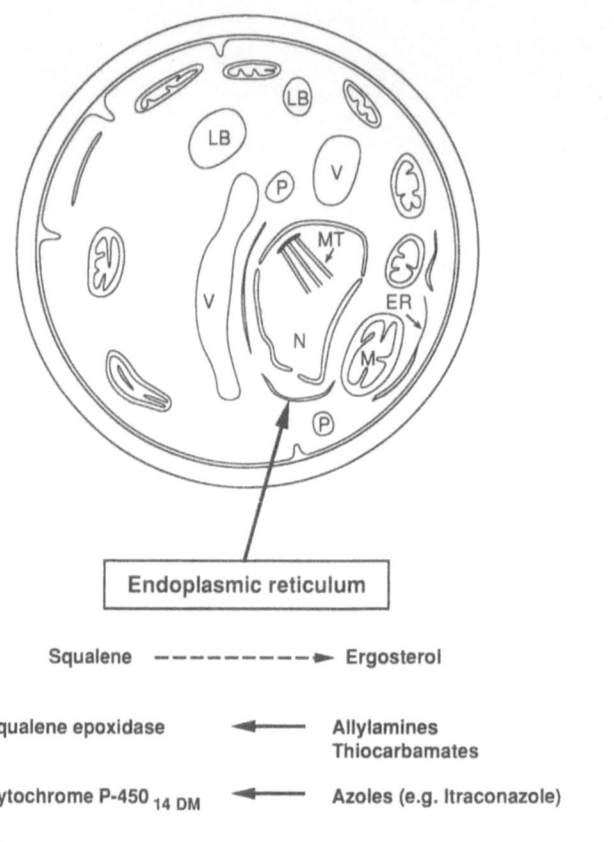

Squalene $\quad$ – – – – – – – –► Ergosterol	
Squalene epoxidase $\quad$ ◄———	Allylamines Thiocarbamates
Cytochrome P-450 $_{14\,DM}$ $\quad$ ◄———	Azoles (e.g. Itraconazole)
Δ^{14}-Reductase $\quad$ ◄———	Morpholines (e.g. Amorolfine)
Δ^8 ——► Δ^7-Isomerase $\quad$ ◄———	Morpholines

Fig.1- Antifungals that interfere with enzymes of the endoplasmic reticulum. ER: endoplasmic reticulum, LB: lipid body, M: mitochondria, MT: microtubules, N: nucleus, P: peroxisome, V: vacuole.

the effects of azoles on ergosterol synthesis in *Candida* (*Torulopsis*) *glabrata* and in *Histoplasma capsulatum*.

COMPOUNDS AFFECTING ENZYMES OF THE SMOOTH ENDOPLASMIC RETICULUM

An important group of modern antifungals interfere with enzymes of the smooth endoplasmic reticulum that are involved in ergosterol synthesis (Fig. 1).

As shown in Fig. 1 these ergosterol biosynthesis inhibitors can be divided into three groups:

1. Squalene epoxidase inhibitors
2. Compounds that interfere with the cytochrome P-450-dependent 14α-demethylase
3. Inhibitors of the 14-reductase and/or Δ^{8-7} isomerase.

The last group of inhibitors comprises the morpholine derivatives of which only one, amorolfine, is currently in clinical evaluation as a topical antifungal for human infections (Polak, 1988b). Since after oral application the drug did not show remarkable activity against systemic infections (Polak, 1988b), this group of inhibitors will not be discussed in detail.

Squalene Epoxidase Inhibitors

Squalene epoxidase is the target for the 2nd group of inhibitors, the allylamines and thiocarbamates. Both naftifine and terbinafine inhibit this endoplasmic enzyme in *C. albicans*. Fifty % inhibition is achieved at 1.2×10^{-6} M and 2.7×10^{-8} M respectively; 95% inhibition is obtained at 2.6×10^{-5} M and 6.86×10^{-7} M (Ryder, 1988). These allylamines are weak inhibitors of the rat liver microsomal squalene epoxidase. The latter enzyme differs in many aspects from the *C. albicans* enzyme. For example, the liver enzyme has a strong requirement for a factor of the cytosol, it has a preference for NADPH instead of NADH, the preferred cofactor of the *Candida* enzyme. Furthermore, allylamines are noncompetitive inhibitors of the *Candida* enzyme and competitive inhibitors of the liver enzyme, and the squalene analogue, 2-aza-2,3-dihydrosqualene, is a much better inhibitor of the liver enzyme (Ryder,1988). Major differences exist not only between the epoxidase of *Candida* and the rat enzyme: important differences in sensitivity to terbinafine are also found between the enzyme present in rat and guinea pig liver, the latter being 23 times more sensitive (Ryder, 1988). Tolnaftate and the closely related tolciclate, both active against dermatophytes, also inhibit the *Candida* squalene epoxidase. The lack of activity of these thiocarbamates against *Candida* might originate from difficulties in crossing the plasma membrane barrier (Ryder et al., 1986). However, this does not explain the weaker effects of allylamines on *C. albicans* as compared with their effects on *Trichophyton* spp.(Ryder, 1988). Indeed, the MICs found for *T. mentagrophytes* and *C.albicans* are 3 µg/ml and 3100 µg/ml, respectively (Ryder, 1986). However, fifty % inhibition of ergosterol synthesis by intact cells of *T. mentagrophytes* and *C. albicans* is achieved at 2 and 8 ng/ml (Ryder, 1986).

14α-Demethylase Inhibitors

The largest group of antifungals, in terms of number of commercial compounds, is that of the azole antifungals. This group contains imidazole (e.g. miconazole, econazole and ketoconazole) as well as triazole derivatives (e.g.fluconazole, itraconazole, saperconazole and terconazole). All these compounds inhibit, in fungal cells, the 14α-demethylation of 24-methylenedihydrolanosterol, a pathogen-specific precursor of ergosterol (Berg, 1986, Vanden Bossche, 1985, 1988a, Vanden Bossche et al., 1983, 1987, 1988, 1990, Yoshida, 1988).

The 14α-demethylation is a cytochrome P-450-dependent process. Cytochrome P-450 (P450) is the terminal enzyme of the pathways involved in the metabolism of a wide variety of foreign (xenobiotics) and endogenous compounds (endobiotics). The broad substrate specificity is partly a result of various forms of P450. P450 isozymes are present in procaryotes such as *Pseudomonas putida* and *Bacillus megaterium*, in protozoa, yeast, fungi, plants, insects, invertebrates, fish, fowls and mammals. In bacteria the P450s are soluble, whereas in all eukaryotes they are membrane-bound (smooth endoplasmic reticulum or inner mitochondrial membrane).

The name of this haemoprotein describes a carbon pigment which absorbs at about 450nm when reduced and complexed with carbon monoxide. The active site of P450 contains an iron protoporphyrin IX moiety (Fig. 2). The iron is penta- or hexacoordinate, four of the ligands being contributed by the tetradentate porphyrin ring. The fifth ligand is a thiolate anion contributed by a cysteine residue of the C-terminal half of the apoprotein. The 6th ligand is still not identified. However, evidence is available that it might be a hydroxyl group from a seryl, threonyl or tyrosine residue (Ruckpaul and Bernhardt, 1984).

P450 enzymes generally insert an atom of oxygen into their substrates. The ability to activate oxygen for insertion into a substrate is also the key property of the P450 (P450$_{14DM}$) involved in the 14α-demethylation of lanosterol (in *Saccharomyces cerevisiae* and mammalian liver) or of 24-methylenedihydrolanosterol in a number of *C. albicans* isolates and all filamentous fungi studied so far. P450$_{14DM}$ (belongs to P450 family LI, Nebert et al., 1989), purified to homogenity from rat liver (Trzaskos et al.,1986) and *S. cerevisiae* (Yoshida, 1988) catalyses three oxidative steps:

1. the hydroxylation of the C-32-methyl (14α-methyl) group of lanosterol with the formation of a C-32 alcohol
2. the oxidation of the C-32 alcohol to a C-32 aldehyde
3. the oxidative elimination of the aldehyde as formic acid, resulting in the formation of and 4,4-dimethyl-$\Delta^{8,14,24}$-cholestatrienol in *S. cerevisiae* and liver. In fungi the product of the oxidative elimination of the C-32-methyl is 4,4-dimethyl-$\Delta^{8,14,24(28)}$-ergostatrienol.

The entire dealkylation sequence is catalysed by a single P450 species.

The amino acid sequences of the P450$_{14DM}$ from *S. cerevisiae* (Kalb et al., 1986, 1988, Ishida et al., 1988), *C.albicans* (Lai and Kirsch, 1989) and *C. tropicalis* (Chen et al., 1988) are known. Using a VGAP (alignment with a variable penalty) alignment programme (Moereels, De Bie, Tollenaere, J.Comp.-aided Mol.Des., in press) the *C.albicans* and *C. tropicalis* P450$_{14DM}$ are found to share with the *S. cerevisiae* P450$_{14DM}$ 64.2 and 65.2% identical amino acids, respectively. P450$_{14DM}$ from *C.albicans* and *C. tropicalis* share 83% identical amino acids. The amino acid sequence of the P450$_{14DM}$ from liver is still not published. Since liver and *S. cerevisiae* P450$_{14DM}$ use lanosterol as substrate it would be of interest to see whether their sequences share more identical amino acids than when compared with P450$_{14DM}$ that preferentially uses 24-methylenedihydro-lanosterol as substrate.

Wilkinson et al. (1972) described a long list of imidazole derivatives as potent inhibitors of P450-dependent reactions in liver microsomes. Each of the imidazoles investigated exhibited a type II difference spectrum with a peak at 430-431 nm and a trough at 390-393 nm. This suggests that the unhindered nitrogen (N$_3$ in the imidazole ring) binds to the catalytic heme iron atom at the site occupied by the exchangeable sixth ligand. Similar type II difference spectra have been obtained with triazole derivatives such as itraconazole, indicating that the N4 in the triazole ring coordinates to the haem iron. Fig. 2 shows a space-filling representation of the itraconazole-haem iron complex. The fifth ligand is a thiolate ligand of the cysteine residue and the 6th ligand is the unhindered nitrogen of itraconazole's triazole moiety.

In contrast with the imidazole derivatives used by Wilkinson et al. (1972) most of the azole antifungals, imidazole as well as triazole derivatives, are highly selective inhibitors of the P450$_{14DM}$ catalysed reaction in fungal cells (Vanden Bossche et al.,1987, 1989). There is no doubt that the azole group of these antifungals is important for binding to the haem iron and thus for inhibiting oxygen binding and activation. However, the binding of these heterocycles alone does not explain differences in activity; the N$_1$ substituent also plays a role. Indeed, imidazole as well as triazole derivatives are found between the potent and much less potent affectors of the *C. albicans* P450. For example, a 50% decrease in the absorption at 448 nm (absorption maximum for the reduced carbon monoxide *C. albicans* P450 complex) is observed at 3.1x10^{-8}M itraconazole (an hydrophobic triazole derivative) whereas 2.5x10^{-7}M fluconazole (a more polar triazole derivative) is needed (Vanden Bossche et al., 1987). This corresponds with fluconazole's low activity in vitro (Troke et al., 1988). Itraconazole forms a stable complex with P450$_{14DM}$ purified from *S. cerevisiae* microsomes (Yoshida and Aoyama, 1987) and with P450 in *C. albicans* microsomes (Vanden Bossche et al., 1988b).

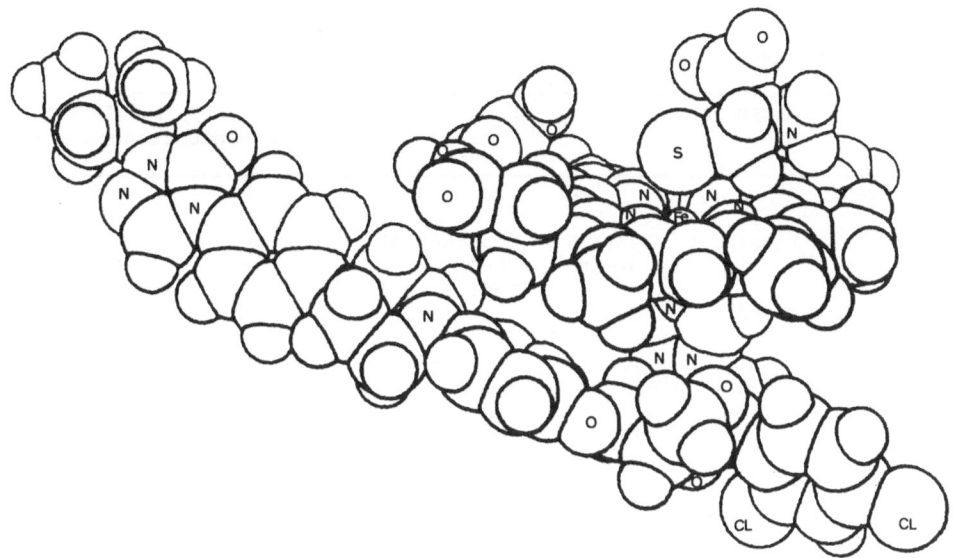

Fig. 2- A space-filling representation of the P450-haem iron-itraconazole complex.

When the triazole ring of itraconazole was replaced by an imidazole ring, a complex with P450 from *C. albicans microsomes* as stable as that seen with itraconazole was obtained (Vanden Bossche et al., 1988b). Compared with itraconazole, ketoconazole forms a less stable P450-azole complex (Yoshida and Aoyama ,1986; 1987, Vanden Bossche et al.,1986) Replacement of the imidazole moiety of ketoconazole by a triazole ring did not increase the binding affinity (Vanden Bossche et al, 1988b). These studies point to the role of the nonligand part (the substituent at N_1 of the azole moiety) of the antifungal compound.

Selectivity

The selectivity of the azole antifungals might also be determined by both the azole and the N-1 substituent of the molecule. For example, replacing the imidazole ring of ketoconazole by a triazole ring (R42164) decreases the interaction with the microsomal P450(s) from piglet testis. IC$_{50}$-values of 3.9×10^{-7}M (ketoconazole) and 4.9×10^{-6}M (R42164) are found. However, it is obvious that the azole group alone cannot be responsible for the differences in selectivity. Major differences in selectivity exist between azole antifungals with an imidazole as nitrogen heterocycle (Vanden Bossche et al.,1989). The agrochemical, imazalil, shows high selectivity: it has almost 80 and 98 times more affinity for the *Candida* P450(s) than for those of the piglet testis microsomes and bovine adrenal mitochondria, respectively. Among the topically active antifungals, clotrimazole and miconazole interact preferentially with the *Candida* P450, whereas bifonazole has highest affinity for P450(s) of the testicular microsomes from piglets (Vanden Bossche et al.,1989). Replacing the triazole of itraconazole by an imidazole ring slightly increased effects on adrenal mitochondrial P450 (from 0 to 25% inhibition at 10^{-6}M) but did not enhance the effects on testicular microsomal P450 (Vanden Bossche et al.,1989). Comparing the effects of ketoconazole with those of nor-ketoconazole (deacylated ketoconazole) indicates that minor structural changes in the non-ligand part affects the interaction with P450. Nor-ketoconazole has 2 and 3 times lower affinity for the microsomal P450 from piglet testis and *C. albicans* , respectively (Vanden Bossche et al.,1989).

These results suggest that the non-ligand hydrophobic part of the azoles has a greater impact on the selective interaction with P450 enzymes

Table 1. Effects of Ketoconazole and Itraconazole on P450-dependent Reactions in Microsomal or Mitochondrial Membranes*

Species	Organ Gland	P450	Product Formed	IC_{50} $(\times 10^{-7}M)$ Ketoconazole	Itraconazole
C. albicans		14-DM	Ergosterol	0.7	0.8
Rat	Liver	14-DM	Cholesterol	20	70
		7α-OH	7α-OH-Cholesterol	2.7	>100
	Testis	17-OH/ C-17,20	Androgens	2	>50**
	Skin	RA	4-OH-RA	6.5	>100
Bovine	Adrenal	SCC	Pregnenolone	17	>50
		C-11ß	Cortisol	3.8	>50
		17-OH/ C-17,20	Androgens	4	>50
		17-OH	11-Deoxycor.	28	>50
		C-21	11-DOC	>100	>100
Chicks	Kidney	C-1	1,25-DiOH-vitD	14	>50
Humans	Placenta	AROM	Estrogens	600[#]	>100

*14-DM = 14α-demethylase; 7α-OH = cholesterol-7α-hydroxylase; 17-OH = 17α-hydroxylase; C-17,20 = 17,20-lyase; RA = retinoic acid; 4-OH-RA = 4-hydroxyretinoic acid; SCC = cholesterol side-chain cleavage; C-11ß = 11ß-hydroxylase; 11-deoxycor. = 11-deoxycortisol; C-21 = 21-hydroxylase; 11-DOC = 11-deoxycorticosterone; C-1 = 1-hydroxylase; 1,25-DiOH-vitD = 1,25-dihydroxyvitamin D3; AROM = aromatase. Methods used are as previously described (Vanden Bossche et al., 1985, 1987, 1988c, 1990; Willemsens and Vanden Bossche, 1985, Willemsens et al., 1980; Princen et al., 1986; Mitropoulos and Balasubramaniam, 1972). **Limit of solubility. [#]Taken from Mason et al., 1985.

than the azole moiety. Thus, the differences in effects of ketoconazole and itraconazole on the P450-dependent reactions summarized in Table 1 might originate more from differences in their non-ligand hydrophobic part.

The data presented in Table 1 prove that itraconazole is a highly selective inhibitor of ergosterol synthesis in *C. albicans*. Of course itraconazole's effects are not limited to *C. albicans*, it is also a potent inhibitor of the P450-dependent ergosterol synthesis in for example *Aspergillus* spp. (Marichal et al., 1985, 1989; Vanden Bossche et al., 1988a.b), *Pityrosporum ovale* (Marichal et al., 1986) and *Leishmania mexicana mexicana* (Hart et al., 1989). Itraconazole is also an inhibitor of the 14α-demethylase of *C. glabrata* and *Histoplasma capsulatum*.

Candida glabrata

In vitro *C. glabrata* is, as compared with *C. albicans*, somewhat less sensitive to azole antifungals (Plempel et al., 1988). It is also difficult to eliminate *C. glabrata* from the vagina with the currently available polyenes and azole antimycotics (Mendling, 1988). As shown in Fig. 3 the *C. glabrata* isolate used here is indeed less sensitive to both itraconazole and fluconazole than the *C. albicans* isolate.

The results shown in Figs. 3 and 4 indicate that itraconazole is a more potent inhibitor of growth than fluconazole. However, as compared with itraconazole, miconazole is a more effective inhibitor of the growth of *C. glabrata*.

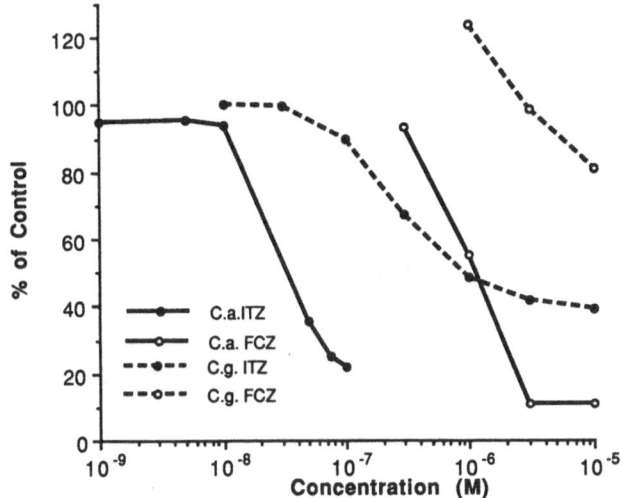

Fig. 3- Effects of itraconazole (ITZ) and fluconazole (FCZ) on the number
of *C. albicans* (C.a., ATCC 28516) and *C. glabrata* (C.g., B 16205) cells.
Cells were grown (at 37°C in a reciprocating shaker) for 24 hours in CYG-
medium (caseinhydrolysate-yeast extract-glucose, each 5 g/l). Azoles
and/or DMSO were added to the medium before inoculation. The number of
cells was obtained by counting a sample diluted with isoton to 20 ml in a
Coulter counter® with a 50μm probe and settings aperture 1, amplification
1/16., lower threshold 4 and upper threshold 105.

As shown in Figs. 4 and 5 the effects of the azole derivatives on
growth of *C. glabrata* might originate from inhibition of ergosterol
biosynthesis. The three azole derivatives lower the ergosterol content
(Fig. 4) and inhibit the incorporation of [^{14}C] acetate into ergosterol
(Fig. 5). This inhibition coincides with an accumulation of 14α-
methylsterols (Fig. 6), indicating that the inhibition of ergosterol
synthesis in *C. glabrata* originates from interaction with P450$_{14DM}$.

Of interest is the accumulation of lanosterol and 4,14-
dimethylzymosterol instead of 24-methylenedihydrolanosterol, found in e.g.
C. albicans and *A. fumigatus*. This indicates that in *C. glabrata* side
chain alkylation proceeds after the demethylation reaction (Margalith,
1986). As mentioned before, incubation of *S. cerevisiae* with azole
antifungals also resulted in the accumulation of lanosterol. This
indicates that, in contrast to the P450$_{14DM}$ of the isolates of *C. albicans*
we used, the *S. cerevisiae* and *C. glabrata* P450$_{14DM}$ prefer lanosterol
instead of 24-methylenedihydrolanosterol as substrate. We also showed that
ketoconazole is a much more potent inhibitor of ergosterol synthesis in *C.
albicans* than in *S. cerevisiae*, both grown for 24h in a casein hydrolysate-
yeast extract-glucose medium. Fifty % inhibition was achieved at 2.1 nM
and 330 nM, respectively.

The results presented in Table 1 indicate that both ketoconazole and
itraconazole are also more potent inhibitors of ergosterol than of
cholesterol synthesis by subcellular fractions of *C. albicans* and rat
liver, respectively. The P450$_{14DM}$ of mammalian microsomes also uses
lanosterol as substrate. Therefore it is possible that P450s using
lanosterol instead of 24-methylenedihydrolanosterol are less sensitive to
azoles. This might explain the differences in sensitivity between *C.
albicans* and *C. glabrata*.

However, there are other possibilities:

1. Recent studies of Watson et al. (1988, 1989) suggest that at least in *S.*

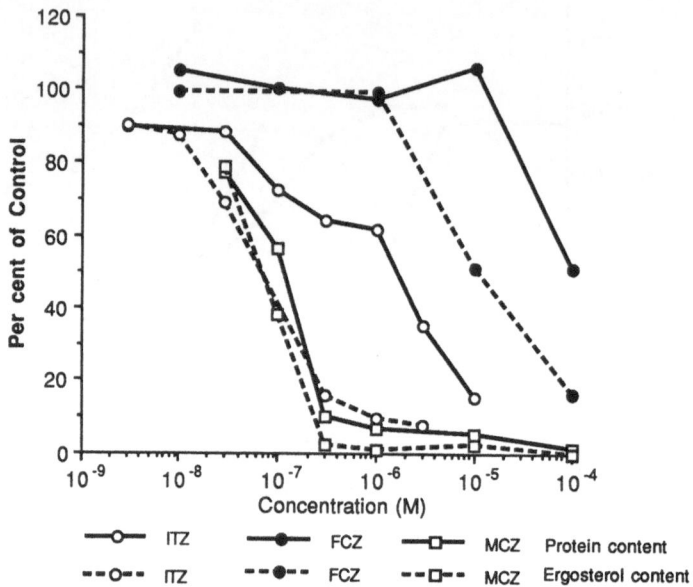

Fig. 4- Effects of itraconazole (ITZ), fluconazole (FCZ) and miconazole
(MCZ) on protein (growth) and ergosterol contents of *C. glabrata* (B 57148
provided by Dr. D.W. Warnock). Cells were grown at 37°C (reciprocating
shaker) for 24 hours in CYG-medium. Azoles and/or DMSO were added to the
medium before inoculation. At the end of the incubation period cells were
collected by centrifugation and washed twice in physiological water. The
pellet was suspended in 3 ml water and added to 2 g of glassbeads (± 0.45
mm). Cells were shaken vigorously (Vanden Bossche et al., 1978) for twenty
cycles of 4 sec with 26 sec intermittent cooling. The homogenates were
separated quantitatively from the glass beads and diluted to 20 ml. The
protein content of the homogenates was determined after appropiate dilution
by the Biorad® method with albumin as the standard. The homogenate was
supplemented with an equal volume of 15% KOH in 90% ethanol. The mixture
was refluxed for 1 hour at 85°C in a waterbath. After cooling sterols were
extracted with one volume of n-heptane (spectrograde). The absorption
difference at 291 and 282 nm (linearly dependent on ergosterol
concentration) was used to determine the ergosterol content.
Concentrations were calculated refering to a standard curve.

cerevisiae azole-induced growth inhibition originates from inhibition of
the 14α-demethylase and the consequent accumulation of 14α-methyl-ergosta-
$\Delta^{8,24(28)}$-dien-3ß,6α-diol (3,6-diol). Their hypothesis is based on the
inability of 3,6-diol to support growth. The amount of the 3,6-diol found
in *C. glabrata* (Fig. 6) incubated in the presence of itraconazole is lower
than that found in *C. albicans* (Vanden Bossche et al., 1989). After 24h of
growth in the presence of 30 nM itraconazole, 69% of the sterols extracted
from the *C. albicans* cells was 3,6-diol whereas 50.5% was found in *C.
glabrata*.

2. Only small amounts of ergosterol are required for yeast cell
proliferation (for a review see Vanden Bossche, 1990). In *C. albicans*
depletion of ergosterol was already found at 30 nM itraconazole (Vanden
Bossche et al., 1989) whereas even at 10 μM still 4.2 % of the ^{14}C-labelled
sterols in *C. glabrata* consisted of ergosterol. This might explain why, at
this high itraconazole concentration, growth (expressed by the number of
cells) was inhibited by about 60% only (Fig. 3).

Although, both itraconazole and fluconazole are less effective growth
inhibitors of *C. glabrata* than of *C. albicans* in vitro (Fig. 3), the
results summarized in Fig. 7 indicate that microsomal P450 from *C. glabrata*

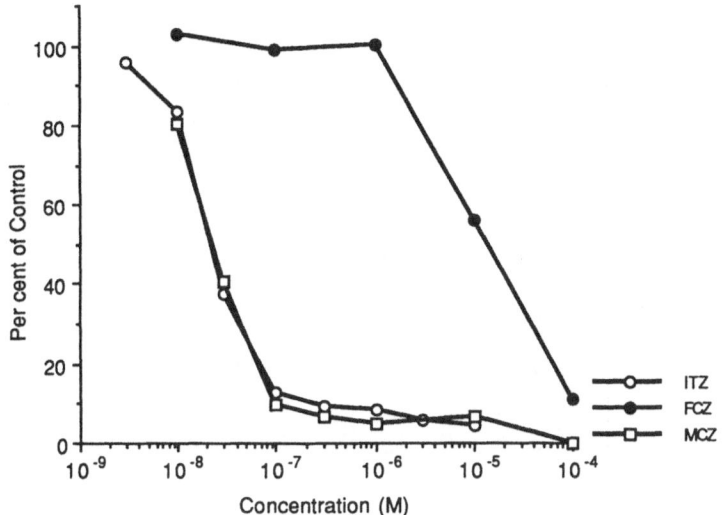

Fig. 5- Effects of itraconazole (ITZ), fluconazole (FCZ) and miconazole
(MCZ) on ergosterol synthesis from [^{14}C] acetate by *C. glabrata* (B 57148,
24h at 37°C in CYG-medium). Total radioactivity in the heptane extract was
determined by liquid scintillation counting. Opti-phase (Pharmacia) was
used as scintillant. The heptane extracts were dried under a nitrogen
stream and dissolved in minimal volumes of methanol/water. Separation of
the sterols was performed by HPLC on a reversed C$_8$ column. Sterols were
eluted with a methanol −H$_2$O mixture (95/5%) for 25 min after which the
column was eluted with pure methanol. Sterols were identified according to
their relative retention time towards standards and/or after GCMS
identification. Results are expressed as % of control. More details are
given in the legend to Fig. 4.

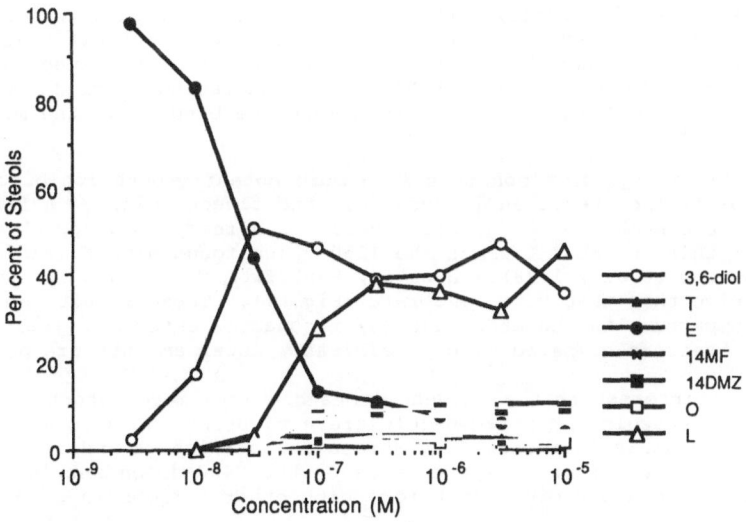

Fig. 6- Effects of itraconazole on sterol synthesis in *C. glabrata*
(B 57148) grown for 24h at 37°C in CYG medium. 3,6-diol=14α-methyl-
ergosta-8,24(28)-dien-3ß,6α-diol; T=14α-methyl-ergosta-5,7,22,24(28)-
tetraen-3ß-ol; E=ergosterol; 14MF=14α-methylfecosterol (14α-methyl-ergosta-
8,24(28)-dien-3ß-ol); 14DMZ=4,14-dimethylzymosterol (4,14-dimethyl-
cholesta-8,24-dien-3ß-ol); O=obtusifoliol; L=lanosterol. More details are
given in the legends to Figs. 4 and 5.

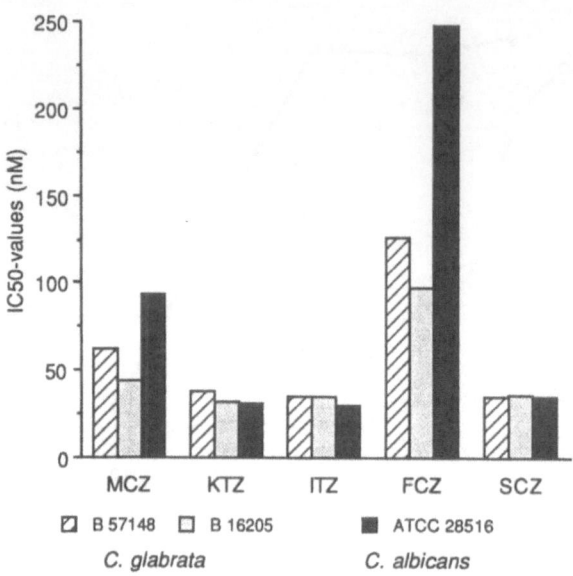

Fig. 7- Effects of miconazole (MCZ), ketoconazole (KTZ), itraconazole
(ITZ), fluconazole (FCZ), and saperconazole (SCZ) on P450 in microsomal
fractions of *C. glabrata* and *C. albicans* isolates. The microsomal
fractions were diluted with buffer to obtain a P450 content of 0.1
nmole/ml. Azoles and/or solvent were added to the P450 suspensions before
reduction with dithionite and saturation with CO (Vanden Bossche et al.,
1984, 1987). IC50-value: drug concentration needed to obtain 50% decrease
in the Soret peak height of the reduced carbon monoxide compound of P450.

is almost as sensitive to itraconazole (and ketoconazole or saperconazole)
as the P450 from *C. albicans*; the latter is even much less sensitive to
fluconazole. The discrepancy between the results obtained with azoles at
the microsomal level and those found with intact cells suggest that these
antifungals are less absorbed by *C. glabrata*. As could be expected from
its lower effect on growth, a much higher concentration of fluconazole is
needed to reach 50% inhibition of carbon monoxide binding to the microsomal
C. glabrata P450s.

As shown in Fig. 4, miconazole is a more potent growth inhibitor of
the *C. glabrata* isolates than itraconazole and fluconazole, . A fifty %
decrease in the number of *C. glabrata* cells is already obtained at 0.17 μM
miconazole: this is about 4 times the IC50 value found with *C. albicans*
(Vanden Bossche et al., 1978). However, P450 from *C.albicans* is somewhat
less sensitive than that of *C. glabrata* (Fig. 7). These results suggest
that, as compared with the other azoles, miconazole enters *C. glabrata* more
easily but that, as compared with *C. albicans*, lower amounts are absorbed.

It is of interest that although miconazole is a more potent growth
inhibitor, miconazole and itraconazole are equipotent inhibitors of
ergosterol synthesis in intact *C. glabrata* cells (Figs. 4 and 5). This
suggests that in addition to its effects on the P450-dependent 14α-
demethylase, miconazole might interfere with other targets in *C. glabrata*.

A number of hypotheses for miconazole's higher growth inhibitory
effects can be formulated. For example, it would be of interest to study
the effects of azoles on the nature of fatty acids present in control and
treated *C. glabrata* cells. Indeed, at concentrations higher than those
needed to affect ergosterol synthesis in *C. albicans*, miconazole also
affects the nature of free and esterified fatty acids (Vanden Bossche et
al., 1982). Furthermore, miconazole's fungicidal effect can partly be

antagonized by oleic acid, whereas the effect of ketoconazole on the viability of *C. albicans* was not affected by this fatty acid (Vanden Bossche et al., 1982, Vanden Bossche, 1985).

It has also been suggested that, at concentrations higher than those needed to affect ergosterol biosynthesis and the nature of fatty acids, miconazole (but not ketoconazole nor itraconazole) destabilizes membranes (Vanden Bossche et al., 1982; Brasseur et al., 1983, Vanden Bossche, 1985). It should be of interest to study the direct interaction of the azole antifungals with the membranes of both *C. albicans* and *C. glabrata*.

Histoplasma capsulatum

Histoplasmosis is a major endemic mycosis associated with AIDS (see Graybill et al., this book). Efficacy of ketoconazole and itraconazole has been demonstrated in histoplasmosis (Stevens, 1988). In guinea pigs itraconazole has been found more effective than ketoconazole against *H. capsulatum var. duboisii* (Van Cutsem and Janssen, 1987). Based on limited studies, itraconazole appears the most promising among the azole antifungals studied to date (Graybill et al., this book).

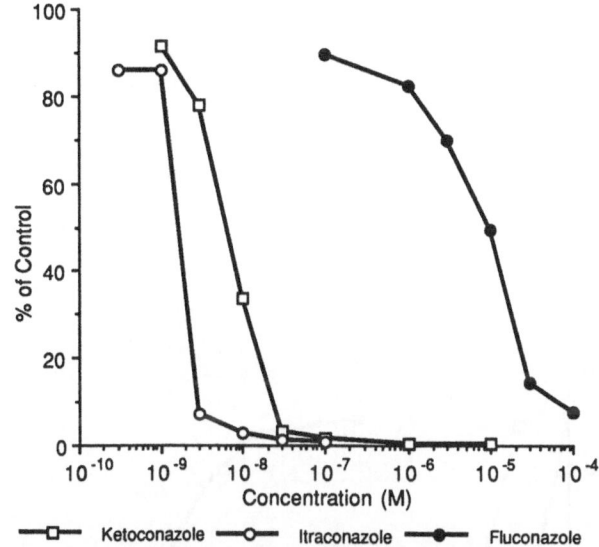

Fig. 8- Effects of ketoconazole, itraconazole and fluconazole on growth (mg protein/100 ml culture) of *H. capsulatum* (yeast form; G217-B, B 55,006, obtained from Prof. G.S. Kobayashi). *H. capsulatum* isolates were maintained on GYA (2% glucose- 1% yeast extract- 0.06% L-cysteine- 2% agar)slants at 37°C. A loopfull from a week old slant was inoculated in 100 ml of GY medium (2% glucose- 1% yeast extract- 0.03% L-cysteine) containing 20 units of penicillin and 40 μg streptomycin per ml and cultivated at 37°C for one week in a reciprocal shaker . Half ml aliquots of this culture were used to inoculate 100 ml GY medium in 500 ml erlenmeyer flasks. Cells were grown at 37°C aerobically in a reciprocating shaker for 72h. These cultures were used as inoculum for inhibition studies. Compounds and/or solvent (DMSO final 0.1%) were added to the medium prior to inoculation. At the end of the incubation period (48h), Cells were collected by centrifugation and washed twice in physiological saline. The pellet was suspended in 3 ml of water and added to 2 g of glassbeads (± 0.45 mm). Cells were shaken vigorously (see legend to Fig. 4) for twenty cycles of 4 sec with 26 sec intermittent cooling. The homogenates were separated quantitatively from the glass beads and diluted to 15 ml.The protein content of the homogenates was determined by the Biorad® method with albumin as the standard. Results are expressed as per cent of control.

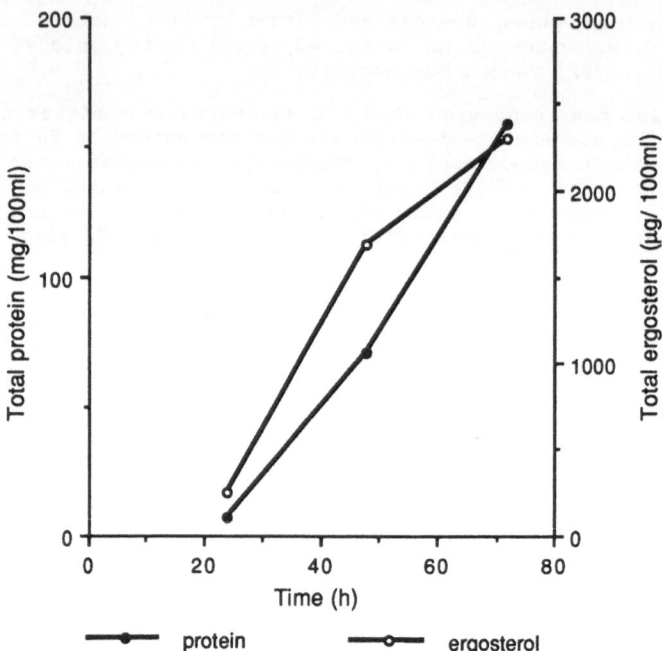

Fig. 9- Ergosterol and protein content of *H. capsulatum* (G217-B) grown for 24, 48, and 72 hours in GY medium. More details can be found in the legend of Fig. 8. The ergosterol content was determined as described in the legend to Fig. 4.

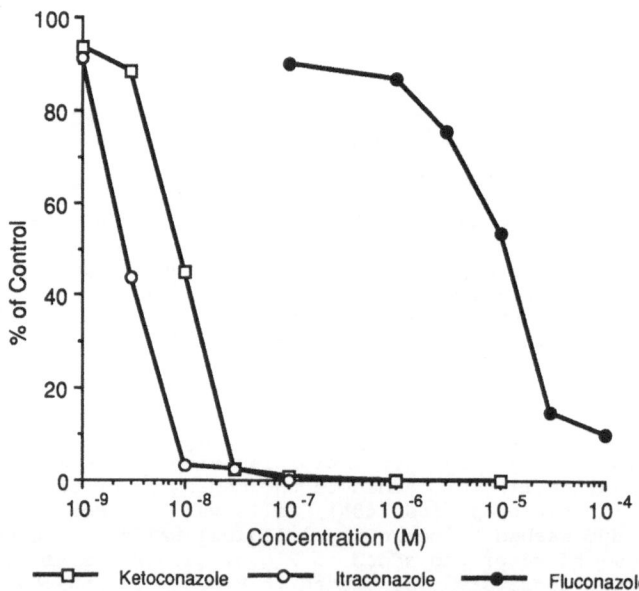

Fig. 10- Effects of ketoconazole, itraconazole and fluconazole on the ergosterol content of *H. capsulatum* (G217-B) grown for 48 hours in GY medium. More details can be found in the legend of Fig. 8. The ergosterol content was determined as described in the legend to Fig. 4.

As shown in Fig. 8, both ketoconazole and itraconazole are potent inhibitors, whereas high concentrations of fluconazole are needed to inhibit the growth of *H. capsulatum* (yeast form). Fifty per cent inhibition of growth (decrease in protein content) is achieved at 1.7 nM itraconazole, 7.1 nM ketoconazole and 9.6 μM fluconazole.

H. capsulatum synthesizes ergosterol (Fig. 9). The amount present correlates with the protein content of the culture and might be used as a measure of growth.

As in other fungi, azole antifungals interfere with ergosterol synthesis by *H. capsulatum*. After 48 h. of growth a 50% decrease in the ergosterol content was observed at 2.6 nM, 8.7 nM or 11 μM itraconazole, ketoconazole or fluconazole, respectively (Fig. 10). After 48 h of

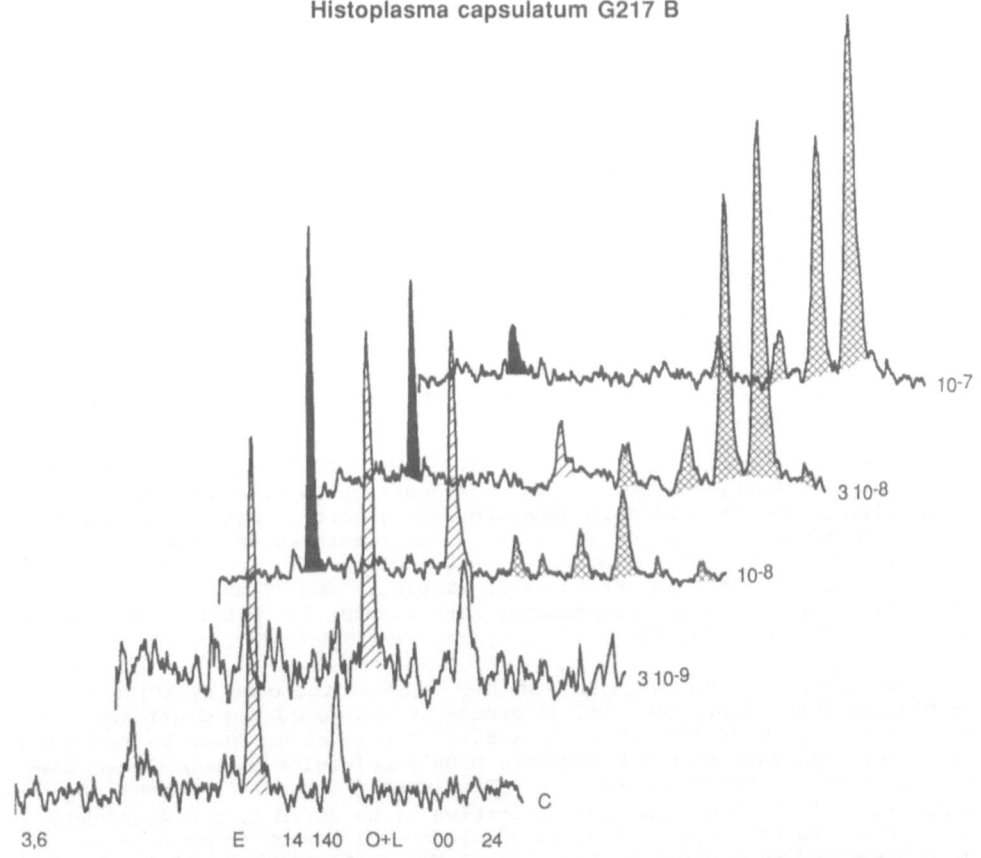

Fig. 11- Effects of ketoconazole on sterol synthesis in *H. capsulatum* grown 48 h in GY medium supplemented with [^{14}C]-acetate and ketoconazole and/or DMSO (C). Homogenisation was as described in Fig. 8. The homogenate was supplemented with an equal volume of 15% KOH in 90% ethanol. The mixture was refluxed for 1 hour at 85°C in a waterbath. After cooling, sterols were extracted with one volume n-heptane (spectrograde). The heptane extracts were dried under a stream of nitrogen and dissolved in minimal volumes of methanol/water (95/5%). Separation of the sterols was performed by HPLC on a reversed C$_8$ column. Sterols were eluted with a methanol –H2O mixture (95/5%) for 25 min after which the column was eluted with pure methanol. Sterols were identified as previously described (Vanden Bossche et al., 1978, 1987). 3,6= 14α-methyl-ergosta-8,24(28)dien-3ß,6α-diol; E= ergosterol; 14= 14-methylfecosterol; 14O= 14-methylfecosterone; O+L= obtusifoliol; OO= obtusifolione; 24= 24-methylenedihydrolanosterol.

Table 2. Sterols and ketones formed by *H. capsulatum* (G217-B)
after 48 h of growth in the presence of 3 nM or 30 nM itraconazole*

Sterol class	Sterols and 3-Ketosteroids	% of Total	
		3nM	30nM
14-Desmethylsterol	Ergosterol	7	0
14-Methylsterols	14-Methyl-$\Delta^{8,24(28)}$-ergosta-diene-3ß, 6-diol	4	4
4,14-Dimethylsteroids	Obtusifolione	37	44
4,4',14-Trimethylsterols	24-Methylenedihydro-lanosterol	43	52

*Details are given in the legend to Fig. 11

incubation of *H. capsulatum* in the presence of ketoconazole concentrations
between 3 nM and 0.1 µM, important shifts in the nature of the sterols
formed from [^{14}C] acetate could be noted (Fig. 11). Fifty per cent
inhibition was achieved at 11±6.7 nM ketoconazole. Complete inhibition was
reached at 0.1 µM. With itraconazole an IC_{50}-value of 2.9±1.2 nM was
found. A complete block of ergosterol synthesis was reached at 30 nM
itraconazole (Table 2).

A block in ergosterol synthesis coincided with the accumulation of
sterols with a methyl group at C-14. In contrast to *C. glabrata*, *H
capsulatum* accumulates 24-methylenedihydrolanosterol. Next to this 24-
alkylated sterol a number of interesting intermediates have been found
(Fig. 11, Table 2). High amounts of obtusifolione are accumulating after
48 h of contact with 3 nM itraconazole (Table 2) and 30 nM ketoconazole
(Fig. 11). Next to this 4,14-dimethylketosteroid, 14-methylfecosterone is
also accumulating (Fig. 11).

These results indicate that the last step in the demethylation at C-4
is blocked (Fig. 12). Both methyl groups at C-4 (C-30 and C-31) are
removed as carbon dioxide (Mercer, 1984). The first group to be removed is
the 4α-methyl. This reaction requires NADPH, molecular oxygen, a cyanide-
sensitive factor and cytochrome b$_5$. The first intermediate formed is a C-4
alcohol. This 4α-hydroxymethyl derivative is oxidated to a C-4 aldehyde
and further to 4α-carboxy-4ß-methyl-3-ketosterol, which is then
decarboxylated by a NAD-dependent sterol 4-decarboxylase to yield the 4α-
methyl-3-ketosteroid, obtusifolione. The final step in the 4-demethylation
reaction is the reduction of the 3-keto to a 3ß-hydroxyl group by the
NADPH-dependent 3-ketosteroid reductase. The second demethylation at C-4
is mechanically identical and catalysed by the same microsomal enzymes
(Mercer, 1984). The intermediate formed is 14-methylfecosterone (Fig. 13).

The accumulation of sterols with a methyl at C-14 proves that both
azole antifungals inhibit the P450-dependent 14α-demethylase in *H.
capsulatum*. C-14 methylated sterols are known to have detrimental effects
on membrane properties (for reviews see Bloch, 1983, Vanden Bossche, 1988,
1990). However, as discussed previously (Vanden Bossche, 1988) ergosterol
depletion might contribute more to the antifungal effects of azole
derivatives. Indeed, even at low ergosterol concentrations some yeast
sterol auxotrophs are able to grow in media supplemented with lanosterol

(Rodriguez et al., 1985). Ergosterol depletion in *H. capsulatum* is achieved at 0.1 µM ketoconazole and 0.03 µM itraconazole.

The accumulation of 3-ketosteroids points to an inhibition of the 3-ketosteroid reductase (Fig. 13).

In *C. albicans*, *C. glabrata*, and *A. fumigatus* we found trace amounts only of these 3-ketosteroids. However, Kerkenaar et al. (1986) reported that imazalil caused in *Penicillium italicum*, in addition of 14-methylsterols, an accumulation of obtusifolione and 14α-methylfecosterone. They suggest that the high sensitivity of *P. italicum* to imazalil results from an additional site of inhibition in the scheme of sterol biosynthesis i.e. the 3-ketosteroid reductase.

The interaction of ketoconazole and itraconazole with two targets, the 14α-demethylase and the 3-ketosteroid reductase (Fig. 13), in the ergosterol biosynthetic pathway might also explain the high sensitivity of *H. capsulatum* to these azole antifungals. Indeed, prerequisites for both the condensing (increase in the chain order of the phospholipids in the liquid crystalline state of the membrane) and liquefying effect (decrease

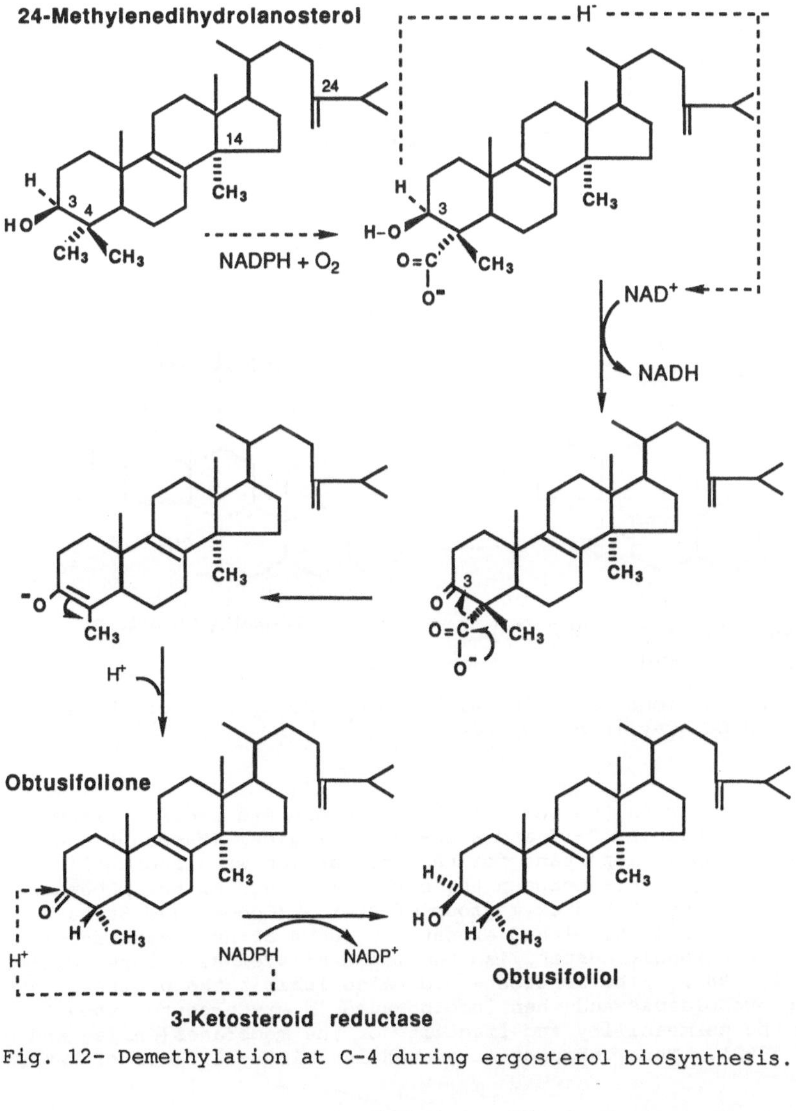

Fig. 12- Demethylation at C-4 during ergosterol biosynthesis.

237

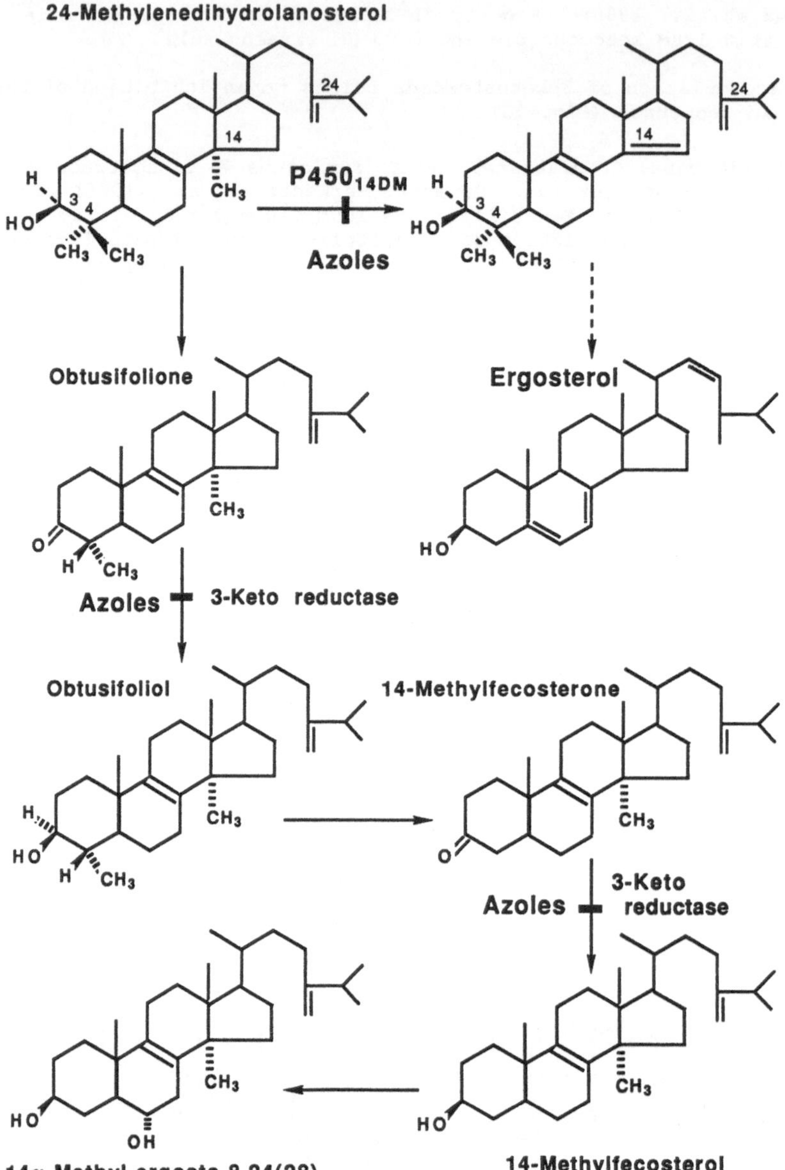

Fig. 13- Interactions of azoles with two enzyme systems of the ergosterol biosynthesis pathway in *H. capsulatum*.

in the chain order in the gel state) of sterols are a planar ring system, a long flexible chain at C-17 and a 3ß-hydroxyl group (Vanden Bossche, 1990). The latter group is important for the interaction with phospholipids: hydrogen bonding occurs between the carbonyl oxygen of the phospholipid-acyl chain and the 3ß-hydroxyl group of sterol (Cooper and Strauss, 1984). In contrast to cholesterol and ergosterol (both 3ß-hydroxysterols), 3-ketosteroids strongly destabilize the lipid bilayer structure (Gallay and De Kruijff, 1982). The 3-ketosteroids also inhibit the growth of sterol-requiring mycoplasmas and when incorporated in erythrocytes they greatly increase the permeability and fragility of the membranes(Gallay and De Kruijff, 1982). Thus, the potent growth-inhibitory effects of ketoconazole and itraconazole on *H. capsulatum* might be the result of interaction with the P450-dependent 14α-demethylase and the NADPH-dependent 3-ketosteroid reductase leading to accumulation of membrane disturbing 14-methylated sterols and 3-ketosteroids, and ergosterol depletion. Since a complete block of ergosterol synthesis is reached at lower itraconazole concentrations, this triazole might prove to be more effective in the treatment of histoplasmosis than ketoconazole.

CONCLUSION

At the subcellular and cellular level, itraconazole is a highly potent and selective inhibitor of ergosterol synthesis in *C. albicans* and *C. glabrata*. It shares with ketoconazole and saperconazole high affinity for P450 of *C. albicans* and *C. glabrata* microsomes. Fluconazole shows much lower affinity for *C. albicans* and *C. glabrata* P450 and is a poor inhibitor of ergosterol synthesis and growth of both *Candida* species in vitro.

Although miconazole has a slighly lower affinity for *C. glabrata* P450 as compared with itraconazole, it is an equipotent inhibitor of ergosterol synthesis and a more potent inhibitor of growth. This suggests that miconazole may interact with another target(s) as well as the P450-dependent 14α-demethylase ($P450_{14DM}$).

In *C. glabrata* grown in the presence of azole antifungals, lanosterol instead of 24-methylenedihydrolanosterol accumulates. This indicates that in *C. glabrata* , as in *S. cerevisiae*, (but unlike for example, *C. albicans* and *A. fumigatus*) side chain alkylation proceeds after the 14-demethylation reaction. It is tempting to speculate that the *C. glabrata* $P450_{14DM}$ is more similar to that of *S. cerevisiae* than of *C. albicans*. This might explain, at least partly, why azole antifungals such as itraconazole and fluconazole need higher concentrations to inhibit growth of *C. glabrata* than of *C. albicans*.

Azole antifungals also inhibit ergosterol synthesis in *H. capsulatum*. Fifty per cent decrease in ergosterol content is achieved at 2.6 nM, 8.7 nM and 11 µM itraconazole, ketoconazole and fluconazole.
The greater sensitivity of ergosterol synthesis to itraconazole correlates with its higher growth-inhibitory effects. This inhibition results in an accumulation of 14-methylsterols (indicative of inhibition of the 14α-demethylase) and 3-ketosteroids (obtusifolione and 14-methylfecosterone). The accumulation of the 3-ketosteroids indicates that itraconazole and ketoconazole interact with two targets, the 14α-demethylase and the last step in the demethylation at C-4 catalysed by the 3-ketosteroid reductase.

Both 3-ketosteroids and 14-methylsterols destabilize membranes. This, together with the observed ergosterol depletion might explain the high sensitivity of this *H. capsulatum* isolate to ketoconazole and itraconazole.

ACKNOWLEDGEMENT

We are grateful to Dr. D.W. Warnock (Bristol Royal Infirmary, Bristol, UK) for providing us with the *C. glabrata* isolate B 57,148 and to Prof. G.S. Kobayashi (Washington University School of Medicine, St. Louis, USA) for the *H. capsulatum* isolate B 55.006 (G217-B).

REFERENCES

Berg, D., 1986, Biochemical mode of action of fungicides. Ergosterol biosynthesis inhibitors, in: "Fungicide chemistry. Advances and practical applications", M.B. Green and D.A. Spilker, eds., American Cemical Society, Washington D.C..
Bloch, K.E., 1983, Sterol structure and membrane function, Crit. Rev. Biochem., 14: 47.
Brasseur, R., Vandenbosch, C., Vanden Bossche, H. and Ruysschaert, J.M., 1983, Mode of insertion of miconazole, ketoconazole and deacylated

ketoconazole in lipid layers. A conformational analysis, Biochem. Pharmacol., 32: 2175.

Chen, C., Kalb, V.F., Turi, T.G.and Loper, J.C., 1988, Primary structure of the cytochrome P-450 lanosterol 14α-demethylase gene from *Candida tropicalis*, DNA, 7: 617.

Cooper, R.A. and Strauss, J.F., 1984, Regulation of cell membrane cholesterol, in: " Physiology of membrane fluidity. Vol.I", M. Shinitzky, ed., CRC Press, Inc., Boca Raton, Florida.

Gallay, J. and Kruijff, B., 1982, Correlation between molecular shape and hexagonal H_{II} phase promoting ability of sterols, FEBS Lett., 143: 133.

Hart, D.T., Lauwers, W.J., Willemsens, G., Vanden Bossche, H. and. Opperdoes, F.R., 1989, Perturbation of sterol biosynthesis by itraconazole and ketoconazole in *Leishmania mexicana mexicana* macrophages, Mol. Biochem. Parasitol., 33: 123.

Ishida, N.Y., Aoyama, Y., Hatanaka, R., Oyama, Y., Imajo, S., Ishiguro, M., Oshima,T., Nakazato,H., Noguchi, T., Maitra, U.S.. Mohan, V.P., Sprinson, D.B., and Yoshida, Y., 1988, A single amino acid substitution converts cytochrome $P450_{14\,DM}$ to an inactive form, cytochrome $P450_{SG1}$: complete primary structures deduced from cloned DNAs, Biochem. Biophys. Res. Commun., 155: 317.

Kalb, V.F., Loper, J.C., Dey, C.R., Woods, C.W. and Sutter, T.R., 1986, Isolation of a cytochrome P-450 structural gene from *Saccharomyces cerevisiae*, Gene, 45: 237.

Kalb, V.F., Woods, C.W., Turi, T.G., Dey, C.R., Sutter, T.R. and Loper, L.C., 1988, Primary structure of the P450 lanosterol demethylase gene from *Saccharomyces cerevisiae*, DNA, 6: 529.

Kerkenaar, A., Jansen, G.G. and Costet, M-F., 1986, Special effects of imazalil on sterol biosynthesis of *Penicillium italicum*,in: "The sixth international congress of pesticide chemistry", Abstract book, IUPAC, Ottawa, Canada.

Kerridge, D. and Whelan, W.L., 1984, The polyene macrolide antibiotics and 5-fluorocytosine: molecular actions and interactions, in: "Mode of action of antifungal agents", A.P.J. Trinci and J.F. Ryley, eds., Cambridge University Press, Cambridge.

Kerridge, D., 1988, Polyene macrolide antibiotics, in: "*Aspergillus* and aspergillosis", H. Vanden Bossche, D.W.R. Mackenzie and G. Cauwenbergh, eds., Plenum Press, New York.

Lai, M.H. and D.R. Kirsch, D.R., 1989, Nucleotide sequence oc cytochrome P450LIA1 (lanosterol 14α-demethylase) from *Candida albicans*, Nucleic Acids Res., 17: 804.

Margalith, P.Z., 1986, "Steroid microbiology", Charles C. Thomas Publishers, Springfield.

Marichal, P., Gorrens, J. and Vanden Bossche, H., 1985, The action of itraconazole and ketoconazole on growth and sterol synthesis in *Aspergillus fumigatus* and *Aspergillus niger*, Sabouraudia: J. Med. Vet. Mycol., 23: 13.

Marichal, P., Gorrens J., Van Cutsem, J., Van Gerven, F. and Vanden Bossche, H., 1986, Effects of ketoconazole and itraconazole on growth and sterol synthesis in *Pityrosporum ovale*, J. Med. Vet. Mycol., 24: 487.

Marichal, P., Vanden Bossche, H., Gorrens, J., Bellens, D. and Janssen, P.A.J., 1989, Cytochrome P-450 of *Aspergillus fumigatus*– Effects of itraconazole and ketoconazole, in: "Cytochrome P-450: biochemistry and biophysics, I. Schuster, ed., Taylor and Francis, London.

Mason J.I., Murry B.A. Olcott M.and Sheets J.J., 1985, Imidazole antimycotics: inhibitors of steroid aromatase, Biochem. Pharm., 34: 1087.

Mendling, W., 1988, Azoles in the therapy of vaginal mycoses, in: "Sterol biosynthesis inhibitors. Pharmaceutical and agrochemical aspects", D. Berg and M. Plempel, eds., Ellis Horwood Ltd., Chichester, England.

Mercer, E.I., 1984, The biosynthesis of ergosterol, Pestic. Sci., 15: 133.

Mitropoulos, K.A. and Balasubramanian, S., 1972, Cholesterol 7α-
hydroxylase in rat liver microsomal preparations, Biochem. J., 128:
1.

Nebert, D.W., Nelson, D.R., Adesnik, M., Coon, M.J., Estabrook, R.W.,
Gonzalez, F.J., Guengerich, F.P., Gunsalus, I.C., Johnson, E.F.,
Kemper, B., Levin, W., Phillips, I.R., Sato, R., and Waterman,
M.R., 1989, The P450 superfamily: updated listing of all genes and
recommended nomenclature for the chromosomal loci, DNA 8: 1.

Plempel, M., Berg, D., Büchel, K-H. and Ritter, W., 1988, Experimental
antimycotic activity of azole derivatives- experience, knowledge
and questions, in: "Sterol biosynthesis inhibitors. Pharmaceutical
and agrochemical aspects", D. Berg and M. Plempel, eds., Ellis
Horwood Ltd., Chichester, England.

Polak, A., 1988a, Mode of action of 5-fluorocytosine in Aspergillus
fumigatus, in: "Aspergillus and aspergillosis", H. Vanden
Bossche, D.W.R. Mackenzie and G. Cauwenbergh, eds., Plenum Press,
New York.

Polak, A., 1988b, Morpholines in clinical use, in: "Sterol biosynthesis
inhibitors. Pharmaceutical and agrochemical aspects", D. Berg and
M. Plempel, eds., Ellis Horwood Ltd., Chichester, England.

Princen, H.M.G., Huysmans, C.M.G., Kuipers, F., Vonk, R.J. and Kempen,
H.J.M., 1986, Ketoconazole blocks bile acid synthesis in
hepatocyte monolayer cultures and in vivo in rat by inhibiting 7α-
hydroxylase, J. Clin. Invest., 78: 1064.

Rodriguez, R.J., Low, C., Bottema, C.D.K. and Parks, L.W., 1985, Multiple
functions for sterols in Saccharomyces cerevisiae, Biochim.
Biophys. Acta, 837: 336.

Ruckpaul, K. and Bernhardt, R., 1984, Biochemical aspects of the
monooxygenase system in the endoplasmic reticulum of mammalian
liver, in: "Cytochrome P-450", K.Ruckpaul and H. Rein, eds.,
Akademie-Verlag, Berlin.

Ryder, N.S., 1986, Biochemical mode of action of the allylamine
antimycotic agents naftifine and SF 86-327, in: "In vitro and in
vivo evaluation of antifungal agents", K. Iwata and H. Vanden
Bossche, eds, Elsevier Science Publishers B.V., Amsterdam.

Ryder, N.S., 1988, Mechanism of action and biochemical selectivity of
allylamine agents, Ann. N.Y. Acad. Sci, 544: 208.

Ryder,N.S., Dupont, M.C., Frank, I., 1986, Ergosterol biosynthesis
inhibition by the thiocarbamate antifungal agents tolnaftate and
tolciclate, Antimicrob Ag Chemother, 29: 858.

Sokol-Anderson, M.L., Brajtburg, J. and Medoff, G., 1986, Amphotericin B-
induced oxidative damage and killing of Candida albicans, J. Inf.
Dis., 154: 76.

Stevens, D.A., 1988, The new generation of antifungal drugs, Eur. J. Clin.
Microbiol. Dis., 7: 732.

Troke, P.F., Marriott, M.S., Richardson, K.,Tarbit, M.H.,1988, In vitro
potency and in vivo activty of azoles, Ann. N.Y. Acad.Sci., 544:
284.

Trzaskos, J., Kawata, S. and Gaylor, J.L., 1986, Microsomal enzymes of
cholesterol biosynthesis. Purification of lanosterol 14α-methyl
demethylase cytochrome P-450 from hepatic microsomes, J. Biol.
Chem., 261: 14651

Van Cutsem, J. and Janssen, P.A.J., 1987, Actvité in vitro et in vivo des
azoles à large spectre sur les champignons resposables de mycoses
d'importation, Bull. Soc. Mycol. Méd., 16: 67.

Vanden Bossche, H., 1985, Biochemical targets for antifungal azole
derivatives: hypothesis on the mode of action, in: "Current topics
in Medical Mycology", Vol 1, M.K. McGinnis, ed., Springer Verlag,
New York.

Vanden Bossche, H., 1987, Itraconazole: a selective inhibitor of the
cytochrome P-450 dependent ergosterol biosynthesis, in: "Recent
trends in the discovery, development and evaluation of
antifungals", R.A. Fromtling, ed., J.R. Prous Science Publishers,
S.A., Barcelona.

Vanden Bossche, H., 1988, Mode of action of pyridine, pyrimidine and azole antifungals, in: "Sterol biosynthesis inhibitors. Pharmaceutical and agrochemical aspects", D. Berg and M. Plempel, eds., Ellis Horwood Ltd., Chichester, England.

Vanden Bossche, H., 1990, Importance and role of sterols in fungal membranes, in: "Biochemistry of cell walls and membranes in fungi", P.J. Kuhn, A.P. Trinci, M.J. Jung, M.W. Goosey and L.G. Copping, eds., Springer-Verlag, Berlin.

Vanden Bossche, H., Willemsens, G., Cools,W., Lauwers, W.F.J. and Le Jeune, L., 1978, Biochemical effects of miconazole on fungi. II. Inhibition of ergosterol biosynthesis in *Candida albicans*, Chem.-Biol. Inter., 21: 59.

Vanden Bossche, H., Ruysschaert, J.M., Defrise-Quertain, F., Willemsens, G., Cornelissen, F., Marichal, P., Cools, W. and Van Cutsem, J., 1982, The interaction of miconazole and ketoconazole with lipids, Biochem. Pharmacol., 31: 2609.

Vanden Bossche, H., Willemsens, G., Cools, W., Marichal, P. and Lauwers, W., 1983, Hypothesis on the molecular basis of the antifungal activity of *N*-substituted imidazoles and triazoles, Biochem Soc. Trans.,.11: 665.

Vanden Bossche, H., Willemsens, G., Marichal, P., Cools, W. and Lauwers, W., 1984, The molecular basis for the antifungal activities of *N*-substituted azole derivatives. Focus on R 51211, in: "Mode of action of antifungal agents", A.P.J. Trinci and J.F. Ryley, eds., Cambridge University Press, Cambridge.

Vanden Bossche, H., Lauwers, W., Willemsens, G. and Cools, W., 1985, The cytochrome P-450 dependent C17,20-lyase in subcellular fractions of the rat testis: differences in sensitivity to ketoconazole and itraconazole, in: "Microsomes and drug oxidations", A.R. Boobis, J. Caldwell, F. de Matteis and C.R. Elcombe, eds., Taylor & Francis, London.

Vanden Bossche, H., Marichal, P., Gorrens, J., Bellens, D., Verhoeven, H., Coene, M.-C., W. Lauwers, W. and P.A.J. Janssen, P.A.J., 1987, Interaction of azole derivatives with cytochrome P-450 isozymes in yeast, fungi, plants and mammalian cells, Pestic. Sci., 21: 289.

Vanden Bossche, H., Marichal, P., Geerts, H.and Janssen, P.A.J., 1988a, The molecular basis for itraconazole's activity against *Aspergillus fumigatus*, in: "*Aspergillus* and aspergillosis", H.Vanden Bossche, D.W.R. Mackenzie and G. Cauwenbergh, eds., Plenum Press, New York.

Vanden Bossche, H., Marichal, P., Gorrens, J., Geerts, H and Janssen, P.A.J., 1988b, Mode of action studies. Basis for the search of new antifungal drugs, Ann. N.Y. Acad. Sci., 544: 191.

Vanden Bossche, H., Willemsens, G. and Janssen, P.A.J., 1988c, Cytochrome P-450-dependent metabolism of retinoic acid in rat skin microsomes: inhibition by ketoconazole, Skin Pharmacol., 1: 176.

Vanden Bossche, H., Marichal, P., Gorrens, J., Coen, M.-C., Willemsens, G., Bellens, D., Roels, I., Moereels, H. and Janssen, P.A.J., 1989, Biochemical approaches to selective antifungal activity. Focus on azole antifungals, Mycoses, 32:35.

Vanden Bossche, H., Willemsens, G. Bellens, D., Roels, I. and Janssen, P.A.J.,1990, From 14α-demethylase inhibitors in fungal cells to androgen and oestrogen inhibitors in mammalian cells, Biochem. Soc. Trans.,.18: 10.

Watson, P.F., Rose M.E. and Kelly S.L., 1988, Isolation and analysis of ketoconazole resistant mutants of Saccharomyces cerevisiae, J. Med. Vet. Mycol., 26: 153.

Watson, P.F., Rose M.E., Ellis S.W., England H. and Kelly S.L., 1989, Defective sterol C5-6 desaturation and azole resistance: a new hypothesis for the mode of action of azole antifungals, Biochem. Biophys. Res. Commun., 164: 1170.

Wilkinson, C.F., Hetnarski K. and Yellin, T.O., 1972, Imidazole derivatives- A new class of microsomal enzyme inhibitors, Biochem. Pharmacol., 21: 3187.

Willemsens, G., Cools, W. and Vanden Bossche, H., 1980, Effects of miconazole and ketoconazole on sterol synthesis in a subcellular fraction of yeast and mammalian cells, in: "The host-invader interplay", H. Vanden Bossche, ed., Elsevier/North Holland, Amsterdam.

Willemsens, G.and Vanden Bossche, H., 1985, effects of ketoconazole on the testicular and adrenal cholesterol side-chain cleavage, in: "Cytochrome P-450: biochemistry, biophysics and induction", L. Vereczkey and K. Magyar, eds., Akadémiaia Kiadó, Budapest.

Yoshida, Y., 1988, Cytochrome P450 of fungi: primary target for azole antifungal agents, in: "Current topics in medical mycology", Vol.2, M.R. McGinnis, ed., Springer Verlag, New York.

Yoshida, Y and Aoyama, Y., 1986, Interaction of azole fungicides with yeast cytochrome P-450 which catalyzes lanosterol 14α-demethylase, in: "In vitro and in vivo evaluation of antifungal agents", K. Iwata and H. Vanden Bossche, eds., Elsevier Science Publishers B.V., Amsterdam.

Yoshida, Y. and Aoyama, Y., 1987, Interaction of azole antifungal agents with cytochrome P-450$_{14DM}$ purified from *Saccharomyces cerevisiae* microsomes, Biochem. Pharmacol., 36: 229.

PHARMACOKINETICS OF ANTIFUNGALS

Geert Cauwenbergh and Jos Heykants

Janssen Research Foundation
B2340 Beerse, Belgium

INTRODUCTION

During the last 10 years, the awareness for the pharmacokinetic
properties of antifungal drugs has gradually increased and, as a
consequence, pharmacokinetics have received much more attention during the
different development phases of antimycotic agents. Indeed, 20 years ago
the antifungal armamentarium existed predominantly of topical and
intravenous remedies. For topical agents the general opinion was that the
active ingredient arrived in any case at the site of the infection. For
intravenous drugs this was not always the case, however pharmacokinetics of
IV antifungals in those days focused mainly on the possibilities to avoid
side effects.

Even for the only 2 oral agents available, griseofulvin and 5-
fluorocytosine, pharmacokinetics played a secondary role. The best example
to illustrate the relatively primitive view on kinetics of antifungals in
those days is the fact that the minimal inhibitory concentrations were
consistently linked to plasma levels rather than tissue levels, although
the importance of tissue levels had already been accepted by the
microbiologists working with antibacterial antibiotics.

PHARMACOKINETIC PARAMETERS AND DRUG EFFECTS

With the development of oral ketoconazole, but even more with the
development of itraconazole and fluconazole, the importance of a variety of
pharmacokinetic properties has become a completely accepted fact (Heeres et
al., 1979; Heeres et al.; 1984; Troke, 1987).

Obviously, the intrinsic antifungal activity of a drug is important in
determining its clinical effect; but even antifungals such as fluconazole,
that are relatively weakly active *in vitro*, become interesting and potent
agents as a result of certain pharmacokinetic properties. Therefore,
factors which are important include:

* Absorption, peak levels, half-life, AUC in plasma;
* Tissue distribution and tissue levels;
* Elimination rate from plasma vs elimination rate from tissues;
* Metabolism and metabolites;
* Protein binding;

Mycoses in AIDS Patients, Edited by
H. Vanden Bossche *et al.,* Plenum Press, New York, 1990

* Biological activity of the absorbed drug;
* Interaction with endogenous enzyme systems and xenobiotics.

In view of the importance of these factors, and because of the existence of bioavailability problems in certain specific patient groups including AIDS-patients (Boogaerts et al., 1989; Cauwenbergh, 1988a), intensive pharmaceutical research has started to focus on optimization of the formulations in an attempt to obtain the maximum therapeutic benefit from an antifungal without increasing its risk for toxic effects. The liposomal formulation of amphotericin B is an interesting example of such an exercise (Lopez-Berestein et al., 1985). Before going in depth into this discussion, it is useful to elaborate somewhat further on the different kinetic aspects listed above.

1. Absorption, Peak Levels, Half-life and AUC's in Plasma

These classic parameters play a major role in explaining the potent clinical antifungal effect of drugs that are *in vitro* relatively weakly active or that of drugs that achieve relatively low plasma levels but are extremely potent *in vitro*. An illustration for this is given in Table 1.

Table 1. Pharmacokinetic parameters of ketoconazole, itraconazole and fluconazole

Parameter	Ketoconazole 200 mg	Itraconazole 100 mg	Fluconazole 50 mg
C_{max} (μg/ml)	4.2	0.13	1.4
MIC_{90} (μg/ml)	2	0.1	5
T_{max} (h)	1.7	4	1
$AUC_{0-\infty}$ (μg.h/ml)	14.7	1.9	37.5
$t_{1/2}$ (h)	7.5	17	24.4

The total bioavailability of ketoconazole, expressed in the AUC-value is 2.5 times lower than that of fluconazole given at a 4 times lower doses. Theoretical extrapolation would mean that at equal doses, this AUC would be 10 times higher with fluconazole compared to ketoconazole. This situation is even more extreme for itraconazole. Indeed, following the same approach, we reach a 40-fold higher AUC for fluconazole, compared to itraconazole at equal doses.

These facts can now be linked to the above mentioned MIC_{90} values. Fluconazole's AUC may be 10 times higher than that of ketoconazole, but its MIC_{90} is 2 times higher as well, giving a factor 5 in favour of fluconazole. Itraconazole's AUC is 40 times lower than that of fluconazole, but its MIC_{90} is 50 times lower giving a slight superiority (factor of 0.8) to itraconazole (Table 2).

Table 2. Theoretical correlation between kinetics and activity of oral
         antifungals (supposed linear kinetics) against the most common
         fungal pathogens (yeasts and dermatophytes)

Parameter	Ketoconazole 200 mg	Itraconazole 200 mg	Fluconazole 200 mg
AUC rel	3.9	1	39.5
MIC90 rel	2.0	1	50
Relative efficacy	3.9/20 = 0.2	1/1 = 1	39.5/50 = 0.8

Lower *in vitro* activity may be compensated by some kinetic features.

2. Tissue Distribution and Tissue Levels

This nice theoretical approach would be acceptable if fungal
infections were to occur in plasma. However, it is common knowledge that
fungal infections predominantly involve tissues; and although fungal
elements are often found in urine, blood, cerebrospinal and other body
fluids, they rarely cause life-threatening situations in those fluids.
These fluids should in the first place be considered as means of
transportation (dissemination) for fungal elements from one organ to the
other or as a source for reinfection of tissues once those tissues have
been healed. The real life-threatening situations with fungal infections
occur when fungi invade organs and inhibit or destroy their normal
function. For this reason it is of primary importance to learn more about
the tissue distribution and tissue levels of an antifungal.

Extensive work on these aspects has been performed with itraconazole,
and it has become clear that in contrast to the situation with
ketoconazole, the levels of itraconazole obtained in the tissues are
consistently higher than the corresponding plasma levels (Table 3) (Van
Cauteren et al., 1987). In addition to this direct phenomenon, it has also
been demonstrated that itraconazole persists in the skin at therapeutic
levels up to 4 weeks after stopping therapy (Fig. 1) (Cauwenbergh et al.,
1988). After a single day treatment of 400 mg (7.5 mg/kg), therapeutic
itraconazole levels persist for 3 days longer in the vaginal epithelium
(Fig. 2) (Van Der Pas and Van Cutsem, 1989). Although the drug cannot be
detected in the cerebrospinal fluid, the actual levels in the brain and in
the meninges are up to 2 times higher than the corresponding plasma levels.
The true therapeutic benefit of a drug will therefore not depend only on
the plasma levels that can be obtained, but also, and maybe more
importantly, on the drug levels that can be obtained in the tissues
involved in the infection (Heykants et al., 1987).

Table 3. Itraconazole plasma and tissue levels after 4 weeks administration of 10 mg/kg/day to rats

	Itraconazole concentrations in µg/ml or g wet tissue								
Time after last dose	Plasma	Brain	Liver	Lung	Kidney	Pancreas	Skin	Fat	Muscle
4 hours	0.16	0.30	4.95	0.78	0.84	4.0	0.37	0.4	0.11
24 hours	0.05	0.11	1.71	0.25	0.47	0.98	0.40	0.15	0.08

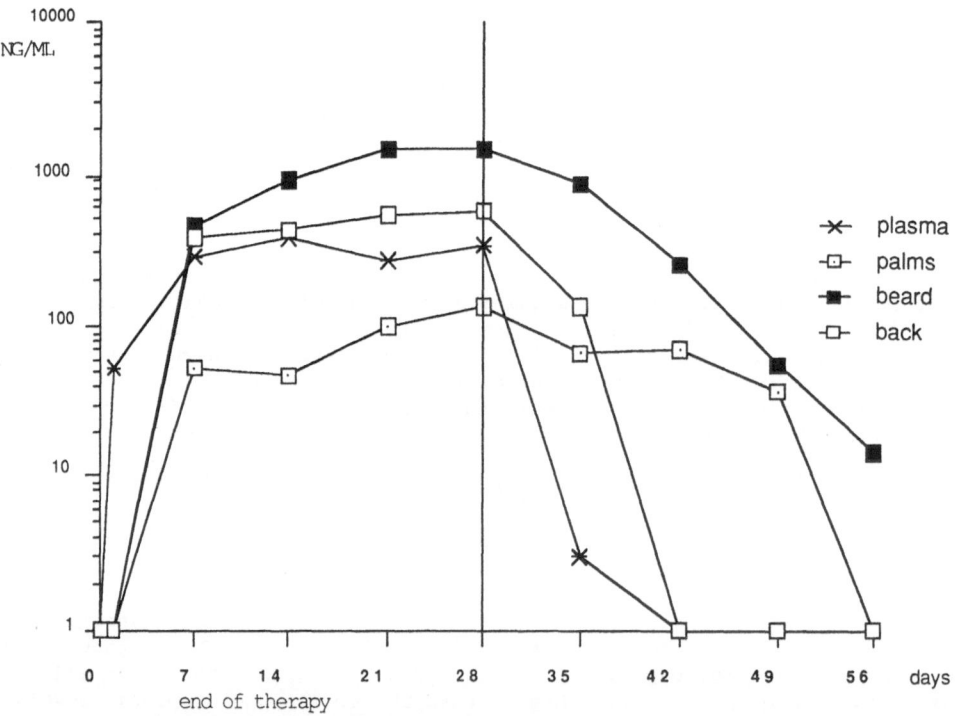

Fig. 1. Mean itraconazole plasma levels and tissue levels after 4 weeks' administration of 100 mg daily.

248

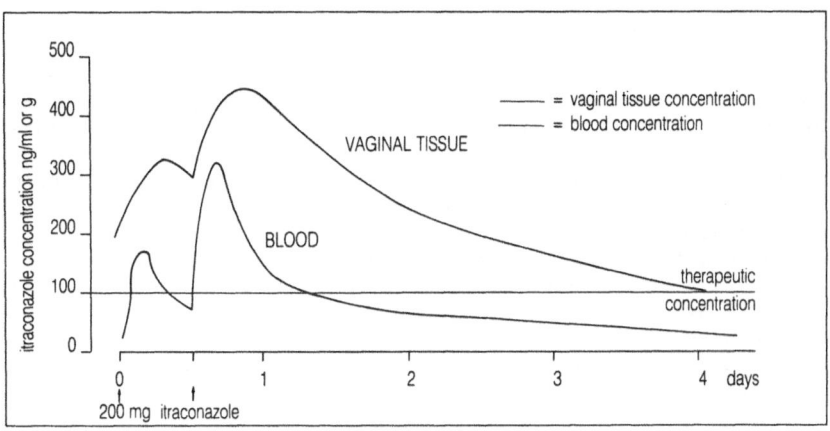

Fig. 2: Uptake and elimination kinetics of itraconazole in vaginal tissue after single-day therapy (400 mg itraconazole total dose, divided over two intakes 12 hours apart). Schematic representations of blood and vaginal tissue levels as a function of time.

3. Elimination Rate from Plasma versus Elimination Rate from Tissues

For most of the internal organs the elimination from the plasma and from the tissues will have a parallel evolution. This situation could be completely different when talking about more "external" organs such as the skin and appendages, oral and vaginal mucosa.

Studies in the past with griseofulvin have indicated that the elimination of this drug from the skin runs in parallel with the elimination from plasma (Wallace et al., 1977). With ketoconazole, reports of a certain degree of drug retention in the skin have appeared, but in vaginal epithelium the elimination of ketoconazole runs parallel to that from plasma (Artis, 1985; Merkus and Bisschop, 1984). Extensive investigations with itraconazole have shown that both in the skin and in the vaginal epithelium the drug persists at therapeutic concentrations for days (vagina) and weeks (skin) after stopping therapy (Cauwenbergh, 1988b). That this phenomenon is not just a pharmacokinetic gimmick without clinical relevance is illustrated by the fact that patients with superficial skin infections can discontinue treatment with itraconazole after 2 weeks, at a moment when clinical lesions are still visible. When these patients are seen for follow-up 2 to 3 weeks after stopping therapy, all lesions have disappeared (Degreef et al., 1987). A similar approach with other oral or topical antifungals would lead to a significant percentage of relapses.

When discussing elimination, something should be said about excretion. Amphotericin B and 5-fluorocytosine can easily be detected in their bioactive form in urine, and as a consequence, they are kinetically suitable for treatment of urinary infections (Borgers et al., 1986). In the azole family of systemic drugs, almost all compounds are 100% metabolized and very little substance is excreted in the urine in a bioactive form (Heykants et al., 1982). The big exception is fluconazole which has a high proportion of unchanged drug being excreted in the urine, suggesting that this molecule should be well active in treatment of mycoses of the urinary tract (Cauwenbergh, 1989).

4. Metabolism and Metabolites

Metabolic clearance is obviously a determining factor for the presence of antifungal drug levels. It is this velocity which makes miconazole inadequate for systemic use, and which makes it necessary to administer this drug in intravenous infusions at least 3 times daily. It is this velocity, or more precisely, an increased metabolic clearance, which made the initially very promising oral clotrimazole useless for systemic administration (Jones, 1982).

Usually metabolites of antimycotics have very little if any antifungal effect. However it is conceivable that some metabolites could still have some antifungal effect (Heykants et al., 1989). Since such metabolites would be more water soluble, it has to be expected that antifungal effects may be observed in organs or body fluids where the parent compound is not necesarilly very active.

5. Protein Binding

Very often plasma protein binding is referred to as inactivation of a molecule's pharmacologic activity (Schaefer-Korting et al., 1989). It is believed that only the free drug concentration is the active one. However, for drugs very extensively bound to plasma proteins, e.g. itraconazole with a plasma protein binding of 99.8%, it is hard to imagine that only the free drug concentration (for itraconazole less than 1 ng/ml) is responsible for the antifungal activity (Van Cutsem et al., 1987). Indeed many of the drugs currently available are bound to plasma proteins, and nobody doubts their therapeutic effect (Brammer and Tarbit, 1987). Also ketoconazole, the reference oral antifungal has a more than 99% plasma protein binding (Michiels, 1982). Therefore plasma protein binding, provided it is reversible, should be regarded as a depot and buffer system for drugs rather than inactivation.

It is even possible to go one step further. Numerous drugs require protein binding for optimal activity. Azole molecules bind to cyt P-450 and two aspects are important in this binding (Vanden Bossche et al., 1986):
* The link between the nitrogen atom in the azole structure and the iron atom in the protoporphyrin moiety of the cyt P-450.
* The link between the rest (tail) of the azole molecule and the apoprotein of the cyt P 450.

In this case it is the binding to the apoprotein which will, in part, determine the activity and the selectivity of the molecule. A low protein binding capacity for a drug may therefore suggest that much higher concentrations of the drug are needed to obtain a therapeutic effect. A high protein binding may be part of the explanation for high *in vitro* activity.

6. Biological Activity of the Absorbed Drug

In some way this aspect relates to the previous point. Indeed, a drug may well be absorbed, but it may not be available in the body in a biologically active form. The reason for this may be an irreversible binding to plasma proteins or a selective accumulation in an organ which is not relevant to the infection. Certain fungal cells can be found intracellularly. If the antifungal does not penetrate in the human cell, it will not have the activity which could be expected from drug level measurements.

The lesson to be learnt from this is that certainly in the early pharmacokinetic studies of an antifungal, bioassay and chromatographic detection methods should be used in combination in order to minimize the risk of unpleasant surprises.

7. Interaction with Endogenous Enzyme Systems and Xenobiotics

This point has been mentioned already briefly when referring to the metabolism of systemic clotrimazole. Certain antifungals have the potential to induce or inhibit endogenous enzyme systems. This is of major importance in the liver since induction could lead to accelerated metabolism of the molecule itself (cf clotrimazole) or to accelerated metabolism of other drugs (cf induction of the ketoconazole metabolism by rifampicin) (Engelhard et al., 1984). Inhibition of metabolism could lead to accumulation of toxic substances leading to increased toxicity (Sheppard et al., 1986).

The potential to induce or inhibit liver enzyme systems can also be regarded as a predictor for drug induced liver toxicity (Cauwenbergh and Van Cutsem, 1988). In this respect, the animal studies usually done to evaluate enzyme induction or inhibition may be valuable tools in predicting certain types of drug induced toxicity.

Enzyme induction and inhibition studies have thus far mainly focussed on the liver. The discovery of the skin as another major organ for metabolism of endobiotics and xenobiotics and the role of cyt P 450 isozymes in this respect opens challenging opportunities for future research into the effects of certain azoles on this metabolic activity in the skin (Janssen et al., 1989). The first studies have already indicated that interesting measurable biochemical effects can be observed.

In the interaction of azoles with the metabolism of xenobiotics, competition can exist for binding sites. Such a competition can also result in an increase in plasma levels of other drugs. The interaction between ketoconazole and cyclosporin A has to be understood in this way (Nolan et al., 1989). Because of a competition in the metabolism of both compounds, cyclosporin levels will gradually increase, which may lead to cyclosporin induced renal toxicity. A careful study of interactions is certainly required when dealing with a patient group such as AIDS-patients who often take multiple drugs at the same time.

8. Formulation Development

A long with an increased awareness of the role of the different pharmacokinetic parameters in determining the antifungal potency of drugs came demands, not only for new substances, but also for improved pharmaceutical formulations of existing drugs.

The liposomal formulation of amphotericin B belongs to this category. It is clear from the presented evidence thus far that the toxicity profile of amphotericin B has been improved dramatically with this new formulation. The evidence to demonstrate equivalent efficacy with the old formulation is however still very limited.

Finally one has to be aware that certain patient populations may have bioavailability problems with certain drugs. Lake-Bakaar has emphasized the point that AIDS-patients occasionally do not absorb ketoconazole adequately (Lake-Bakaar et al., 1988). For treatment of oral candidosis, this may cause problems, and therefore a suspension of ketoconazole has been developed, produces a local as well as a systemic effect when the drug is administered. A similar approach is being followed for itraconazole. The development of such a solution has further improved the clinical results obtained with this drug in AIDS-patients with oropharyngeal candidosis (Desmet et al., 1989).

The special formulation (pellets) of itraconazole has been another succesful attempt to standardize and optimize the absorption obtained in the different patient populations. Other formulation work aiming at improved kinetic profiles includes itraconazole in a cyclodextrin formulation for intravenous use.

CONCLUSION

In conclusion it can be said that the relevance of the pharmacokinetic profile of an antifungal has become evident to those who try to understand why antifungals work the way they do. However, the increased number of inquiries about the pharmacokinetic behaviour of systemically used antifungals has not made the development of an antifungal drug an easier task. Even with the improved equipment and with better methodology, limitations will continue to exist when evaluating the pharmacokinetic profile of antifungal agents. These limitations will be of practical, technical or ethical nature.

ACKNOWLEDGEMENTS

The authors are grateful to Mrs. Hilde Dergent for typing and retyping the manuscripts.

REFERENCES

Artis, W.M., 1985, Final pathway for delivery of oral antifungals to keratinized cornified skin, in: "Oral Therapy in Dermatomycoses: a Step Forward". Proceedings of a Symposium held in Frankfurt, W. Meinhof, ed., The Medicine Publishing Foundation, Oxford.

Boogaerts, M.A., Verhoef, G.E., Zachee, P., Demuynck, H., Verbist, L., and De Beule, K, 1989, Antifungal prophylaxis with itraconazole in prolonged neutropenia: correlation with plasma levels, Mycoses 32 (Suppl. 1): 103.

Borgers, M., Vanden Bossche, H., and Cauwenbergh, G., 1986, The pharmacology of agents used in the treatment of pulmonary mycoses, Clin. Chest. Med. 7(3):439.

Brammer, K.W., and Tarbit, M.H., 1987, A review of the pharmacokinetics of fluconazole in laboratory animals and man, in: "Recent Trends in the Discovery, Development and Evaluation of Antifungal Agents", R.A. Fromtling, ed., Prous Science Publishers, S.A.

Cauwenbergh, G., 1988a, Prophylaxis of aspergillosis in immunocompromised patients, in: "Aspergillus and Aspergillosis" H. Vanden Bossche, D.W.R. Mackenzie and G. Cauwenbergh, eds., Plenum Press, New York.

Cauwenbergh, G., 1988b, Skin Mycoses: Effects of Ketoconazole and Itraconazole in Relation to Drug Distribution in the Skin, Thesis, Catholic University Leuven.

Cauwenbergh, G., 1989, Current drugs for the treatment of deep fungal infections, Dermatologica, in press.

Cauwenbergh, G., Degreef, H., Heykants, J., Woestenborghs, R., Van Rooy, P., and Haeverans, K., 1988, Pharmacokinetic profile of orally administered itraconazole in human skin, J. Am. Acad. Dermatol. 18(2)(Part 5):263.

Cauwenbergh, G., and Van Cutsem, J., 1988, Role of animal and human pharmacology in antifungal drug design, Ann. NY Acad. Sci. 544:264.

Degreef, H., Marien, K., Deveylder, H., Duprez, K., Borghys, A., and Verhoeve, L., 1987, Itraconazole in the treatment of dermatophytoses: a comparison of two daily dosages, Rev. Infect. Dis. 9(1):S104.

Desmet, P., Kayembe, K., Stoffels, P., Mulumba, M.P., De Beule, K., and Cauwenbergh, G., 1989, Treatment of oral candidosis in AIDS patients with itraconazole oral solution, 3rd Symposium on Topics in Mycology - Mycoses in AIDS patients, Paris, France.

Engelhard, D., Stutman, H.R., and Marks, M.I., 1984, Interaction of ketoconazole with rifampin and isoniazid, New Engl. J. Med. 311(26):1681.

Heeres, J., Backx, L.J.J., Mostmans, J.H., and Van Cutsem, J., 1979, Antimycotic imidazoles. Part 4. Synthesis and antifungal activity of ketoconazole, a new potent orally active broad-spectrum antifungal agent, J. Med. Chem. 22:1003.

Heeres, J., Backx, L.J.J., and Van Cutsem, J., 1984, Antimycotic azoles. 7. Synthesis and antifungal properties of a series of novel triazol-3-ones, J. Med. Chem. 27(7):894.

Heykants, J., Michiels, M., Meuldermans, W., Monbaliu, J., Lavrijsen, K., Van Peer, A., Levron, J.C., Woestenborghs, R., and Cauwenbergh, G., 1987, The pharmacokinetics of itraconazole in animals and man: an overview, in: "Recent Trends in the Discovery, Development and Evaluation of Antifungal Agents, R.A. Fromtling, ed., J.R. Prous Science Publishers, S.A.

Heykants, J., Van Peer, A., Van De Velde, V., Van Rooy, P., Meuldermans, W., Lavrijsen, K., Woestenborghs, R., Van Cutsem, J., and Cauwenbergh, G., 1989, The clinical pharmacokinetics of itraconazole: an overview, Mycoses (supplement), in press.

Heykants, J., Woestenborghs, R., Bisschops, M., and Merkus, J., 1982, Distribution of oral ketoconazole to vaginal tissue, Eur. J. Clin. Pharmacol. 23:331.

Janssen, P.A.J., Vanden Bossche, H.F.A., Van Wauwe, J.P., Cauwenbergh, G.F.M.J., and Degreef, H.J., 1989, The role of cytochrome P-450 in dermatology, Int. J. Dermatol. 28(8):493.

Jones, H.E., 1982, Current therapies for mycotic diseases, in: "Ketoconazole in the Management of Fungal Diseases", H.B. Levine, ed., Adis Press, Hong Kong.

Lake-Bakaar, G., Tom, W., Lake-Bakaar, D., Gupta, N., Beidas, S., Elsakr, M., and Straus, E., 1988, Gastropathy and ketoconazole malabsorption in the acquired immunodeficiency syndrome (AIDS), Ann. Intern. Med. 109:471.

Lopez-Berestein, G., Fainstein, V., Hopfer, R., Mehta, K., Sullivan, M.P., Keating, M., Rosenblum, M.G., Mehta, R., Luna, M., Hersch, E.M., Reuben, J., Juliano, R.L., and Bodey, G.P., 1985, Liposomal amphotericin B for the treatment of systemic fungal infections in patients with cancer: a preliminary study, J. Infect. Dis. 151(4):704.

Merkus, J.M.W.M., and Bisschop, M.P.J.M., 1984, Optimum dose of oral ketoconazole in the treatment of vaginal candidosis, in: "Oral Therapy in Vaginal Candidosis" - Proceedings of a Symposium held in Oxford, B.W. Eliot, ed., The Medicine Publishing Foundation, Oxford.

Michiels, M., 1982, Zur Pharmakokinetik von Ketoconazol, in: "Chemotherapie von Oberflaechen-, Organ- und Systemmykosen", H.P.R. Seeliger and H. Hauck, eds., Perimed Fachbuch-Verlagsgesellschaft MBH, Erlangen.

Nolan, P.E., Butman, S.M., Wild, J.C., Finley, P.R., and Copeland, J.G., 1989, Potentially favorable pharmacokinetic interaction between cyclosporine and ketoconazole in cardiac transplant recipients, Pharmacotherapy 9(3):192.

Schaefer-Korting, M., Korting, H.C., and Lukacs, A., 1989, The influence of plasma protein binding on the penetration of oral antifungal agents into skin blister fluid, Eur. J. Clin. Pharmacol. 36:218.

Sheppard, J.H., Canafax, D.M., Simmons, R.L., and Nararian, J.S., 1986, Cyclosporine-ketoconazole: a potentially dangerous drug-drug interaction, Clin. Pharm. 5:468.

Troke, P.F., 1987, Efficacy of fluconazole in animal models of superficial and opportunistic systemic fungal infection, in: "Recent Trends in the Discovery, Development and Evaluation of Antifungal Agents", R.A. Fromtling, ed., Prous Science Publishers, S.A.

Van Cauteren, H., Heykants, J., De Coster, R., and Cauwenbergh, G., 1987, Itraconazole: pharmacologic studies in animals and humans, <u>Rev. Infect. Dis.</u> 9(Suppl 1):S43.

Van Cutsem, J., Van Gerven, F., and Janssen, P.A.J., 1987, Activity of orally, topically and parenterally administered itraconazole in the treatment of superficial and deep mycoses: animal models, <u>Rev. Infect. Dis.</u> 9(1):15.

Vanden Bossche, H., Bellens, D., Cools, W., Gorrens, J., Marichal, P., Verhoeven, H., Willemsens, G., De Coster, R., Beerens, D., Haelterman, C., Coene, M.C., Lauwers, W., and Lejeune, L., 1986, Cytochrome P-450: Target for itraconazole, <u>Drug Dev. Res.</u> 8:287.

Van Der Pas, H., and Van Cutsem, J., 1990, Itraconazole in vaginal candidosis: relation to antifungal levels in the vagina. Satellite Symposium "Therapy of Candidosis: from Screening to Clinical Evaluation" to FEMS-Symposium "Candida and Candidamycosis", Alanya, Turkey.

Wallace, S.M., Shah, V.P., Epstein, W.L., Greenberg, J., and Riegelman, S., 1977, Topically applied antifungal agents. Percutaeous penetration and prophylactic activity against *Trichophyton mentagrophytes* infection, <u>Arch. Dermatol.</u> 113:1539.

SKIN CANDIDOSIS IN AIDS PATIENTS. EFFECTS OF KETOCONAZOLE

AND ITRACONAZOLE. FOCUS ON TISSUE LEVELS

H.C. Korting

Department of Dermatology, Ludwig-Maximilians-Universität,
München, FRG

INTRODUCTION

Mycoses in HIV-infected patients most of all involve the skin and the neighbouring mucous membranes. This in particular applies to the oral cavity. Oral candidosis is considered as the most frequent opportunistic infection in AIDS patients (Holmberg and Meyer, 1986). In HIV-infected patients from Munich, *Candida albicans* was found in the oral cavity in 27.7% at stage I (sero-positive latency), in 76.5% at stage II (lymphadenopathy syndrome/AIDS-related complex (PGL-ARC)), and in 87.5% at stage III (full-blown AIDS) (Korting, 1989). At the same time clinical signs for oral candidosis were found in 0, 35.3 and 54.2 %, respectively. In fact, increasing probability of manifest oral candidosis is linked to decreasing immunocompetence as to be judged by the T4/T8 ratio (Korting et al., 1988). Hence the question how to treat oral candidosis in HIV-infected patients is one of the most frequent questions to be answered by the clinician in charge of pertinent patients.

As in general in antimicrobial chemotherapy therapeutic decisions have to be based on a detailed knowledge of what has been called the magic triangle (Fig. 1).

As to be seen in Fig. 1 the term "tissue levels at the infection site" has to replace the more conventional term "blood levels". Although true in general this in particular applies to the problem of infections of skin and mucous membranes. With systemic antimycotic treatment to be discussed for

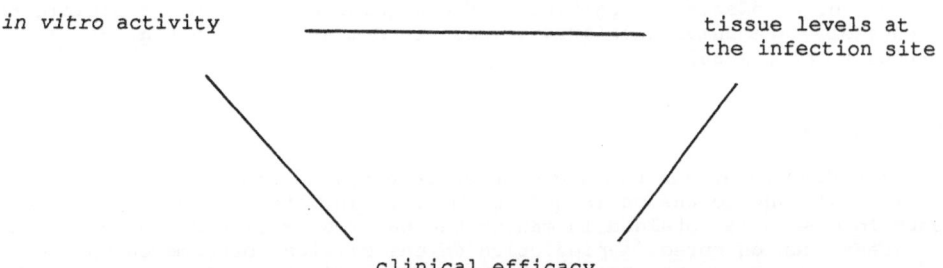

Fig. 1. The magic triangle of antimicrobial therapy.

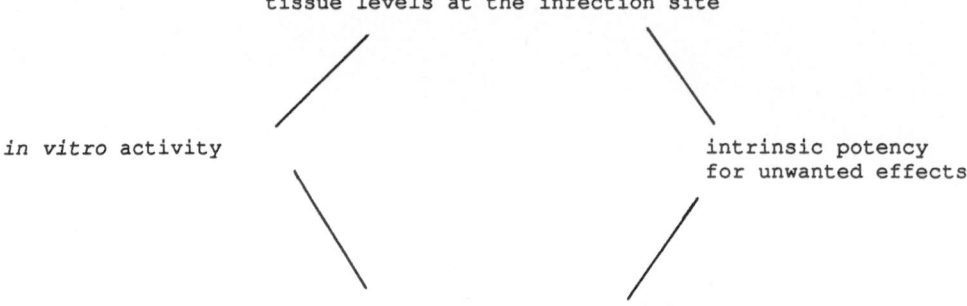

Fig. 2.   The magic pentangle of antimycotic chemotherapy for mucocutaneous candidosis

these diseased states one has also to take into account that an intrinsic potency for unwanted effects might compromise clinical safety and thus also limit clinical efficacy. For clinical mycologists this is conventional wisdom with respect to amphotericin B when given systemically. Safety aspects, however, also are of importance when it comes to azole treatment for oral candidosis (Cauwenbergh, 1989). It seems justified to speak of a magic pentangle instead of just a magic triangle of antimycotic chemotherapy for mucocutaneous candidosis (Fig. 2).

IN VITRO ACTIVITY

The relevance of the *in vitro* evaluation of antimicrobial susceptibility of fungi has been the subject of controversial debate for decades. Yet most recently at least with respect to yeasts there is remarkable evidence that the clinical outcome of antimycotic treatment in fact is linked to *in vitro* activity (for review compare Korting and Georgii, 1988). This especially applies to the modified microdilution test as devised by Johnson et al. (1984) modifying an earlier proposal by Galgiani and Stevens (1976). The method being independent of the inoculum size is easy to be performed and gives reproducible data even with azole antimycotics which pose the most severe problems with other test procedures. Investigating 62 consecutive *Candida albicans*-isolates from the oral cavity of HIV-infected patients from Munich it became evident that most strains are fully susceptible to all antimycotics to be presently taken into account when it comes to the treatment of oral or other muco-cutaneous candidoses. While IC30 values never exceed 8 μg/ml with the polyenes amphotericln B and nystatin this, however, is not the case with flucytosine and both ketoconazole and itraconazole. With all these drugs some of the strains tested have to be considered as clinically resistant the geometric mean inhibitory concentration increasing with the stage of HIV-infection except for nystatin. Strains isolated from patients in the later stages of disease in particular are significantly less susceptible to itraconazole. The cumulative IC30 values are demonstrated in Fig. 3 (Korting et al., 1988).

TISSUE LEVELS

If a *Candida* strain causing disease in a given clinical case is not totally resistant to the envisaged antimycotic *in vitro* it depends on the tissue levels to be obtained in man at the particular site of infection if the disease can be cured. A prediction of the clinical outcome on this base requires an access to the biophase. This is possible applying:

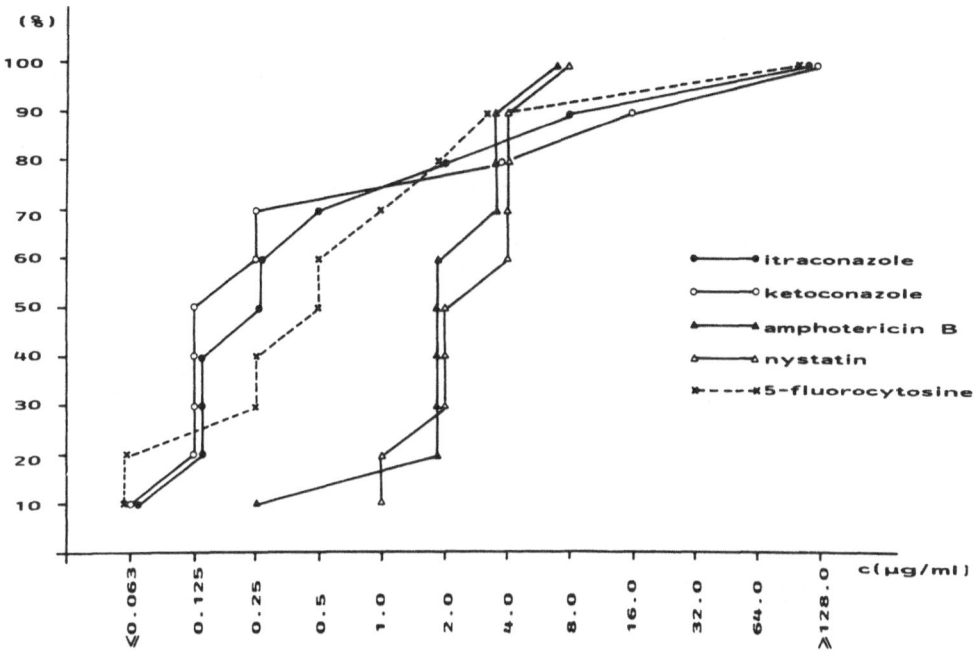

Figure 3. Cumulative IC$_{30}$ values of ketoconazole, itraconazole, amphotericin B, nystatin, flucytosine with *Candida albicans* isolates from the oral cavities of HIV-infected patients from Munich (c, concentration of drug).

1. mathematical models allowing the calculation of the drug amount in a peripheral compartment or
2. experimental methods allowing the collection of tissue fluid (for review compare Schäfer-Korting and Korting, 1989).

Experimental methods allowing mucocutaneous tissue level determinations comprise:

1. implantation of tissue level cages and fibrin clots,
2. skin window technique,
3. implantation of nylon/cotton threads,
4. cantharides blistering,
5. suction blistering.

Skin blistering in man can be based on the method first described by Kiistala (1968). Cantharides blistering can be performed according to the procedure proposed by Simon (1976). The major characteristics of both methods are listed in Table 1.

Both suction and cantharides blistering have been performed in parallel in human volunteers taking ketoconazole once or repeatedly at a dose of 200 mg perorally (Schäfer-Korting et al., 1984; Korting et al., 1989). Suction blister and cantharides blister fluid-as well as serum level profiles after both a single and repeated peroral application of ketoconazole 200 mg are shown in Fig. 4 and Fig. 5.

Table 2 gives the key pharmacokinetic parameters, i.e. maximum concentration (C$_{max}$) and area under the curve (AUC).

Table 1. Major characteristics of suction blisters (SB) and cantharides
blisters (CB)

Characteristic	SB	CB
Induction	mechanical	chemical
Split	subepidermal	intraepidermal
Nature	non-inflammatory	inflammatory
Albumin concentration	30 - 42 %	70 - 80 %

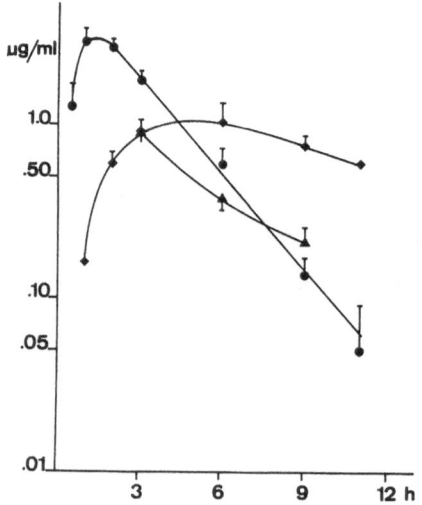

Figure 4. Serum (●); suction blister (▲) and cantharides blister (◆)
fluid level profiles after a single peroral application of
ketoconazole 200 mg (from Schäfer-Korting et al., 1984).

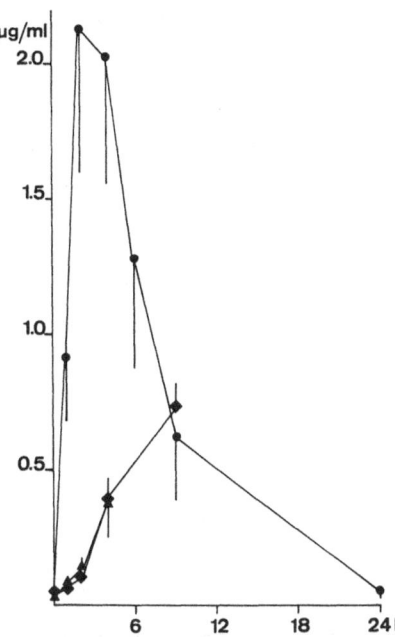

Figure 5.   Serum (●), suction blister (▲) and cantharides blister (◆)
             fluid level profiles after a repeated peroral application of
             ketoconazole 200 mg once daily (from Korting et al., 1989).

Table 2.   Key pharmacokinetic parameters of ketoconazole after the single
            and repeated peroral application of 200 mg.

Material*	Parameter		
	$c_{max}[\mu g.ml^{-1}]$	$c_{min}[\mu g.ml^{-1}]$	$AUC_{0-9}[\mu g.ml^{-1}.h]$
SBF	0.914/0.376	-/0.028	4.09/4.69
CBF	1.047/0.74	-/0.058	6.84/9.44
Serum	3.24/2.32	-/0.052	11.19/17.36

*SBF: suction blister fluid; CBF: cantharides blister fluid

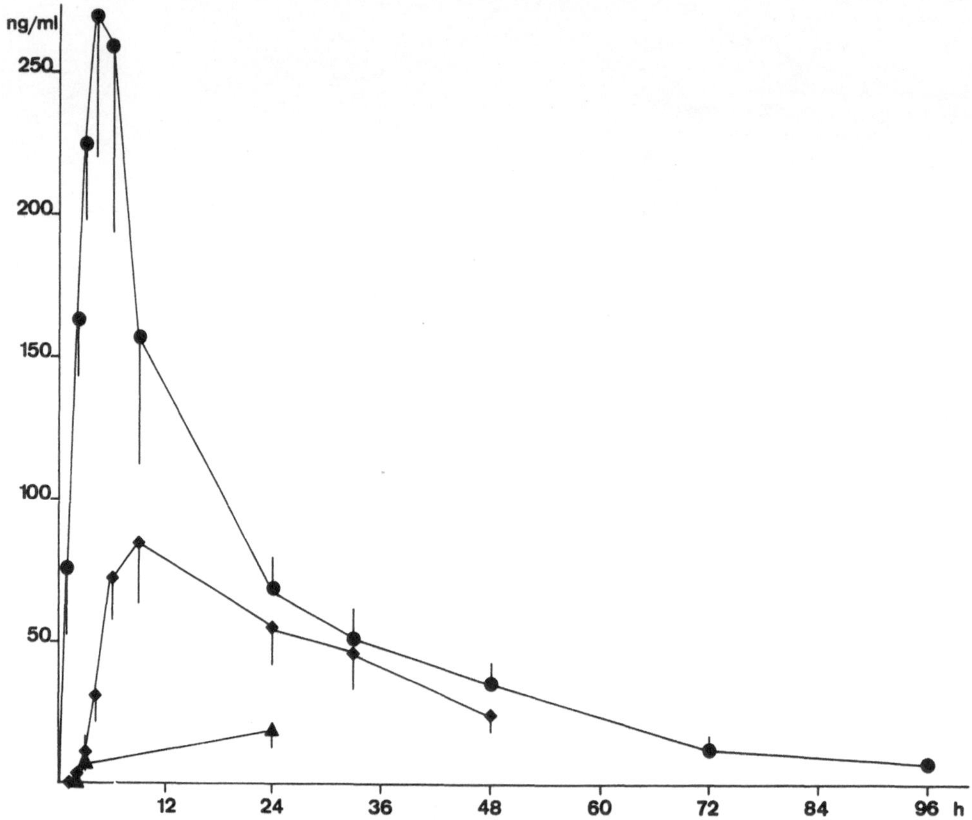

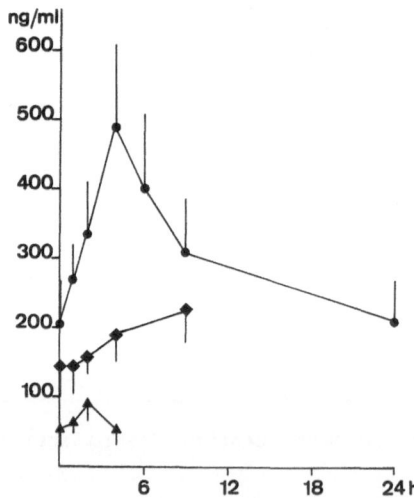

Figure 6. Serum (●), suction blister (▲) and cantharides blister (♦) fluid level profiles after the single peroral application of itraconazole 200 mg (from Schäfer-Korting et al., 1990).

Figure 7. Serum (●), suction blister (▲) and cantharides blister (♦) fluid level profiles after repeated peroral application of itraconazole 100 mg once daily (from Schäfer-Korting et al., 1990).

Corresponding data for itraconazole 200 mg administered once and itraconazole 100 mg administered repeatedly once a day are shown in Figures 6 and 7.

Table 3 shows the key pharmacokinetic parameters. Azole levels found in the skin using the blistering techniques can be discussed before the background of those obtained from the analysis of stratum corneum material. According to Faergemann and Maibach (1983) 0.4 to 1.3 µg ketoconazole per gram are encountered after the single peroral application of 200 mg. After the repeated peroral administration of 200 mg, 3.1 µg per gram are found according to Haneke (1987). While analysis of stratum corneum material does not allow to differentiate between drug bound to protein and unbound drug this is the case with blister fluid. As it is a common belief in clinical pharmacology that only the fraction of drug not bound to protein is the active one it deserves interest to see that the unbound fraction of ketoconazole and itraconazole is comparatively small. In SBF it amounts to 2.3 and 0.64% resp., in CBF to 1.2 and 0.23 and in serum to 1.0 and 0.2% resp. (Schäfer-Korting, 1989).

Table 3. Key pharmacokinetic parameters of itraconazole after the single peroral application of 200 mg and after the repeated peroral application of 100 mg.

Material	Parameter		
	$c_{max}[µg.ml^{-1}]$	$c_{min}[µg.ml^{-1}]$	$AUC_{0-00/024}[µg.ml^{-1}.h]$
SBF	0.019/0.094	—/0.15	—/1.30
CBF	0.093/0.23	—/0.052	3.50/4.39
Serum	0.32/0.54	—/0.21	5.76/7.21

IN VITRO SIMULATION OF TISSUE LEVELS

As early as in 1978 Grasso et al. have described a "new *in vitro* model to study the effect of antibiotic concentration and rate of elimination on antibacterial activity". This method can be applied to the simulation of level profiles of antimycotics in the presence of fungi. Although Grasso et al. themselves only concentrated on blood levels the model is appropriate for the simulation of the drug-micro-organism relationship in other tissues (Korting et al., 1986). Simulating cantharides blister fluid levels after the single peroral application of ketoconazole 200 mg in the presence of *Candida albicans* does not lead to a sharp decline of the number of colony forming units (CFU) as found earlier with ceftixozime levels after the single intramuscular application of 1 g in the presence of *Neisseria gonorrhoeae*. In comparison to the control, ketoconazole just reduces yeast growth. This effect is somewhat more marked if a medium is used promoting pseudomycelial growth. As to be expected simulating free drug levels is still less effective than simulating total drug levels. Figure 8 shows the growth of *Candida albicans* both in the presence and in the absence of (total) ketoconazole CBF levels. This supports the hypothesis to be based on the comparison of ketoconazole *in vitro* activity and skin tissue levels according to which ketoconazole (as well as probably itraconazole) might only prove successful in a significant portion but not in all patients suffering from mucocutaneous candidoses, and in particular oral candidosis. Treatment of gonorrhoea with third-generation cephalosporins such as ceftixozime effective clinically in 100 % of cases in fact is based on skin

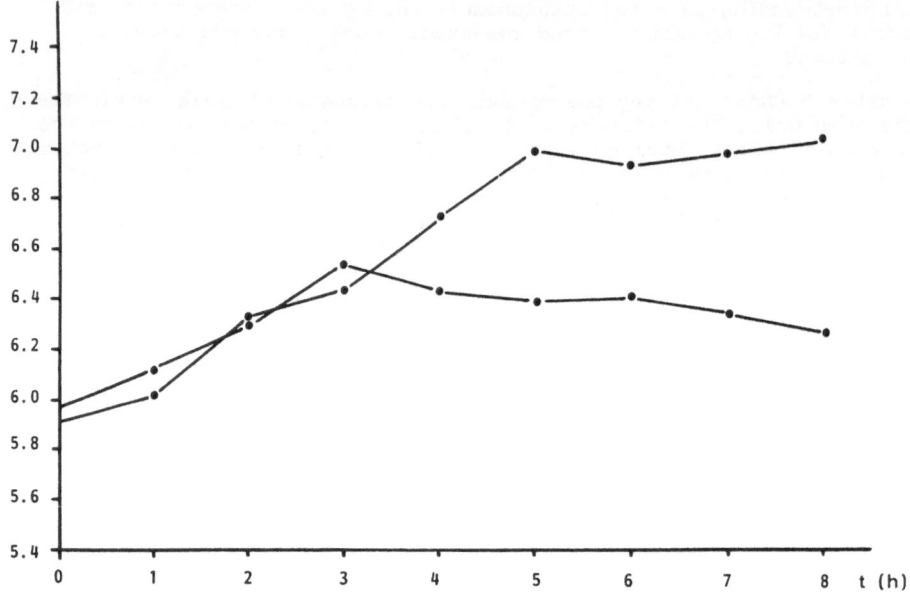

Figure 8. Colony-forming units of *Candida albicans* (pseudomycelial form) in the presence of ketoconazole CBF levels (after a single peroral dose of 200 mg) simulated *in vitro* (upper curve represents control) (from Korting et al., 1986).

tissue levels about 103 times as high as the minimum inhibitory concentration (Korting, 1987).

CLINICAL EFFICACY

In fact, ketoconazole 200 mg once daily for 28 days has been able to cure 75% of ARC- and AIDS-patients in clinical and 69% in mycological terms (De Wit et al., 1989). In an own trial based on a very limited number of patients, corresponding cure rates in the short term after ketoconazole 200 mg a day for 7 to 10 days amounted to 87 and 53% resp. As to be expected results in the medium term were still less satisfactory: 56 and 9% were found. The clinical outcome was in so far linked to the *in vitro* susceptibility of the causative organism as *Candida albicans* strains with IC_{30} values exceeding 256 µg/ml were not eradicated which corresponded to the persistence of typical mucosal lesions. The study based on quantitative assessment of the effect of ketoconazole on colony density in mouthwash fluid moreover backed the hypothesis that patients on zidovudine might profit more from ketoconazole treatment as to be judged from short- and medium-term results (Korting et al., in press). This in a way reflects the correlation between the manifestation of oral candidosis and a decrease of the T4/T8 ratio.

Improvement of treatment of oral candidosis in the short run might be obtained putting the patients on regular intermittent therapy as for example ketoconazole 200 mg a day for 7 days every fourth week. In the long run azoles should be investigated in higher daily doses. Facing its safety profile this might be possible with itraconazole. Conventional doses of itraconazole might not alter the situation fundamentally as both *in vitro* activity and blister fluid level data differ not much from those found with ketoconazole.

REFERENCES

Cauwenbergh, G., 1989, Safety aspects of the most common antifungals for oral and gastrointestinal candidosis, Mycoses, 32 (suppl. II), in press.

De Wit, S., Werts, D., Goossens, H., and Clumeck, N., 1989, Comparison of fluconazole and ketoconazole for oropharyngeal candidiasis in AIDS. Lancet, 1:746.

Faergemann, J., and Maibach, H., 1983, Griseofulvin and ketoconazole in dermatology, Semin. Dermatol., 2:262.

Galgiani, J.N., and Stevens, D.A., 1976, Antimicrobial susceptibility testing of yeasts: A turbidimetric technique independent of inoculum size, Antimicrob. Agents Chemother., 10:721.

Grasso, S., Meinardi, G., De Carneri, J., and Tamassia, V., 1978, New in vitro model to study the effect of antibiotic concentration and rate of elimination on antibacterial activity, Antimicrob. Agents Chemother., 13:570.

Haneke, E., 1987, Verweildauer von Ketoconazol in der Haut nach oraler Behandlung.,Hautarzt, 38:93.

Holmberg, K., and Meyer, R.D., 1986, Fungal infections in patients with AIDS and AIDS-related complex, Scand. J. Infect. Dis., 18:179.

Johnson, E.M., Richardson, M.D., and Warnock, D.W., 1984, In vitro resistance to imidazole antifungals in Candida albicans, J. Antimicrob. Chemother., 13:547.

Kiistala, U., 1968, Suction blister device for separation of viable epidermis from dermis, J. Invest. Dermatol., 50:129.

Korting, H.C., 1967, Cephalosporin-Therapie der Gonorrhoe, Karger, Basel, New York.

Korting, H.C., 1989, Clinical spectrum of oral candidosis and its role in HIV-infected patients, Mycoses, 32 (suppl. II), in press.

Korting, H.C., and Georgii, A., 1988, Antimykotikatestung dermato-venerologisch bedeutsamer Hefen: Methoden, Ergebnisse und klinische Relevanz, Hautarzt, 39:343.

Korting, H.C., Blecher, P., Fröschl, M. and BraunFalco, O., in press, Quantitative assessment of the efficacy of peroral ketoconazole for oral candidosis in HIV-infected patients, Eur. J. Clin. Microbiol. Infect. Dis., in press.

Korting,, H.C., Lukacs, A., Schäfer-Korting, M., Heykants, J., and Behrendt, H., 1989, Skin blister fluid levels of ketoconazole during repetitive administration in healthy man, Mycoses, 32:39.

Korting, H.C., Niederauer, H., and Ollert, M., 1986, Bactericidal activity of cefotiam and ceftizoxime against Neisseria gonorrhoeae in an in vitro model simulating plasma and cantharidal blister fluid levels after the single intramuscular application of one gram, Chemotherapy, 32:260.

Korting, H.C., Ollert, M., Georgii, A., and Fröschl, M., 1988, In vitro susceptibilities and biotypes of Candida albicans isolates from the oral cavities of patients infected with human immunodeficiency virus, J. Clin. Microbiol., 26:2626.

Korting, H.C., Schlaf-Maier, U., and Schäfer-Korting, M., 1986, Antimikrobielle Effekte des in vitro simulierten Ketokonazol-Kantharidinblasenflussigkeitsspiegel-Profils bei Candida albicans. Mykosen, 29:297.

Schäfer-Korting, M., 1989, Pharmacokinetics of topical and oral antifungals, Mycoses, 32 (suppl. II), in press.

Schäfer-Korting, M., and Korting, H.C., 1989, Skin blisters and skin windows: An access to total and free drug concentrations in the skin, in: "Models in Dermatology", Vol. 4, H.I. Maibach and N.J. Lowe, eds., Karger, Basel.

Schäfer-Korting, M., Korting, H.C., Dorn, M., and Mutschler, E., 1984, Ketoconazole concentrations in human skin blister fluid and plasma. Int. J. Clin. Pharmacol. Ther. Toxicol., 22:371.

Schäfer-Korting, M., Korting, H.C., Lukacs, A., Heykants, J., and Behrendt, H., 1990, Levels of itraconazole in skin blister fluid after a single oral dose and during repetitive administration, <u>J. Am. Acad. Dermatol.</u>, 22:211.

Simon, C., 1976, Antibiotika-Spiegel in Serum, Speichel, Tranen und Hautblasenflussigkeit, <u>Infection</u>, 4 (suppl. II):91.

THE MODERN REVOLUTION IN ANTIFUNGAL DRUG THERAPY

John R. Graybill

University of Texas Health Science Center in San Antonio and
Audie L. Murphy Memorial Veterans Hospital, 7400 Merton
Minter Blvd., Infectious Diseases Division (111), San
Antonio, Texas 78240

INTRODUCTION

For a number of years the mycoses and their management have languished
in obscure medical tomes, drawing little attention. Although treatment was
much needed for individual patients, the systemic mycoses were generally
very uncommon and not regarded as a stimulus for development of antifungal
agents. Further, the systemic dimorphic mycoses were mostly problems of the
new world, with highly limited range of endemicity. Since most people
infected with these pathogens had either no illness or only transitory
illness, effective therapy was not high on health care demands. The 1960's
saw the advent of amphotericin B, the only member of the polyenes that was
sufficiently "reduced" in toxicity to permit systemic use. Amphotericin B
provided one broad spectrum antifungal and thus fulfilled the need for
"something that worked". Consequently, very little work was done on new
systemic antifungals for some years thereafter. Despite the much larger
numbers of patients with vaginal candidosis and dermatologic fungal
infections, these were considered by many to be trivial medical problems.
Accordingly, the pharmaceutical industry opted to develop "safe" topical
antifungals rather than systemic agents which might cause significant
toxicity during treatment of largely cosmetic problems.

The stable decade of the 1960's was dominated by amphotericin B,
nystatin, griseofulvin, potassium iodide, and tolnaftate. However,
stability ended with the marked increase of intensive antileukemia
chemotherapy and the use of other cytotoxic drugs in the 1970's. In this
decade the oncologists documented a marked increase in candidiasis, such
that by now it causes upto 25% of all deaths of leukopenic patients (Horn
et al., 1985). In one hospital alone, 135 cases of candida caused 123
cases of fungal sepsis in a 3 year period, with death in 59% and candidemia
contributing to death in 75% of cases (Komshian et al., 1989). *Torulopsis
glabrata* (Aisner et al., 1976; Komshian et al., 1989), *Trichophyton
beigelii* (Walsh et al., 1986), *Fusarium* (Blazar et al., 1984), *Aspergillus*
species (Young et al., 1970), *Zygomycetes* (Rinaldi, 1989) and lesser known
fungi have further enlarged the cast of fungal pathogens. These fungi
could cause lingering illness or death within a few weeks or even days. It
quickly became apparent that treating only confirmed diagnoses of
candidosis or aspergillosis was totally infective (Horn et al., 1985;
Rinaldi, 1989). This lesson in futility gave birth to efforts to diagnose
systemic mycoses on increasingly empiric grounds, the ultimate one being
simply continued fever after treatment with antibacterial agents (Pizzo et

Mycoses in AIDS Patients, Edited by
H. Vanden Bossche *et al.,* Plenum Press, New York, 1990

al., 1982; Aisner et al., 1977; Sinclair et al., 1978; Fisher et al., 1981). Increasing empiric use of amphotericin B was effective in lowering mortality, but at the cost of treating many patients unnecessarily with a highly toxic agent. Increasing numbers of patients with systemic mycoses and dissatisfaction with amphotericin B and its only modestly less toxic companion flucytosine were strong motivations for reconsideration of development of new antifungals. Because many of these patients were acutely ill, and had gastrointestinal mucositis from cancer chemotherapy, parenterally administered agents were sought for treatment of established disease. However, as investigators became intrigued by the concept of prophylaxis of these infections, oral agents also became attractive (Fainstein et al., 1987).

A second major event was the discovery of AIDS and its mycologic stepchildren. These were initially considered to be mucosal candidiasis, which occurred in more than half of the patients, and cryptococcosis, which occurred in 6-8% of patients. However, times have changed, and in the United States we are coming to appreciate histoplasmosis and coccidioidomycosis as significant pathogens in as many as 5% of patients in endemic areas (Johnson et al., 1982; Bronniman et al., 1987). Dermatophyte infections and seborrheic keratosis may also be quite severe in patients with AIDS. Nevertheless, the major problem at the outset was cryptococcosis. Initial studies with the "usual" regimen of flucytosine and amphotericin B reported mortality above 70% (Kovacs et al., 1985). More recently, suppressive regimens with amphotericin B delivered through a long dwelling intravenous catheter have reduced fungal mortality to 23% in some series (Mycoses Study Group, 1989).

Therefore, a sharp rise in patients with depressed humoral immune response and leukopenia has been associated with increased disseminated candidosis, aspergillosis, and multiple less common infections. Conversely, the rise in AIDS, with depression of cell mediated immunity, has been associated with a completely different spectrum consisting of mucosal candida, cryptococcosis, histoplasmosis, and coccidioidomycosis. The management of AIDS associated mycoses will be discussed by Just et al.; Dupont et al.; Denning et al. (this book), while the discussion below will concentrate on management of the mycoses in the normal and the leukopenic host.

KETOCONAZOLE

Although it was not the first drug to be tried, ketoconazole was the first broad spectrum antifungal azole which could be given orally (Van Tyle, 1984). Like other antifungal azoles, ketoconazole inhibits synthesis of ergosterol in fungal cell membranes (Vanden Bossche, 1985). Early studies, ranging from coccidioidomycosis to histoplasmosis and blastomycosis, found ketoconazole effective and simple to take. Ketoconazole was so easy to take that Janssen Pharmaceutica quickly appreciated its potential in serious dermatophyte infections and vaginitis. Suddenly there was an alternative for griseofulvin for topical drug therapy.

However, as with every new drug, increased use saw increasingly conservative reports of efficacy, and also increased reporting of toxicities. Much of the initial experience was gained with coccidioidomycosis, one of the dimorphic endemic mycoses limited to the western hemisphere. Coccidioidomycosis is known for its frequency of failure and relapse after amphotericin B therapy. The initial reports for coccidioidomycosis had varied response rates, up to 88% in some groups (Defelice et al., 1982). These initial studies were followed by studies with more conservative criteria for response and more uniform followup. In the Mycoses Study Group trial, only 23% of 56 patients achieved remission at 400 mg per day, and 32 % of 56 patients at 800 mg per day (Galgiani et al., 1988). Even when dosing was carried as high as 1600 mg per day, less than 50% of patients achieved remission. Of those who achieved a remission, 38% later relapsed. However, at doses of 1200 mg per day and above, there was evidence of clinical improvement in patients with fungal meningitis, especially coccidioidomycosis (Buchsbaum et al., 1985; Craven et al., 1983; Graybill et al., 1988). However, at doses above 800 mg per day as many as half of the patients developed dose limiting nausea and vomiting (Sugar et al., 1987; Craven et al., 1983). Other toxicities included hepatotoxicity (Lewis et al., 1984) and endocrine suppression of adrenal and testicular steroid synthesis (Pont et al., 1984; Tucker et al., 1985). As more patients were treated, interaction with other drugs became an increasingly recognized problem. Rifampin, oral hypoglycemics, dilantin and other agents are associated with lowering ketoconazole blood levels (Van Tyle, 1984), while concurrent cyclosporine A and ketoconazole are associated with increased cyclosporine A levels (Dieperink and Moller, 1982; White et al., 1984; Morgenstern et al., 1982).

Ketoconazole is used at doses ranging from 200 mg per day, for dermatophytes, to 400 mg per day for systemic mycoses. The doses used for meningitis, 1200 to 2000 mg per day, frequently caused nausea and sometimes vomiting in recipients (Craven et al., 1983). Nevertheless, ketoconazole was considered such a success and so well tolerated that its use was broadened to many systemic infections, and also refractory dermatophyte infections. Ketoconazole today occupies a major role in antifungal therapy, and in the United States is the drug of choice for mucosal candidosis, including esophagitis and thrush (Graybill et al., 1980; Petersen et al., 1980; Fazio et al., 1983), paracoccidioidomycosis (Restrepo et al., 1983), histoplasmosis (National Institute of Allergy and Infectious Diseases Mycoses Study Group, 1985) and blastomycosis (National Institute of Allergy and Infectious Diseases Mycoses Study Group, 1985; Bradsher et al., 1985), and is a reasonable alternative to amphotericin B in treatment of coccidioidomycosis (Galgiani et al., 1988; Graybill et al., 1988). Ketoconazole is also used in infections caused by *Pseudallescheria boydii*, an uncommon pathogen which is characteristically resistant to amphotericin B (Galgiani et al., 1984).

Despite its considerable success, the lack of effect in aspergillosis (Tricot et al., 1987; Shale et al., 1987) and limitation to the oral route of administration curtailed the use of ketoconazole in leukopenic patients. Gastrointestinal mucosal toxicity from cancer chemotherapy made drug absorption unreliable, and physicians were hesitant to prescribe it (Stockley et al., 1988). An acid milieu was critical for optimal absorption (Lelawongs et al., 1988). The importance of acid for absorption was also found critically important in AIDS patients, who frequently poorly absorbed ketoconazole because of achlorhydria (Lake-Bakaar et al., 1988). Ketoconazole did find an increasing role as a prophylactic agent given to prevent fungal infections in leukopenic patients. Several studies suggested that ketoconazole had considerable value over placebo and nystatin or amphotericin B given orally (Hahn et al., 1982; Meunier-Carpentier, 1984; Shepp et al., 1985; Fainstein et al., 1987). However, the benefits were limited by overgrowth of *Torulopsis glabrata* (*Candida glabrata*) and by limitation of the benefit in most studies to depressed *Candida* colonization of mucosal surfaces. Days of empirical amphotericin B use, febrile days, or number of documented fungal infections were not altered, and the value of ketoconazole prophylaxis in the neutropenic patient is not completely clear.

Just as the limitations of ketoconazole were becoming well defined, the antifungal triazoles became available (Saag and Dismukes, 1988; Dismukes, 1988). These agents are now well into clinical trials, and both itraconazole and fluconazole have been licensed in several countries. SCH39304 is much earlier in clinical evaluation. Although the triazoles act by mechanisms similar to ketoconazole, their differences in pharmacokinetics and fungal spectrum are significant, and may ultimately place each of them in different primary clinical roles. These differences are summarized in Table 1.

Table 1. Characteristics of antifungal azole drugs

	KETO	FLU	ITRA	SCH39304
Aspergillus	no	no	yes	animals
Sporothrix schenckii	+/−	unknown	yes	unknown
Fonsecaea pedrosoi	no	unknown	yes	animals
T1/2	4−9h	30h	20−40+h	>60h
Acid for oral absorp	yes	no	yes	no
IV Administration	no	yes	soon	soon
Clearance	liver	renal	liver	renal
Used for meningitis	yes	yes	yes	animals
Toxicity	++++	+++	++	unknown

KETO=ketoconazole; FLU=fluconazole; ITRA=itraconazole; h=hours
T1/2=serum half live

ITRACONAZOLE

This was the first drug to enter clinical trials. Itraconazole is characterized by poor water solubility which is somewhat increased by an acid environment, wide distribution into tissues, poor penetration into cerebrospinal fluid, and slow clearance by hepatic degradation. The serum half life ranges from 20 to more than 40 hours, depending on both dose and duration of treatment (Hardin et al., 1988). About 2 weeks of daily dosing are required to achieve steady state. Doses range from 50 mg per day for mild superficial infections through 400 mg per day, for severe systemic infections. Itraconazole clearance becomes slower over time and at higher doses, suggesting that clearance mechanisms are saturable. Itraconazole interacts adversely with other drugs, including oral hypoglycemics, rifampin and cyclosporine A, but the problem may be less than with ketoconazole. Itraconazole effects on the endocrine system are less severe and different from those of ketoconazole (Phillips et al., 1987). Treatment is occasionally associated with hypokalemia and increased blood pressure (Sharkey et al., 1988). The mechanism may be due to increased mineralocorticoids, but has not yet been worked out in detail. It is probably dose related, and occurs only at doses well above those used for dermatophyte infections. Hepatotoxicity also occurs with itraconazole, but abnormalities in liver functions occur in less than 5% of patients, and are about half as frequently seen as with ketoconazole. At present itraconazole is available only orally, and should not be used with antacids or H2 blockers, which decrease absorption.

The most important distinguishing characteristics of itraconazole may not be the above, but rather is its tremendous *in vitro* and *in vivo* potency. Itraconazole is markedly more effective than ketoconazole in virtually all experimental infections tested, and has a significantly increased spectrum over ketoconazole, with clear efficacy in aspergillosis, sporotrichosis, and agents of chromoblastomycosis. If efforts to solubilize itraconazole in cyclodextrins are clinically successful, itraconazole may offer the first useful alternative to amphotericin B in the patient with leukopenia and aspergillosis.

Although the data are not yet published, the Mycoses Study Group has found itraconazole highly effective in both histoplasmosis and blastomycosis (Saag et al., 1988; Negroni et al., 1987). Because it is better tolerated than ketoconazole, and also because of equal or superior efficacy, 80% or more, itraconazole should become the drug of choice for either of these infections as soon as it is licensed in a country. For coccidioidomycosis, the Mycoses Study Group has also found itraconazole effective, with probably more frequent responses and definitely fewer relapses than with ketoconazole (Graybill et al., 1986). The dose is 400 mg per day for 6 months or, in the case of coccidioidomycosis, as long as a year. Remission rates with itraconazole are above 50% for coccidioidomycosis, as compared to less than one third of patients treated with ketoconazole. Others have confirmed the good responses of coccidioidomycosis to itraconazole (Tucker et al., 1988). Further, the relapse rate with itraconazole is about 12%, versus 38% with ketoconazole (Graybill et al., 1986).

Several infections more common in tropical climates respond well to itraconazole. Perhaps the most dramatic is chromoblastomycosis, which traditionally has been poorly responsive to antifungal therapy. Restrepo, Lavalle, Borelli and their collegues have independently found that itraconazole is highly effective in chromoblastomycosis caused by *Fonsecaea pedrosoi* and *Cladosporium* (Restrepo et al., 1988; Lavalle et al., 1987; Borelli, 1987). Of 10 patients reported by Restrepo et al., 8 responded to treatment (Restrepo et al., 1988). This is extremely welcome news, because these infections of the subcutaneous tissues are terribly disfiguring and have been refractory to most efforts at medical therapy. For paracoccidioidomycosis, itraconazole is highly effective in a treatment period as short as 6 months, versus a year required for ketoconazole, again making itraconazole the drug of choice (Negroni et al., 1987; Restrepo et al., 1987). Studies are few with sporotrichosis, but those reported by Restrepo indicate that fluconazole is very effective in lymphocutaneous disease, with a dose of 200 mg per day for 6 months clearing virtually all patients of their disease (Restrepo et al., 1986). Fourteen patients of a total 17 treated were evaluable, with a mean post treatment observations of 115 days. There were no relapses. The Mycoses Study Group has had a small but encouraging experience with itraconazole in sporotrichosis (Saag et al., 1988). Another group of rare but increasingly frequent mycoses is caused by the dematiaceous or melanin pigment producing fungi. These cause a very slowly progressing pulmonary cutaneous or osseous disease called phaeohyphomycosis. This frequently follows traumatic inoculation, and has responded poorly to ketoconazole or amphotericin B. The United States experience with 17 evaluable cases has been very encouraging for itraconazole, with 3 patients cured and 8 patients improved by treatment with itraconazole. Many had been refractory to prior treatment (Graybill, J.R., unpublished observations). The ideal duration of treatment is not clear.

Itraconazole may also be effective in life threatening disasters like invasive aspergillosis. The initial experiences with aspergillosis have involved patients with cardiac transplants, some with leukemia, others with immunosuppression, and some with *Aspergillus* fungus ball. Both failures and successes have been reported (Viviani et al., 1987; Viviani et al., 1989; Dupont and Drouhet, 1988; Phillips et al., 1987; Ganer et al., 1987). Among the largest experiences is that of Viviani et al. (1989). Eight of ten patients with invasive aspergillosis in the presence of leukemia or heart transplant patients responded. In five of these the diagnosis was confirmed by biopsy. Therefore, enough patients have dramatically improved or completely resolved their aspergillosis to make itraconazole an attractive candidate for treatment of the leukopenic patient, if drug delivery to involved tissues can be assured. On the other hand, the patients with fungus ball are the most difficult to interpret. While their disease is unquestioned, the extent of pulmonary fibrosis and destruction, and poor drug penetration, and slow resolution of symptoms all prevent reaching definite conclusions. However, some of Restrepo's patients clearly improved with treatment, and itraconazole may be worth considering, especially in the patients who cannot tolerate surgical resection (Restrepo et al., 1988).

There is a major caveat which the physician must consider when using itraconazole, and this is drug absorption. In the United States patients receiving corticosteroids and most physicians dogmatically add H2 blockers and antacids to such a regimen. We have had a tragic failure treating a marrow transplant recipient with itraconazole in such an environment, where the primary physicians wanted itraconazole but would not stop the H2 blockers. The widely disseminated cutaneous aspergillosis, slowly worsening on amphotericin B, became fulminating when amphotericin B was changed for itraconazole. Drug absorption was nil. Our other most impressive failure was a patient with aspergillosis of the spine. He had no improvement with months of treatment, and his serum concentrations were less than 0.22 mcg/ml after 2 weeks of 200 mg per day (Phillips et al., 1987). Therefore, if there is any question about absorption, serum concentrations should be measured after 2-3 days of therapy. Absorption may be increased by the addition of acid, commonly in the form of Acidulin, 2-3 tablets with the dose, and giving itraconazole with a lipid containing meal. Coca-Cola may be useful as well.

Success in aspergillosis has prompted consideration of itraconazole for use in prophylaxis of the neutropenic patient. The most dramatic result is that of Tricot et al. (1987). Their study is the first one published, and is a historical comparison. Rising deaths from fungal infections were noted with ketoconazole prophylaxis, and these were sharply reduced when itraconazole was used. This impressive result in a cancer hospital has led to several trials in the United States, with unpublished and less impressive results. One problem may be impaired drug absorption in these patients with severe mucosal damage from chemotherapy. To avoid this problem, a new formulation has been developed for itraconazole. A new pellet formulation was recently compared with the polyethylene glycol formulation in patients undergoing remission induction for acute myelocytic anemia (Prentice et al., 1989). There was a trend to higher absorption with the new form, with mean maximum serum concentration 412+/-227 ng/ml vs 317+/-177 ng/ml for the standard formulation at 14 days of treatment.

Absorption of itraconazole, generally good in the non-AIDS patient and still sufficient in the AIDS patient, remains an impediment to use in the severely leukopenic patient. The use of cyclodextrins to solubilize itraconazole for parenteral administration should markedly increase the usefulness of this potent antifungal azole.

FLUCONAZOLE

Fluconazole is much more water soluble than itraconazole, is well absorbed with or without an acid milieu, penetrates the cerebrospinal fluid at more than 70% of serum concentration (Arndt et al., 1988; Hanson and Stevens, 1988b), and is excreted largely unchanged by the renal route. The serum half life is about 30 hours in man. The greatest advantage of fluconazole may be that it can be administered readily intravenously. Fluconazole is active against a broad range of fungi. It has also been found effective in an immunosuppressed rabbit model of invasive aspergillosis (George et al., 1989). However, in animal models of aspergillosis fluconazole is only effective at high concentrations, correlating to 800 mg per day or more in man (a dose largely unstudied thus far). Data on sporotrichosis, blastomycosis, and histoplasmosis are few and do not suggest great efficacy, but they reflect doses of 50 to 100 mg per day, which are known to be suboptimal (Mycoses Study Group, unpublished observations). The greatest clinical experience with fluconazole has been collected in infections caused by *Candida* species and *Cryptococcus neoformans*. Mucosal forms of candidiasis respond readily to fluconazole, as little as 50 mg per day, both in AIDS and non-AIDS patients. In a small study fluconazole has been found superior to clotrimazole in treatment of thrush (Koletar et al., 1989).

Fluconazole has also been found equal or superior to polyenes in prophylaxis of fungal infections in neutropenic patients (Rozenberg-Arska et al., 1988). In a recent large study comparing fluconazole with oral polyenes, a dose of 50 mg per day fluconazole was associated with proven or suspected fungal infection in 33% of 122 evaluable patients, versus 45% of

119 patients treated with polyenes (Feczko et al., 1989). There were 17 proven fungal infections in the polyene group versus only 1 in the fluconazole group. Of the 17 mycotic infections, 12 were mucosal and 5 were invasive infections, with candidemia in 4. Other studies compared fluconazole with ketoconazole, and found that fluconazole was quite effective, but the numbers are small (De Wit et al., 1988). There is also preliminary experience suggesting that fluconazole may be of value in disseminated candidiasis (8/10 cured or improved) and candida urinary tract infection (10/10 cured or improved) (Rubin et al., 1989). Other preliminary studies also found fluconazole useful in urinary tract infections (Graybill et al., 1988; Bru et al., 1988). However, there was little uniformity in treatment regimens used. Large studies are ongoing with both thrush and with disseminated candidosis in leukopenic patients, but results are not yet available. Although fluconazole has not been used for treatment of aspergillosis in man, thus far there does not appear to be any predisposition for *Aspergillus* overgrowth in leukopenic patients treated with fluconazole.

In the United States, by far the largest experience with fluconazole has been accumulated in patients with AIDS, where it may be equally effective as amphotericin B (Mycoses Study Group, 1989). More recent studies are commencing with 200-400 mg doses in coccidioidomycosis, histoplasmosis, sporotrichosis, and blastomycosis. The only results available from these studies are from a series of patients treated with fluconazole at up to 400 mg per day for coccidioidal meningitis (Tucker et al., in press). The response rate was 64% and was similar in patients who the 9 patients who were treated only with fluconazole and the 7 who had received concurrent amphotericin B intrathecally.

SCH39304

This drug has several distinguishing features, among them good absorption from the stomach, not being dependent on acid, extremely slow clearance by the renal route, with a half life of more than 60 hours, penetration of the cerebrospinal fluid at more than 30% of serum concentrations, a broad antifungal spectrum that is similar to itraconazole, and evidence suggesting greater potency in animal models than other azoles (Lin et al., 1989; Walsh et al., 1989; DeFaveri and Graybill, 1989; Kramer et al., 1988; Schmitt et al., 1988). An intravenous form is under development, but is not yet available. SCH39304 has just been introduced into clinical trials in the United States, and in a very few cases appears effective in thrush, candidal esophagitis, and coccidioidomycosis (Graybill, J.R., unpublished observations). SCH39304 also appears well tolerated in these few patients, but a toxicity profile in man has yet to be defined. Large multicenter trials are in development for cryptococcosis, histoplasmosis in AIDS, and in prophylaxis/empiric treatment of fungal infections in leukopenic patients.

REPACKAGING AMPHOTERICIN B

Ampholiposomes, in the hands of Lopez-Berestein and Meunier, have been found to be extremely well tolerated at doses up to more than 5 mg/kg by rapid intravenous infusion (Lopez-Berestein, 1989; Wiebe and deGregorio, 1988). There is virtually no fever, nephrotoxicity, nausea, vomiting, or the other effects usually attributed to amphotericin B. Ampholiposomes have been used successfully to treat particularly prolonged and refractory mycoses caused by *Candida* or *Aspergillus* in leukopenic patients, and also hepatosplenic candidiasis. This last has been especially frustrating, with a mortality above 30% even in patients treated for prolonged courses of amphotericin B (Haron et al., 1987; Thaler et al., 1988). At present, Squibb has developed a preparation called amphotericin lipid complex (ABLC) in which the lipid form is homogeneous rather than in the multilamellar form of liposomes. ABLC is effective in mice with cryptococcosis, and is now commencing clinical trials in the United States (Whitney et al., 1989). Ultimately, these preparations may find use in patients severely ill with acute systemic mycoses, particularly the group with leukopenia and fulminating disease.

However, even for hepatic candidiasis there may be a role in the future for antifungal azoles. We have had several patients treated successfully with itraconazole and fluconazole. Our experience with just a few patients is insufficient for generalization.

TERBINAFINE

The other direction has been away from the systemic mycoses, toward dealing with dermatophytes. Terbinafine has been developed by Sandoz, and has a broad spectrum of activity against superficial mycelial infections. It appears effective in refractory fungal onychomycosis. Despite high activity against filamentous dermatophytes, terbinafine has less *in vivo* and *in vitro* activity against yeasts, and little *in vivo* activity in a few trials with systemic mycoses.

SUMMARY

With all of the activity noted above, and numerous ongoing studies, current recommendations for antifungal therapy have greatly changed over the past decade. For systemic mycoses, amphotericin B has retreated from its unique position as the "only effective drug". The initial challenge by ketoconazole has been followed by itraconazole, which appears to have excellent potency against all of the endemic systemic mycoses of the western hemisphere. Itraconazole also has great efficacy in sporotrichosis and chromoblastomycosis, and has promise for treatment of aspergillosis. The improved tolerance of this drug, benign enough to use in superficial mycotic infections, makes it a very attractive alternative to ketoconazole, and it may become the drug of choice for these mycoses.

Fluconazole appears highly effective in mucosal forms of candidosis, and may be as effective as amphotericin B in cryptococcal meningitis in AIDS patients. There are insufficient data to suggest a role for fluconazole in endemic mycoses. At doses currently recommended, fluconazole is not likely to be effective in aspergillosis.

Based on animal studies, SCH39304 has great promise, but it is too early in clinical trials to make any assessment of its role.

The aggressive *Candida* and *Aspergillus* infections associated with leukemia are best still treated very early using protocols for empiric therapy with amphotericin B. Fluconazole is being evaluated at present, and when intravenously administrable forms of itraconazole and SCH39304 become available, there will be trials with these, both for empiric treatment and for prophylaxis. However, the clearest alternative to amphotericin B at present appears to be ampholiposomes or ABLC, which offers a much less toxic form of amphotericin B, ability to infuse large amounts in a short time, and apparently superior clinical results to traditional amphotericin B. However, no controlled trials have been done yet, and only a few centers are conducting studies.

REFERENCES

Aisner, J., Schimpff, S.C., Sutherland, J.C., Young, V.M., and Wiernik, P.H., 1976, *Torulopsis glabrata* infections in patients with cancer, Am. J. Med., 61:23.

Aisner, J., Schimpff, S.C., and Wiernik, P.H., 1977, Treatment of invasive
    aspergillosis: relation of early diagnosis and treatment to
    response, Ann. Intern. Med., 86:539.
Arndt, C.A., Walsh, T., McCully, C.L., Balis, F.M., Pizzo, P., and Poplack,
    D., 1988, Fluconazole penetration into cerebrospinal fluid:
    implications for treating fungal infections of the central nervous
    system, J. Infect. Dis., 157:178.
Blazar, B.R., Hurd, D.H., Snover, D.C., Alexander, J.W., and McGlave, P.B.,
    1984, Invasive Fusarium infections in bone marrow transplant
    recipients, Am. J. Med., 77:645.
Borelli, D., 1987, A clinical trial of itraconazole in the treatment of
    deep mycoses and leishmaniasis, Rev. Infect. Dis., 9 (Suppl 1):S57.
Bradsher, R.W., Rice, D.C., and Abernathy, R.S., 1985, Ketoconazole therapy
    for endemic blastomycosis, Ann. Intern. Med., 103:872.
Bronnimann, D.A., Adam, R.D., Gagliani, J.N., Habib, M.P., Peterson, E.A.,
    Porter, B., and Bloom, J.W., 1987, Coccidioidomycosis in the
    acquired immunodeficiency syndrome, Ann. Intern. Med., 106:372.
Bru, J.P., Lebeau, B., Stahl, J.P., and Micoud, M., 1988, Oral fluconazole
    treatment of urinary mycoses. Rev. Iber. Micol., 5 (1): abstract 0-
    94:34.
Buchsbaum, H.W., Fahs, J.J., and Friedman, B.A., 1985, Coccidioidal
    meningitis treated with high-dose ketoconazole, in:
    "Coccidioidomycosis. Proceedings of the 4th International
    Conference", H.E. Einstein, and A. Catanzaro, eds., National
    Foundation for Infectious Diseases, Wash, D.C.
Craven, P.C., Graybill, J.R., Jorgensen, J.H., Dismukes, W.E., and Levine,
    B.E., 1983, High-dose ketoconazole for treatment of fungal
    infections of the central nervous system, Ann. Intern. Med., 98:160.
DeFaveri, J., and Graybill, J.R., 1989, SCH39304 in treatment of murine
    pulmonary aspergillosis. Abstract 822. Twenty-ninth Interscience
    Conference on Antimicrobial Agents and Chemotherapy, Houston, TX.
DeFelice, R., Galgiani, J.N., Campbell, S.C., Palpant, S.D., Friedman,
    B.A., Dodge, R.R., Weinberg, M.G., Lincoln, L.J., Tennican, P.O.,
    and Barbee, R.A., 1982, Ketoconazole treatment of nonprimary
    coccidioidomycosis, Am. J. Med., 72:681.
De Wit, S., Weeris, D., Hermans, P., DeCock, F., and Clumeck, N., 1988,
    Double-blind randomized study of oral fluconazole (F) versus
    ketoconazole (K) in 40 episodes of oropharyngeal candidiasis (OC) in
    ARC/AIDS patients. Abstract 572. Twenty Eighth Interscience
    Conference on Antimicrobial Agents and Chemotherapy. Los Angeles,
    CA.
Dieperink, H., and Moller, J., 1982, Ketoconazole and cyclosporin A,
    Lancet, 2:1217.
Dismukes, W., 1988, Azole antifungals: old and new, Ann. Intern. Med.,
    109:177.
Dupont, B., and Drouhet, E., 1988, The treatment of aspergillosis with
    azole derivatives, in: "Aspergillus and Aspergillosis", H. Vanden
    Bossche, D. Mackenzie, and G. Cauwenbergh, eds., Plenum, New York.
Fainstein, V., Bodey, G.P., Elting, L., Maksymiuk, A., Keating, M., and
    McCredie, K.B., 1987, Amphotericin B or ketoconazole therapy of
    fungal infections in neutropenic cancer patients, Antimicrob. Agents
    Chemother., 31:11.
Fazio, R.A., Wickremesinghe, P.C., and Arsura E.L., 1983, Ketoconazole
    treatment of Candida esophagitis: a prospective study of 12 cases,
    Am. J. Gastroenterol., 78:261.
Feczko, J.M., Brammer, K.W., and Buell, D.N., 1989, Superiority of oral
    fluconazole (FLU) over oral non-absorbable polyenes (POL) as fungal
    prophylaxis in neutropenic patients. Abstract 70. Twenty-ninth
    Interscience Conference on Antimicrobial Agents and Chemotherapy,
    Houston, TX.

Fisher, B.D., Armstrong, D., Yu, B., and Gold, J.W.M., 1981, Invasive aspergillosis: progress in early diagnosis and treatment, Am. J. Med., 71:571.

Galgiani, J.N., Stevens, D.A., Graybill, J.R., Dismukes, W.E., and Cloud, G.A., 1988, Ketoconazole therapy of progressive coccidioidomycosis. Comparison of 400 and 800 mg doses and observations at higher doses. Am. J. Med., 84:603.

Galgiani, J.N., Stevens, D.A., Graybill, J.R., Stevens, D.L., Tillinghast A.J., and Levine H.B., 1984, Pseudallescheria boydii infections treated with ketoconazole. Clinical evaluations of seven patients and in vitro susceptibility results, Chest, 86:219.

Ganer, A., Arathoon, E., and Stevens, D.A., 1987, Initial experience in therapy for progressive mycoses with itraconazole, the first clinically studied triazole, Rev. Infect. Dis., 9 (Suppl 1):577.

George, D., Miniter, P., and Andriole, V.T., 1989, Fluconazole prophylaxis in experimental invasive aspergillosis. Abstract 815. Twenty-ninth Intersciencee Conference on Antimicrobial Agents and Chemotherapy, Houston, TX.

Graybill, J.R., Herndon, J.H., Kniker, W.T., and Levine, H.B., 1980, Ketoconazole treatment of chronic mucocutaneous candidiasis, Arch. Dermatol., 116:1137.

Graybill, J.R., Sharkey, P.K., and Watson, C., 1988, Fluconazole treatment of funguria, Rev. Iber. Micol., 5 (1):abstract 0-93:34.

Graybill, J.R., Stevens, D.A., Galgiani, J.N., Dismukes, W.E., Cloud, G., Ganer, A., Arathoon, E., Fetchick, R., and Diaz, M., 1986, Itraconazole treatment of coccidioidomycosis. Abstract 788. Twenty Sixth Interscience Conference on Antimicrobial Agents and Chemotherapy, New Orleans, LA.

Graybill, J.R., Stevens, D.A., Galgiani, J.N., Sugar, A.M., Craven, P.C., Gregg, C., Huppert, M., Cloud, G., and Dismukes, W.E., 1988, Ketoconazole treatment of coccidioidal meningitis, Ann. N.Y. Acad. Sci., 544:488.

Hahn, I.M., Prentice, H.G., Corringham, R., Blacklock, H.A., Keany, M., Shannon, M., Noone, P., Gascoigne, E., Fox, J., Boesen, E., Szawatkowski, M., and Hoffbrand, A.V., 1982, Ketoconazole versus nystatin plus amphotericin B for fungal prophylaxis in severely immunocompromised patients, Lancet, 1:826.

Hanson, L.H., and Stevens, D.A., 1988b, Pharmacokinetics of fluconazole in cerebrospinal fluid and serum in human coccidioidal meningitis, Antimicrob. Agents Chemother., 32:369.

Hardin, T.C., Graybill, J.R., Fetchick, R., Woestenborghs, R., Rinaldi, M., and Kuhn, J., 1988, Pharmacokinetics of itraconazole following oral administration to normal volunteers, Antimicrob. Agents Chemother., 32:1310.

Haron, E., Feld, R., Tuffnell, P., Patterson, B., Hasselback, R., and Matlow, A., 1987, Hepatic candidiasis: an increasing problem in immunocompromised patients, Am. J. Med., 83:17.

Horn, R., Wong, B., Kiehn, T.E., and Armstrong, D., 1985, Fungemia in a cancer hospital: changing frequency, earlier onset, and results of therapy, Rev. Infect. Dis., 7:646.

Johnson, P.C., Khardori, N., Najjar, A.F., Butt, F., Mansell, P.W.A., and Sarosi, G.A., 1988, Progressive disseminated histoplasmosis in patients with acquired immunodeficiency syndrome, Am. J. Med., 85:152.

Koletar, S.L., Russell, J.A., Fass, R.J., and Plouffe, J.F., 1989, Comparison of fluconazole with clotrimazole troches as treatment for oral candidiasis in HIV-infected patients. Abstract 1064. Twenty-Ninth Interscience Conference on Antimicrobial Agents and Chemotherapy, Houston, TX.

Komshian, S.V., Uwaydah, A.K., Sobel, J.D., and Crane, L.R., 1989, Fungemia caused by Candida species and Torulopsis glabrata in the hospitalized patient: frequency, characteristics, and evaluation of factors influencing outcome, Rev. Infect. Dis., 11:379.

Kovacs, J.A., Kovacs, A.A., Polis, M., Wright, W.C., Gill, V.J., Tuazon, C.U., Gelmann, E.P., Lane, H.C., Longfield, R., Overturf, G., Macher, A.M., Fauci, A.S., Parillo, J.E., Bennett, J.E., and Masur, M., 1985, Cryptococcosis in the acquired immunodeficiency syndrome, Ann. Intern. Med., 103:533.

Kramer, W., Kim, H., Symchowics, S., Perentesis, G., Affrime, N., and Lin, C., 1988, Pharmacodynamics of SCH39304 in man. Abstract 165. Twenty Eighth Interscience Conference on Antimicrobial Agents and Chemotherapy, Los Angeles.

Lake-Bakaar, G., Tom, W., Lake-Bakaar, D., Gupta, N., Beidas, S., Elsakr, M., and Straus, E., 1988, Gastropathy and ketoconazole malabsorption in the acquired immunodeficiency syndrome (AIDS), Ann. Intern. Med., 109:471

Lavalle, P., Suchil, P., de Ovando, F., and Reynoso, S., 1987, Itraconazole for deep mycoses: preliminary experience in Mexico, Rev. Infect. Dis., 9 (Suppl 1):S64.

Lelawongs, P., Barone, J.A., Colaizzi, J.L., Hsuan, A.T.M., Mechlinski, W., Legendre, R., and Guarnieri, J., 1988, Effect of food and gastric acidity on absorption of orally administered ketoconazole, Clin. Pharm., 7:228.

Lewis, J.H., Zimmerman, H.J., Benson, G.D., and Ishak, K.G., 1984, Hepatic injury associated with ketoconazole therapy, Gastroenterology, 86:503.

Lin, C., Kim, H., Radwanski, E., Affrime, M., and Symchowicz, S., 1989, Steady-state pharmacokinetics of SCH39304 in man. Abstract 1359. Twenty-ninth Interscience Conference on Antimicrobial Agents and Chemotherapy, Houston, TX.

Lopez-Berestein, G., 1989, Liposomal-amphotericin B in the treatment of systemic mycoses in patients with cancer, Chapter 14, in: "Diagnosis and Therapy of Systemic Fungal Infections", K. Holmberg and R. Meyer, eds , Raven Press, New York.

Meunier-Carpentier, F., 1984, Chemoprophylaxis of fungal infections, Am. J. Med., 76:652.

Morgenstern, G.R., Powles, R., Robinson, B., and McElwain, T.J., 1982, Cyclosporin interaction with ketoconazole and melphalan, Lancet, 2:1342.

Mycoses Study Group: Dismukes, W.E., Cloud, G., Thompson, S., Sugar, A., and Tuazon, C., 1989, Fluconazole (FLU) versus amphotericin B (AMB) therapy (Rx) of acute cryptococcal meningitis. Abstract 1065. Twentyninth Interscience Conference on Antimicrobial Agents and Chemotherapy, Houston, TX.

National Institute of Allergy and Infectious Diseases Mycoses Study Group, 1985, Treatment of blastomycosis and histoplasmosis with ketoconazole, Ann. Intern. Med., 103:861.

Negroni, R., Palmieri, O., Koren, K., Tiraboschi, I.N., and Galimberti, R.L., 1985, Oral treatment of paracoccidioidomycosis and histoplasmosis with itraconazole, Rev. Infect. Dis., 9 (Suppl 1):47.

Petersen, E.A., Alling, D.W., and Kirkpatrick, C.H., 1980, Treatment of chronic mucocutaneous candidiasis with ketoconazole, Ann. Intern. Med., 93:791.

Phillips, P., Fetchick, R., Weisman, I., Foshee, S., and Graybill, J.R., 1987, Tolerance to and efficacy of itraconazole in treatment of systemic mycoses: preliminary results, Rev. Infect. Dis., 9 (Suppl 1):587.

Phillips, P., Graybill, J.R., Fetchick, R., and Dunn, J.F., 1987, Adrenal response to corticotropin during therapy with itraconazole, Antimicrob. Agents Chemother., 31:647.

Pizzo, P.A., Robichaud, K.J., Gill, F.A., and Witebsky, F.G., 1982, Empiric antibiotic and antifungal therapy for cancer patients with prolonged fever and granulocytopenia, Am. J. Med., 72:101.

Pont, A., Graybill, J.R., Craven, P.C., Galgiani, J.N., Dismukes, W.E., and Reitz, R.E., 1984, High-dose ketoconazole and adrenal and testicular function in man, Arch. Intern. Med., 144:2150.

Prentice, A.G., Bradford, C.R., and Warnock, D.W., 1989, A double-blind cross over pharmacokinetic study of two formulations of itraconazole (ITRA) during remission induction for acute myeloblastic leukemia. Abstract 1358. Twenty-ninth Interscience Conference on Antimicrobial Agents and Chemotherapy, Houston, TX.

Restrepo. A., Gomez, I., Canio, L.E., Arango, M.D., Gutierrez, F., and Sanin, A., 1983, Treatment of paracoccidioidomycosis with ketoconazole: a three year experience, Am. J. Med., 74(1B):48.

Restrepo, A., Gomez, I., Robledo, J., Patino, M.M., and Cano, L.E., 1987, Itraconazole in the treatment of paracoccidioidomycosis: a preliminary report, Rev. Infect. Dis., 9:S51.

Restrepo, A., Gonzalez, A., Gomez, I., Arango, M., and deBedout, C., 1988, Treatment of chromoblastomycosis with itraconazole, Ann. N.Y. Acad. Sci., 544:504.

Restrepo, A., Munera, M.I., Arteaga, I.D., Gomez, I., Tabares, A.M., Patino, M.M., and Arango, M., 1988, Itraconazole in the treatment of pulmonary aspergilloma and chronic pulmonary aspergillosis, in: "Aspergillus and Aspergillosis", H. Vanden Bossche, D.W.R. Mackenzie, and G. Cauwenbergh, eds., Plenum, New York.

Restrepo A., Robledo, J., Gomez, I., Tabares, A.M., and Gutierrez, R., 1986, Itraconazole therapy in lymphangitic and cutaneous sporotrichosis, Arch. Dermatol., 122:413.

Rinaldi, M.G., 1989, Zygomycosis. Infect. Dis. Clin. North Am., 3:20.

Rozenberg-Arska, M., Dekker, A.W., and Verhoff, J., 1988, Efficacy of fluconazole (FL) for prevention of fungal infections (FI) in comparison to amphotericin B (AmB) in granulocytopenic patients (pts). Abstract 577. Twenty Eighth Interscience Conference on Antimicrobial Agents and Chemotherapy. Los Angeles, CA.

Rubin, R.H., Debruin, M.F., and Knirsch, A.K., 1989, Fluconazole therapy for patients with serious Candida infections who have failed standard therapies. Abstract 71. Twenty-ninth Interscience Conference on Antimicrobial Agents and Chemotherapy, Houston, TX.

Saag, M., Bradsher, R., Chapman, S., Kauffman, C., Girard, W., Stevens, D., Bowles-Patton, C., Dismukes, W.E., and the NIAID Mycoses Study Group, 1988, Itraconazole (I) therapy (Rx) for blastomycosis (B), histoplasmosis (H), and sporotrichosis (5). Abstract 574. Twenty Eighth Interscience Conference on Antimicrobial Agents and Chemotherapy. Los Angeles, CA.

Saag, M.S., and Dismukes, W.E., 1988, Azole antifungal agents: emphasis on new triazoles, Antimicrob. Agents Chemother., 32:1.

Schmitt, H.J., Barnard, E.M., Hauser, M., and Armstrong, D., 1988, Comparison of antifungal agents in a rat model of aspergillosis. Abstract 171. Twenty Eighth Interscience Conference on Antimicrobial Agents and Chemotherapy, Los Angeles.

Shale, D.J., Faux, J.A., and Lane, D.J., 1987, Trial of ketoconazole in noninvasive pulmonary aspergillosis, Thorax, 42:26.

Sharkey, P.K., Rinaldi, M.G., Lerner, C.J., Fetchick, R., Dunn, J.F., and Graybill, J.R., 1988, High dose itraconazole in the treatment of systemic mycoses. Abstract 575. Twenty Eighth Interscience Conference on Antimicrobial Agents and Chemotherapy. Los Angeles, CA.

Shepp, D.H., Klosterman, A., Siegel, M.S., and Meyers, J.D., 1985, Comparative trial of ketoconazole and nystatin for prevention of fungal infection in neutropenic patients treated in a protective environment, J. Infect. Dis., 152:1257.

Sinclair, A.J., Rossof, A.H., and Coltman, C.A., 1978, Recognition and successful management in pulmonary aspergillosis in leukemia, Cancer, 42:2019.

Stockley, R.J., Daneshmend, T.K., Bredow, M.T., Warnock, D.W., Richardson M.D., and Slade, R.R., 1988, Ketoconazole pharmacokinetics during chronic dosing in adults with hematological malignancy, Eur. J. Clin. Microbiol., 5:513.

Sugar, A.M., Alsip, S., Galgiani, J.N., Graybill, J.R., Dismukes, W.E., Cloud, G., Craven, P.C., and Stevens, D.A., 1987, Pharmacology and toxicity of high dose ketoconazole, Antimicrob. Agents Chemother., 31:1874.

Thaler, M., Pastikia, B., Shawker, T.H., O'Leary, T., and Pizzo, P.A., 1988, Hepatic candidiasis in cancer patients: the evolving picture of the syndrome, Ann. Intern. Med., 108:88.

Tricot, G., Joosten, E., Boogaerts, M.A., Vande Pitte, J., and Cauwenbergh, G., 1987, Ketoconazole vs itraconazole for antifungal prophylaxis in patients with severe granulocytopenia: preliminary results of two non-randomized studies, Rev. Infect. Dis., 9 (Suppl 1):594.

Tucker, R.M., Denning, D.W., Rinaldi, M.G., and Hanson, L.H., 1988, Itraconazole (IZ) therapy (Rx) of progressive coccidioidomycosis (C). Abstract 573. Twenty Eighth Interscience Conference on Antimicrobial Agents and Chemotherapy. Los Angeles, CA.

Tucker, R.M., Galgiani, J.N., Graybill, J.R., Denning, D.W., Hanson, L.H., Eckman, D., Salemi, C., and Stevens, D.A., 1989, Treatment of coccidioidal meningitis with fluconazole, Rev. Infect. Dis., in press.

Tucker, W.S., Snell, B.B., Island, D.P., and Gregg, C.R., Reversible adrenal insufficiency induced by ketoconazole, J.A.M.A., 253:2413.

Vanden Bossche, H., 1985, Biochemical targets for antifungal azole derivatives: hypothesis on the mode of action, in: "Current Topics in Medical Mycology", M. McGinnis, ed., Springer-Verlag, New York.

Van Tyle, J.H., 1984, Ketoconazole: mechanism of action, spectrum of activity, pharmacodynamics, drug interactions, adverse reactions, and therapeutic use, Pharmacotherapy, 4:343.

Viviani, M.A., Tortorano, A.M., Langner, M., Almaviva, M., Negri, C., Cristina, S., Scoccia, S., deMaria, R., Fuiocchi, R., Ferrazzi, P., Goglio, A., Gavazzeni, G., Faggian, G., Rinaldi, R., and Cadrobbi, P., 1989, Experience with itraconazole in cryptococcosis and aspergillosis. J. Infect., 18:151.

Viviani, M.A., Tortorano, A.M., Woenstenborghs, R., and Cauwenbergh, G., 1987, Experience with itraconazole in deep mycoses in Northern Italy, Mykosen, 30:233.

Walsh, T.J., McCully, C., Rinaldi, M.G., Balis, F., Lee, J., Pizzo, P., and Poplack, D., 1989, Pharmacokinetics and penetration of SCH39304 into cerebrospinal fluid of rhesus monkeys. Abstract 1361. Twenty-ninth Interscience Conference on Antimicrobial Agents and Chemotherapy, Houston, TX.

Walsh, T.J., Newman, K.R., Moody, M. Wharton, R.C., and Wade, J.C., 1986, Trichosporonosis in patients with neoplastic disease, Medicine, 65:268.

White, G.J.G., Blatchford, N.R., and Cauwenbergh, G., 1984, Cyclosporine and ketoconazole, Transplantation, 37:214.

Whitney, R.R., Kunselman, L., Clark, J.M., and Bonner, D.P., 1989, Efficacy of amphotericin B lipid complex (ABLC) in cryptococcal meningitis in normal or immunocompromised mice. Abstract 166. Twenty-ninth Interscience Conference on Antimicrobial Agents and Chemotherapy, Houston, TX.

Wiebe, V.J., and deGregorio, M.W., 1988, Liposome-encapsulated amphotericin B: a promising new treatment for disseminated fungal infection, Rev. Infect. Dis., 10:1097.

Young, R.C., Bennett, J.E., Vogel, C.L., Carbone, P.P., and DeVita, V.T., 1970, Aspergillosis: the spectrum of the disease in 98 patients, Medicine, 49:147.

TREATMENT OF CANDIDOSIS IN AIDS PATIENTS

G. Just, D. Steinheimer, M. Schnellbach, C. Böttinger,
E.B. Helm and W. Stille

Frankfurt University Hospital
Dept. of Internal Medicine/Infectiology
Theodor-Stern-Kai 7
6000 Frankfurt/M 70

Fungal infections are the most common manifestations of infections by facultative pathogens in AIDS. The clinical manifestations range from mild infections of the mucous membrane to life-threatening systemic mycoses which are difficult to control. *Candida* oropharyngitis and *Candida* esophagitis are the most common fungal infections (Glatt et al., 1988; Holmberg and Meyer, 1986). In the latest CDC definition of AIDS, candidiasis of the esophagus, trachea, bronchi and lungs are indicative of AIDS in case of absence of serological evidence of HIV. In case of positive HIV serology, *Candida* infections of the esophagus indicate AIDS (CDC, 1987). The colonisation of the mucous membrane with *Candida* largely depends on the immune function and can be an early hint for HIV infection (Brodt et al., 1986; Klein et al., 1984). With the progression of T-cell defect, long-term or intermittent antimycotic therapy becomes necessary - even more so in cases receiving antibiotics or cytostatics at the same time. The dissemination of *Candida*., e.g. into the lung, CNS or other organs, is quite rare and only seen in more advanced stages of AIDS (Holmberg and Meyer, 1986, Langford et al., 1988).

At Frankfurt University Hospital, only one case of invasive *Candida* infection was observed in 130 autopsied patients. In this patient, severe necrotic *Aspergillus* pneumonia was observed at the same time.

Oropharyngeal *Candida* infections show a diversity of clinical pictures (Table 1).

Acute pseudomembraneous candidiasis with creamy white plaques is the most common type, followed by atrophic types (with enanthema) and perlèche (Levy et al., 1989).

Oral candidiasis is a marker for concomitant esophageal involvement, especially when symptoms like painful swallowing and retrosternal chest pain are present (Porro et al., 1989).

However, absence of symptoms and oral candidiasis does not mean absence of esophageal involvement. Endoscopy with brushing of the esophagus and biopsy samples is indicated in cases of non-response to antifungal therapy (Tavitan et al., 1986; Porro et al., 1989).

Table 1. Distribution of oral *Candida* manifestations

1. Acute pseudomembraneous candidiasis (oral thrush)

2. Acute atrophic candidiasis (midline glossitis)

3. Chronic atrophic candidiasis

4. Chronic hyperplastic candidiasis (*Candida* leukoplakia)

5. Perlèche (angular cheilitis)

At Frankfurt University Hospital, HIV-infected patients have been examined for *Candida* since 1985. Clinical and cultural courses of antimycotic therapy were monitored. Beside examination of the oral cavity and registration of complaints, samples of oral wash-out (5 ml NaCl) for quantification and identification (API 20C and AT B32 Merieux) of *Candida* spp. have been shown to be useful.

In addition, cultural examination of oral wash-out is helpful for differential diagnosis, especially to differentiate leukoplakia caused by Epstein Barr Virus.

The number of colony-forming units (CFU) per ml oral wash-out and the clinical evaluation is as following:

CFU $<10^2$/ml = low colonisation
CFU $10^2 - 10^3$/ml = medium colonisation
CFU $> 10^3$/ml = significant colonisation

As the efficacy of topical antimycotics, e.g. nystatin, amphotericin B, miconazole, clotrimazole, are insufficient in the treatment of oral *Candida* infection in progressed stages of HIV disease, orally absorbed azoles have been successfully used.

Since its development in the late 1970s, ketoconazole as the first orally active azole was initially the most important compound. In recent years, the high efficacy of ketoconazole in the treatment of local and systemic fungal infections has been described in many studies (Drouhet and Dupont, 1987; Jones, 1987; Saag and Dismukes et al., 1988; Tricot et al., 1987; Van Tyle, 1984).

Our study group investigated the efficacy of ketoconazole with a new form of application, as suspension.

In 52 treatment courses, 33 patients in advanced stages of HIV infection with cultural and clinical signs of *Candida* infection were treated with ketoconazole suspension at a dosage of 200 to 600 mg/day (200 mg: n=1; 400 mg: n=43; 600 mg: n=8). Duration of therapy was 1-12 days. Clinical symptoms were burning (n=14), furry sensation (n=39), painful swallowing (n=27), dry mouth (n=5), and substernal chest pain (n=6). No symptoms were observed in 4 patients. Thrush was found in 45 out of 52 treatment episodes.

The treatment with ketoconazole suspension at a dosage of 400-600 mg/day led to a rapid improvement and disappearance of clinical symptoms and to a simultaneous reduction of *Candida* in the culture in 33 treatment courses.

Table 2 shows the reason for termination of therapy in the study.

Table 2. Reasons for cessation of the therapy with ketoconazole suspension in 52 treatment courses

	Treatment courses: n = 52
Disappearance of clinical symptoms and reduction of *Candida* in the culture	37
Beginning of therapy with rifampicin	2
Taste of the drug not accepted (lack of compliance)	9
Absent absorption after resection of the stomach, persistence of thrush and clinical symptoms	2
Selection of *Candida glabrata*, persistence of the thrush and clinical symptoms	2

The compound was well-tolerated without severe side effects; only in 9 cases therapy was stopped because patients did not tolerate the taste.

Due to its good therapeutic results and both local and systemic efficacy, ketoconazole suspension will remain a useful concept for short-term treatment of oropharyngeal *Candida* infections.

Ketoconazole, however, has certain disadvantages which restrict its use in AIDS patients. Hepatopathy and interactions with rifampicin and other drugs as well as unfavourable pharmacokinetic properties are the most important ones. Furthermore, ketoconazole is largely ineffective against *Torulopsis* spp., *Aspergillus* spp. and *Cryptococcus neoformans* (Drouhet and Dupont, 1987; Graybill, 1989; Tavitan et al., 1986). Within the last few years, two novel orally absorbed azole derivatives with a broad spectrum of activity itraconazole and fluconazole have been developed and are now subjected to clinical trials.

Itraconazole has been shown to be effective in the treatment of oral candidiasis as well as other severe mycoses in several studies (Cauwenbergh et al., 1987b; Smith et al., 1989; Tricot et al., 1987). In a small number of patients, good results were achieved in the treatment of oral *Candida* infection with itraconazole solution at our department.

Our team also collected substantial experience in the treatment of oral thrush respectively esophagitis with fluconazole.

140 HIV-positive patients with clinical and cultural signs of oropharyngeal *Candida* infection were treated in 152 treatment episodes. Usually the patients were in more advanced stages of HIV infection (9 patients PGL, 25 patients ARC, 106 patients AIDS). Initial dosage was 200 mg/d, followed by a daily dosage of 100 mg.

Duration of therapy is shown in Table 3. Therapy lasted between 3 and 149 days. 79 patients were treated for less than 3 weeks. In a few patients with increased relapse risk, treatment was continued after clinical and mycological cure.

Table 3. Number of treatment episodes (n=152) in relation to treatment days with fluconazole

Treatment days	Treatment courses
0 - 5	16
6 - 10	27
11 - 15	27
16 - 20	9
21 - 25	21
26 - 30	11
31 - 50	20
51 - 90	16
>90	5

In the majority of patients (more than 80% of all treatment episodes), elimination of thrush and enanthema, and accordingly recession of complaints, were achieved within 10 days (Table 4). Disappearance of the complaints reported by the patients is shown in Table 5. The group of nonresponders consisted of patients in advanced stages of HIV disease, patients with cytostatic respectively antibiotic therapy, or patients with a selection of *Candida spp.* known to be less sensitive to azole compounds.

Table 6 shows the reduction of colony-forming units in the oral wash-out. Within 10 days, CFU > 10^3/ml could be detected in 41 out of 152 treatment courses. Obviously the elimination of clinical symptoms is not

Table 4: Course of clinical signs and complaints according to the duration of treatment with fluconazole (100-200 mg/d) in 140 HIV-positive patients with a total of 152 treatment episodes.

Treatment days	Thrush	Enanthema	Complaints
Before	138	71	115
5 days	57	54	41
7 days	34	25	35
8-10 days	26	13	29
End	13	7	17

Table 5. Disappearance of clinical signs in the treatment of fluconazole 100-200 mg/d in 152 treatment courses (HlV-positive patients)

Treatment days	Furry sensation	Burning	Painful swallowing	Distaste
Before	89	5	5	5
5 days	22	7	10	18
7 days	18	4	7	19
8-10 days	16	6	7	12
End	8	4	6	3

Table 6: Reduction of colony-forming units/ml (CFU) in the oral wash-out in the treatment of 140 HIV-positive patients in 152 treatment courses.

Treatment	No. episodes	CFU 10^2	CFU 10^2–10^3	CFU $>10^3$
0	152	0	0	152
5	143	27	58	58
7	126	39	39	48
10	104	31	32	41
14	57	43	10	39
21	57	25	10	22
>21	34	12	8	14

always accompanied by concomitant cultural reduction. On the whole, clinical results of fluconazole treatment were better than the microbiological parameters.

The most frequently isolated *Candida* spp. was *C. albicans*, and the two most selected species during therapy were *C. glabrata*, *C. krusei*, and *C. inconspicua* (Table 7). It is remarkable that *C. glabrata*, a low-grade pathogen, caused enanthema, but not thrush. This corresponds with the findings of Tom and Aaron (1987) who described *C. glabrata* as a cause of esophagitis with ulcerations in an AIDS patient.

However, the influence on the spectrum of *Candida* spp. as well as the impact on clinical course during the treatment with azoles is not clear and will have to be considered. Fluconazole was well tolerated, and apart from mild gastrointestinal symptoms in a small number of patients, no adverse effects were seen even after long-term treatment.

The new azoles present a major step forward in the treatment of fungal infections. At present, 3 compounds are available for the treatment of candidiasis: ketoconazole, itraconazole, and fluconazole.

Ketoconazole is available as tablets and as suspension, itraconazole is available as capsules and solution, and fluconazole is available as capsules, syrup, and as vials for i.v. application. The choice of drug

Table 7: Distribution of *Candida* spp. in oral wash-outs, 100–200 mg/d fluconazole, 152 treatment courses in 140 HIV-positive patients. Only samples with CFU > 103/ml were evaluated.

Candida species	Before treatment	During treatment (Selection)	After treatment
C. albicans	142	0	13
C. tropicalis	3	1	1
C. pseudotropicalis	2	0	0
T. candida	0	1	1
C. glabrata	7	10	5
C. inconspicua	4	5	3
C. krusei	2	7	9

depends on the extent of candidiasis. In mild cases, local application of nystatin or amphotericin B can be prescribed.

However, in cases of extensive candidiasis, systemic therapy is preferred. In severe cases, especially with septic spread of candidiasis, parenteral application of amphotericin B and flucytosine may be necessary.

When choosing a drug, other factors, e.g. patient's condition, absorption, distribution, drug interaction, hepatic metabolism, and hepatic toxicity, should be considered. However, a number of clinical and pharmacokinetic questions are still open.

Our experience shows that mycoses other than candidiasis, such as *Aspergillus* infections, might present a problem in advanced stages of HIV infection. As a consequence, in patients with predisposing factors for *Aspergillus* infection, antifungal drugs with activity.against aspergillosis should be preferred.

The development of more resistant *C. albicans* strains and the selection of other resistant *Candida spp.*, e.g., *C. glabrata*, which may occur due to increasing application of long-term therapy with azole derivatives will have to be considered.

REFERENCES

Brodt, H.-R., Helm, E.B., Joetten, A., Bergmann, L, Klüver, A., and Stille, W., 1986, Spontanverlauf der LAV/HTLV 111 Infektion, Dtsch. med. Wschr., 111:1175.

Cauwenbergh, G., De Doncker, P., Stoops, K.,Dedier, A.M., and Goyvaerts, H., 1987b, Itraconazole in the treatment of human mycosis: Review of three years of clinical experience, Rev. Infect. Dis., 9 (1):146.

Centers for Disease Control, 1987, Revision of the CDC surveillance case definition for aquired immunodeficiency syndrome, M.M.W.R. (Suppl.), 36:1.

Drouhet, E., and Dupont, B., 1987, Evaluation of antifungal agents: past, present, and future, Rev. Infect. Dis., 9 (1):4.

Glatt, A.E., Chirgwin, K., and Landesmann, S.H., 1988, Treatment of infections associated with human immunodeficiency virus, N. Engl. J. Med., 318:1439.

Graybill, J.R., 1989, New antifungal agents, Eur. J. Clin. Microbiol. Infect. Dis., 402.

Holmberg, K., and Meyer, R.D., 1986, Fungal infections in patients with AIDS and AlDS-related complex, Scan. Infect. Dis., 18:179.

Jones, H.E., 1987, Ketoconazole Today - A Review of Clinical Experience, ADIS Press Limited, Manchester.

Klein, R.S., Harris, C.A., Small, C.B., Moll, B., Lesse, Y.M, and Friedland, G.H., 1984, Oral candidiasis in high risk patients as the initial manifestation of the aquired immunodeficiency syndrome, N. Engl. J. Med., 311:357.

Levy, R.M, Janssen, R.S., Bush, T.J., and Rosenblum, M.L, 1988, Neuroepidemiology of aquired immunodeficiency syndrome, J. of AIDS, 1: 31.

Langford-Kuntz, A.A., Rüchel, R., and Reichart, P.A., 1988, Orale Manifestation der Candidiasis bei HIV infektion, Dtsch. Mund-Kiefer-Gesichts-Chir., 12:28.

Porro, G.B., Parente, F., and Cernuschi, M., 1989, The diagnosis of esophageal candidiasis in patients with aquired immune deficiency syndrome: Is endoscopy always necessary? Am. J. Gastroenterol., 1989, 84:143.

Saag, M.S., and Dismukes, W.E, 1988, Minireview. Azole antifungal agents: emphasis on new triazoles, Antimicrob. Ag. Chemother., 32:1.

Saul, A., Bonifaz, A., and Arias, J., 1987, Itraconazole in the treatment of superficial mycosis: an open trial of 40 cases, Rev. Infect. Dis., 9 (1):100.

Smith, D.E., Allan, M., Migley, J., and Gazzard, B.G., Recurrence rate for buccal candidosis and its relationship to T4-count and P24-antigen, V. International Conference of AIDS, Montreal, Abstract N.B.P. 175:251

Tavitan, A., Raufmann, J.P., Rosenthal, L.E., and Weber J., 1986 Ketoconazole resistant *Candida* esophagitis in a patient with aquired immunodeficiency syndrome, Gastroenterol., 71:395.

Tavitan, A., Raufman, J.-P., Rosenthal, L E., 1986, Oral candidiasis as a marker for esophageal candidiasis in the aquired immunodeficiency syndrome, Ann. Intern. Med., 104:54.

Tom, W., Aaron,J.S., 1987, Esophageal ulcers caused by *Torulopsis glabrata* in a patient with aquired immunodeficiency syndrome, Am. J. Gastroenterol., 82:766.

Tricot, G., Joosten, E., Boogaerts, M.A., Vande Pitte, J., and Cauwenbergh, G., 1987, Ketoconazole vs. itraconzole for antifungal prophylaxis in patients with severe granulocytopenia: preliminary results of two non-randomized studies, Rev. Infect. Dis., 9 (1):94.

Van Tyle, J.H., 1984, Ketoconazole: mechanismen of spectrum of activity, pharmacodynamics, drug reactions, adverse reactions, and therapeutic use, Pharmacotherapy, 4:343.

CRYPTOCOCCAL MENINGITIS IN AIDS PATIENTS

A PILOT STUDY OF FLUCONAZOLE THERAPY IN 52 PATIENTS

B. Dupont*, I. Hilmarsdottir, A. Datry, M. Gentilini, P.
Dellamonica, E. Bernard, S. Lefort, J. Frottier, P. Choutet,
J.L. Vilde and the French study group on fluconazole in
cryptococcal meningitis in AIDS patients**

** (M. Armengaud, J.P. Bleriot, J.P. Brion, G. Calamy, A.
Chapman, J.P. Coulaud, P. de Truchis, G. Humbert, C.
Lafaix, J. Modai, M. Obadia, O. Patey, J. Prinseau, F.
Raffi, W. Rosenbaum, R. Roue, F. Vachon, A. Vergnenegre)

* Pasteur Institute, Hospital and Mycology Unit, 25 rue du
Dr. Roux, F 75015 Paris, France

SUMMARY

Cryptococcal meningitis in AIDS patients is liable to relapse and
carries a high mortality rate. Standard treatment with amphotericin B and
5-fluorocytosine is limited by toxic effects. In the present study 64 AIDS
patients with cryptococcal meningitis were admitted to a multicentre trial
of fluconazole given either orally or intravenously. After a loading dose
of 400 mg, two dosage regimens were adopted: 150-200 mg, or 400 mg daily
for at least 45-60 days; subsequent maintenance therapy was 100 or 200 mg
daily for both groups. Forty seven of the 52 evaluable patients were alive
and improved or asymptomatic at day 60 and respectively 7 of 14 and 27 of
31 of those on 200 mg or 400 mg per day had negative cultures of the
cerebrospinal fluid. Relapse rate under maintenance therapy was 23%.
Fluconazole was well tolerated and free from severe side effects. These
results merit further study of fluconazole, either alone or in combination.

INTRODUCTION

Acquired immunodeficiency syndrome is currently the leading
predisposing factor for cryptococcosis (Eng et al., 1986; Kovacs et al.,
1985; Zuger et al., 1988). In most areas in the USA 6 to 9% of AIDS
patients develop this infection. The frequency may be as low as 1.9 to
3.2% or as high as 20% depending on undefined local epidemiological
features (Dismukes, 1988; Katlama et al., 1984; Piot et al., 1984). Our
epidemiological survey in France showed an incidence of 4% in 204 cases
over a period of four years 1985-1988.

Mycoses in AIDS Patients, Edited by
H. Vanden Bossche *et al.,* Plenum Press, New York, 1990

Despite treatment, prognosis is poor with initial lethality of 15-54% and a relapse rate of 40-50% (Kovacs et al., 1985, Zuger et al., 1988) within 6 months after the end of treatment. In a recent study from Chuck and Sande (1989) mean survival was 4.7 months in patients treated with amphotericin B and 6.1 months in patients treated with the association amphotericin B and flucytosine. The standard antifungal treatment amphotericin B plus flucytosine is toxic to renal function and bone marrow. Patients show a lack of compliance to protracted hospitalization due to amphotericin B infusions whether as initial or preventive therapy.

New approaches to the management of cryptococcosis in AIDS patients include the use of new antifungal agents with a potential activity against *Cryptococcus neoformans* such as itraconazole or fluconazole (Byrne and Wajszczuk, 1988; De Gans, et al., 1988; Graybill, 1987; Perfect et al., 1986; Van Cutsem et al., 1987; Van 't Wout et al., 1988; Viviani et al., 1987; see also Denning et al., this book). Fluconazole is a new triazole differing from other azoles pharmacokinetically. In particular the drug diffuses into CSF and is actively eliminated in the urine (Perfect et al., 1986). Results of MICs on *Cryptococcus neoformans* are questionable since this antifungal agent is inhibited by many compounds present in culture media. However, the efficacy of fluconazole in cryptococcal meningitis was established *in vivo* in several animal models (Graybill, 1987; Perfect et al., 1986).

Early in 1986 we had a dramatic success with fluconazole in the treatment of an AIDS patient with cryptococcal meningitis who had relapsed after a standard course of amphotericin B and 5-fluorocytosine (Dupont and Drouhet, 1987). This, combined with the problems generally encountered in managing these patients persuaded us of the need to set up a clinical trial with fluconazole. We report here the results of an open French multicentre study of treatment of cryptococcal meningitis with fluconazole as the initial therapy.

The objectives of this study were to assess the clinical efficacy of fluconazole on the symptoms of meningitis and localizations outside the central nervous system, and the mycological efficacy on the CSF culture and various other sites which were initially positive. The aim of a dose/efficacy study was to specify more accurately the dosage for initial therapy and maintenance therapy. The antigen values were compared as the therapy progressed in order to establish their prognostic value. Finally, the clinical and biological tolerance and patient compliance were observed. The term relapse applied to patients whose CSF culture became positive again after at least one negative sample under treatment.

PATIENTS AND METHODS

The French multicentre study was of open design and the protocol was approved by local Ethical Committees. The 64 patients admitted to the study gave informed consent in accordance with the Helsinki Declaration. All were adult patients with positive CSF culture for *C. neoformans* prior to fluconazole treatment. Sixty two were human immunodeficiency virus (HIV) positive, two patients were HIV negative. One patient was considered as having AIDS because he had two relapses of cryptococcal meningitis and he had a CD4 T. lymphocyte count of 154/ml. The other patient was also considered as having AIDS because of the cryptococcal infection and a count of CD4 T lymphocytes at 113/ml. However, neither patient had detectable antibodies against HIV1 or HIV2 antigen; HIV could not be cultured from their blood and no risk factors were known (Pts. 29 and 37). Patients with

288

positive cultures for *C. neoformans* were also admitted if they had received amphotericin B with a total dose of less than 1 mg/kg, had relapsed and had had no other therapy in the previous month; had relapsed with meningeal symptoms while on maintenance treatment; or had failed to respond (as evidenced by two positive cerebrospinal fluid cultures) to a full standard course of amphotericin B. In such cases patients were enrolled if they had again a CSF positive culture just before starting fluconazole treatment.

Fluconazole was given once a day orally with a meal or intravenously depending on consciousness and vomiting. After an initial loading dose of 400 mg, treatment was continued at a dosage of 150-200 mg or 400 mg per day for at least 45-60 days. Thereafter, once CSF cultures were negative, maintenance therapy was 100 or 200 mg daily. No other antifungal therapy was administered.

The efficacy of treatment was assessed both clinically (symptoms and signs of meningeal infection) and mycologically. CSF, obtained by lumbar puncture on days 1, 30, 45-60, 90, 150 and every two or three months, was examined by the India ink test and cultured on Sabouraud chloramphenicol glucose agar at 30° and 37°C. The cultures were incubated for at least three weeks, since the growth of cryptococci from patients under treatment may be delayed for this length of time. Latex agglutination tests for *C. neoformans* antigen were also performed (Crypto LA test International Biological Labs. Cranbury, N.J. USA) in CSF and in serum.

Side effects were recorded and routine haematological and a standard range of hepatic and renal function tests were carried out at baseline and at intervals during treatment.

RESULTS

Sixty four patients were enrolled between July 1986 and May 1989: 52 were evaluable (Table 1). Twelve patients were excluded from analysis: three died within one to three days after initiation of treatment, nine violated the protocol: errors in dosing or length of treatment (3 patients), associated systemic antifungals (3 patients), insufficient data, particularly no CSF sampling (3 patients). The demographic characteristics of the 52 patients are shown in Table 2. Twenty nine patients had cryptococcal organisms located at least at one site outside the central nervous system (Table 3). At the time of diagnosis of cryptococcal meningitis all patients were full blown AIDS according to the CDC classification. Twenty-two were already at this stage, 15 converted from ARC (IV and IV C_2) to AIDS (IV C_1) 5 converted from asymptomatic (II) to AIDS and in 8 the previously unknown HIV positivity was discovered because of cryptococcal meningitis (Table 1). The mean CD4 T. lymphocyte count was 66/ml with a range of 3 to 345. Eight patients were treated for mycobacterial infection mainly due to tuberculosis with rifampicin 600 to 900 mg per day associated with other antimycobacterial agents.

CLINICAL RESULTS

By day 60 of treatment, 47 (90%) patients were alive and had improved or were asymptomatic. Five patients died: three patients (Pts. 10, 41, 43) died respectively on day 51, 28, 18 of progressive cryptococcosis; one (Pt. 3) died on day 50 without evidence of cryptococcosis, as no antigen was detected in his CSF and CSF culture was negative on day 30; the fifth (Pt. 2) died on day 18, despite initial clinical improvement, from an acute rise in intracranial pressure. No autopsies were performed.

Table 1: Characteristics of fluconazole treatment in 52 AIDS patients with cryptococcal meningitis

Patient number	HIV class* prior cryptococcosis	Cr. neoformans localizations	Intensity of meningeal signs and symptoms**	Fluconazole dose mg/day***	treatment duration days	CSF culture day of treatment and result	Follow up and outcome since therapy began
1	AIDS	CNS	1	a) 150 b) 150 (twice a week)	1 – 60 61 – 390	pre+ 9– 30– 60– 90–	Cleared, death 13 months after 1st day of treatment. No relapse. Autopsy: encapsulated yeasts in meningeal membranes and brain in paraventricular areas, culture negative.
2	revealed HIV+	CNS	1	a) 200	1 – 18	pre+	Failure, death on day 18 after initial clinical improvement. No autopsy.
3	AIDS	CNS	1	a) 200	1 – 50	pre+ 30–	Cleared, death day 50: no evidence of active cryptococcosis. No autopsy.
4	AIDS	CNS, skin lung, spleen	1	a) 200 b) 400	1 – 30 31 – 60	pre+ 17+ 43+	Failure, splenectomy day 47 with positive culture for Cr. neoformans. Failed then to respond to IV ampho, then to fluconazole (400 mg) + 5 FC, then to itraconazole (400 mg) death at 8.5 months No autopsy.
5	AIDS	CNS, blood urine	1	a) 200	1 – 220	pre+ 15+ 30+ 60+ 120+ 160+	Marked clinical improvement, asymptomatic. Death at 6.5 months: wasting disease. No autopsy.
6	AIDS	CNS, urine	1	a) 200 b) 100 c) 200–400	1 – 60 61 – 87 88 – 189	pre+ 15+ 30– 60– 82+ 97+ 127+ 140+ 200–	Cleared, relapse day 82: CSF and blood culture positive. Cleared with 400 mg. Death at 8 months wasting disease. No autopsy.
7	AIDS	CNS, lung	1	a) 200 b) 100	1 – 60 61 – 72	pre+ 15+ 30– 60–	Cleared, death day 72: diffuse Toxoplasma encephalitis. Autopsy: histology and culture negative for Cr. neoformans.
8	AIDS	CNS	1	a) 200 b) 100	1 – 68 69 – 97	pre+ 30+ 60– 90–	Cleared, death day 97. End stage of AIDS. No autopsy.

Table 1: Continued 1

Patient number	HIV class* prior cryptococcosis	Cr. *neoformans* localizations	Intensity of meningeal signs and symptoms**	Fluconazole dose mg/day***	treatment duration days	CSF culture day of treatment and result	Follow up and outcome since therapy began
9	revealed HIV+	CNS, blood	1	a) 400 b) 100	1 – 60 61 – 270	pre– 15– 30– 60– 90– 150–	Cleared, relapsed at 9 months. Failed to respond to fluconazole + itraconazole. Death at 13 months with evolutive cryptococcosis. No autopsy.
10	AIDS	CNS, blood urine, lung, spleen	1	a) 200 b) 400	1 – 30 31 – 36	pre+ 15+ 30+ 41+	Failure, death day 51 after failure of treatment with ampho + 5FC (15 days). No autopsy.
11	ARC	CNS	2	a) 400 b) 100	1 – 60 61 – 180	pre+ 15– 30– 60–	Cleared, death at 6 months: wasting disease. No relapse. No autopsy.
12	AIDS	CNS	1	a) 150 b) 100	1 – 130 131 – 265	pre+ 30– 60– 120– 150– 210–	Cleared, death day 230: no relapse. Autopsy: histology positive, culture negative.
13	AIDS	CNS, blood lung, urine bone marrow	1	a) 400 200 400 b) 100 c) 400	1 – 7 8 – 19 20 – 63 64 – 194 195 – 208	pre+ 15– 30– 60– 90– 120– 150– 180– 190+	Cleared, blood cultures positive day 1 through day 29. Relapsed at 6.5 months. Autopsy:multivisceral involvement except CNS.
14	AIDS	CNS	1	a) 200 no maintenance	1 – 66	pre+ 15+ 30– 60–	Cleared, relapsed day 140, responded to second course of fluconazole. Death end stage of AIDS at 6.5 months. No autopsy.
15	ARC	CNS, blood	1	a) 200 400	1 – 22 23 – 53	pre+ 30+ 56+	Marked clinical improvement, mycological failure. CSF failed to clear with fluco+5FC, cleared with ampho+5FC. Lost for follow up.
16	ARC	CNS, blood urine	1	a) 400 0 400 b) 100	1 – 57 58 – 67 68 – 86 87 – 102	pre+ 30+ 60– 82–	Cleared, death by suicide at 3.5 months. No relapse. No autopsy.

Table 1: Continued 2

Patient number	HIV class* prior cryptococcosis	Cr. *neoformans* localizations	Intensity of meningeal signs and symptoms**	Fluconazole dose mg/day***	treatment duration days	CSF culture day of treatment and result	Follow up and outcome since therapy began
17	ARC	CNS	1	a) 200	1 - 52	pre+ 30+ 52+	Marked clinical improvement, mycological failure. Switch to ampho. Lost for follow up.
18	ARC	CNS	2	a) 400 200 b) 100	1 - 44 45 - 94 95 - 216	pre+ 30+ 45+ 90- 180- 210-	Cleared, death at 8 months. No relapse. No autopsy.
19	AIDS	CNS	2	a) 200 400	1 - 114 115 - 161	pre+ 30- 60+ 80+ 120+ 150+	Failure, cleared with ampho + 5FC. Alive under maintenance with ampho (21 months).
20	AIDS	CNS, urine	1	a) 400 b) 100	1 - 55 56 - 115	pre+ 30- 60- 90- 450-	Cleared, returned to Africa then lost for follow up after 11 months.
21	revealed HIV+	CNS	1	a) 400 b) 100	1 - 45 46 - 440	pre+ 30- 50- 115- 180- 300- 430-	Cleared, death at 16.5 months. No relapse. No autopsy.
22	ARC	CNS	1	a) 400 b) 100 c) 400	1 - 63 64 - 210 211 - 325	pre+ 30+ 60- 120- 180- 210+ 225- 255-	Cleared, relapsed at 7 months, cleared with 400 mg, death at 11 months. No autopsy.
23	AIDS	CNS, blood	1	a) 400 b) 100	1 - 73 74 - 335	pre+ 30+ 60- 90- 135- 220- 310-	Cleared, death at 11 months. No relapse. No autopsy.
24	AIDS	CNS	3	a) 400 b) 200	1 - 74 75 - 191	pre+ 30- 60- 120- 170-	Cleared, lost for follow up at 6 months.
25	revealed HIV+	CNS, urine lung	1	a) 400 b) 200	1 - 85 86 - 157	pre+ 30+ 60- 100-	Cleared, death day 157 end stage of AIDS. No relapse. No autopsy

Table 1: Continued 3

Patient number	HIV class* prior cryptococcosis	Cr. neoformans localizations	Intensity of meningeal signs and symptoms**	Fluconazole dose mg/day***	treatment duration days	CSF culture day of treatment and result	Follow up and outcome since therapy began
26	revealed HIV+	CNS	1	a) 400	1 – 100	pre+ 30+ 60+ 90+	Mycological failure. Cleared with fluconazole + itraconazole, alive (19.5 months).
27	AIDS	CNS, lung	1	a) 200 b) 100	1 – 78 79 – 295	pre+ 30- 60- 105- 240-	Cleared, death at 10 months (lymphoma). No relapse. No autopsy.
28	AIDS	CNS, urine	2	a) 400 b) 200	1 – 81 82 – 134	pre+ 30+ 60- 90- 130-	Cleared, death day 134, relapsed with acute cryptococcal ventriculitis. Autopsy: positive culture of brain and ventricular fluid.
29	HIV-	CNS, blood	1	a) 200 400 b) 200	1 – 60 61 – 190 191 – 720	pre+ 30+ 60+ 90- 120- 180- 330- 730-	Cleared, under maintenance therapy since 24 months.
30	asymptomatic	CNS, blood urine	1	a) 200	1 – 60	pre+ 30+ 60+	Mycological failure, cleared with ampho. Death at 31 months.
31	AIDS	CNS, liver	3	a) 400 b) 200 c) 400	1 – 129 130 – 258 259 – 274	pre+ 30+ 60+ 90- 150- 220- 250+	Cleared, relapsed at day 220, failed to respond at 400 mg then failed with ampho. Death at day 288. No autopsy.
32	AIDS	CNS	3	a) 400 b) 200	1 – 60 61 – 450	pre+ 30- 60- 120-	Cleared, death at day 450. No relapse. No autopsy.
33	ARC	CNS skin	2	a) 400 b) 200	1 – 66 67 – 520	pre+ 45- 110- 180-	Cleared, under maintenance therapy (17 months).
34	ARC	CNS	2	a) 400 b) 200	1 – 60 61 – 360	pre+ 30- 60- 90- 210- 275-	Cleared, under maintenance therapy (12 months).
35	ARC	CNS urine	2	a) 400 b) 200 c) 400	1 – 61 62 – 343 344 – 396	pre+ 15- 30- 60- 90- 120- 180- 300- 330+	Cleared, relapsed at day 330 but was responding clinically at 400 mg.

Table 1: Continued 4

Patient number	HIV class* prior cryptococcosis	Cr. neoformans localizations	Intensity of meningeal signs and symptoms**	Fluconazole dose mg/day***	treatment duration days	CSF culture day of treatment and result	Follow up and outcome since therapy began
36	revealed HIV+	CNS blood urine	1	a) 400 b) 200 c) 50	1 – 115 116 – 290 291 – 320	pre+ 30+ 60+ 90+ 120– 320+	Cleared, wrongly put on 50 mg at day 29, and relapsed 30 days later; responded then to ampho.
37	HIV–	CNS lungs	1	a) 400 b) 200	1 – 69 70 – 466	pre+ 30+ 60– 90–	Cleared, asymptomatic under maintenance (17 months).
38	asymptomatic	CNS	1	a) 400 b) 200	1 – 76 77 – 512	pre+ 30– 45– 60– 397–	Cleared, asymptomatic under maintenance (16 months).
39	ARC	CNS urine, blood	1	a) 400 b) 200	1 – 45 46 – 541	pre+ 30–	Cleared, asymptomatic under maintenance, patient refused further CSF sampling.
40	AIDS	CNS urine	1	a) 400 b) 200 c) 400	1 – 60 61 – 100 101 – 130	pre+ 30– 60– 100+	Cleared, relapsed at day 100, was clinically responding at 400 mg.
41	ARC	CNS blood, urine	2	a) 400	1 – 28	pre+ 15+ 21+	Death with evolutive cryptococcosis and wasting disease.
42	asymptomatic	CNS	3	a) 400 b) 200	1 – 87 88 – 150	pre+ 30+ 60– 90– 120– 290–	Cleared, fluco. stopped for pregnancy at day 150, still cleared at day 320.
43	asymptomatic	CNS blood	3	a) 400	1 – 18	pre+ 10+	Death due to cryptococcosis.
44	AIDS	CNS blood	1	a) 400 b) 200	1 – 62 63 – 210	pre+ 30– 60– 90– 150– 210–	Cleared, asymptomatic under maintenance (8 months).
45	ARC	CNS blood, urine skin spleen	1	a) 400 b) 200	1 – 81 82 – 168	pre+ 15+ 30+ 60– 105– 150–	Cleared, death at 5.5 months of wasting disease. No relapse. No autopsy.
46	AIDS	CNS	2	a) 400 b) 200	1 – 58 59 – 78	pre+ 30– 60– 208+	Cleared, patient stopped maintenance at day 78 and relapsed at day 208. Fluconazole 400 mg given again.

Table 1: Continued 5

Patient number	HIV class* prior cryptoccocosis	Cr. neoformans localizations	Intensity of meningeal signs and symptoms**	Fluconazole dose mg/day***	treatment duration days	CSF culture day of treatment and result	Follow up and outcome since therapy began
47	ARC	CNS blood, urine	1	a) 400 b) 200	1 - 61 62 - 214	pre+ 15+ 30- 60- 120- 180-	Cleared asymptomatic under maintenance (9 months).
48	ARC	CNS	1	a) 400 b) 200	1 - 60 61 - 202	pre+ 15+ 30-	Cleared, no further CSF sampling (antivitamine K treatment) no clinical relapse at day 340. Stopped maintenance at day 202.
49	revealed HIV+	CNS	3	a) 400 b) 200	1 - 60 61 - 150	pre+ 15+ 30- 60-	Cleared, returned to his country. Asymptomatic, lost for follow up.
50	ARC	CNS	1	a) 400 b) 200 c) 400	1 - 90 91 - 111 112 - 145	pre+ 30+ 60- 95+ 115+ 145+	Cleared, relapsed 10 days after maintenance, failed to respond to fluco 400 mg and then to ampho. Fluco + flucytosine started.
51	asymptomatic	CNS blood, lungs	1	a) 400 b) 200	1 - 60 61 - 159	pre+ 15+ 30- 60- 90- 120-	Cleared, death of brain toxoplasmosis at day 160.
52	revealed HIV+	CNS	2	a) 400 b) 200	1 - 125 126 - 346	pre+ 30- 60- 90- 150- 210-	Cleared, asymptomatic under maintenance (12 months).

*   AIDS means full blown AIDS
** 1: latent or meningismus; 2: obtundation, cranial nerve palsy, seizure; 3: comatous
*** a: means initial therapy, b: means maintenance therapy, c: means treatment for relapse
ampho: amphotericin B
fluco: fluconazole

Table 2. Demographic characteristics of the 52 patients with cryptococcal
meningitis

Sex		
male	44	
female	8	
Age (years)		
mean (range)	34.5	(22-77)
Weight (kg)		
mean (range)	55.9	(39-76)
Ethnic origin		
White	43	
Black and Haitian	9	
Homosexual	20	
Heroin addict	10	
Africans, Haitians	9	
Heterosexual	6	
Others	7	
Past history of cryptococcosis		
meningeal	3	
pulmonary	1	
Full blown AIDS*	22	
Chronic HIV positive	20	
HIV positivity revealed by meningitis	8	
HIV negative	2	

* prior diagnosis of cryptococcosis

Table 3. 47 cryptococcal localizations with positive cultures outside the
central nervous system in 29 patients

Origin	No. of patients	
Blood (hemoculture)	15	(28.8)
Urinary tract (uroculture)	18	(34.6)
Skin (biopsy)	3	(5.7)
Lung (broncho-alveolar lavage)	2	(3.8)
Liver (puncture-biopsy)	5	(9.7)
Spleen	2	(3.8)
Eye*	1	(1.9)
Bone marrow	1	(1.9)

*Diagnosis on examination of the fundus of the eye

In those patients who responded, the efficacy on fever and on meningeal signs and symptoms was rapid, within 3-5 days, and often dramatic. The disappearance of all clinical abnormalities was observed during the third or fourth week of treatment. Most of the patients not otherwise debilitated by other pathological processes were able to return home within the first month of treatment and in time to return to work. Even Pt. 4 who remained culture positive was stable and ambulatory. Three patients who had previous cryptococcal meningitis responded to fluconazole: one patient (Pt. 1) had relapsed after standard treatment with amphotericin B and flucytosine, another patient (Pt. 29) relapsed twice despite apparent clinical and mycological clearance with amphotericin B and flucytosine; a third patient (Pt. 12) relapsed under maintenance therapy with amphotericin B and had failed to respond to higher dose of amphotericin B and flucytosine treatment.

One patient had simultaneously cerebral toxoplasmosis and cryptococcal meningitis, 4 patients developed cerebral toxoplasmosis during fluconazole treatment. Symptoms of toxoplasmosis were mixed with those of cryptococcosis. Patient No. 7 died of acute diffuse *Toxoplasma* encephalitis diagnosed at autopsy, although he had a normal brain CT Scan two days before death.

Up to the time of this report 30 patients have died, 5 are not available for follow up, one is free of symptoms without maintenance therapy (Pt. 48) and 16 are receiving a maintenance therapy. Cryptococcal infection was the direct cause of or contributed to death in 10 patients, in 20 patients the cause of death was a wasting disease and/or an opportunistic infection with no clinical or mycological signs or symptoms of reactivation or relapse of the quiescent or cleared cryptococcal infection.

MYCOLOGICAL RESULTS

The results of the microbiological study of the cerebrospinal fluid are shown in Table 4 where the effects of the two treatment regimens of fluconazole are compared at day 30 and day 45-60. Quite clearly the continuation of therapy in the 400 mg per day group produced the better outcome as only 12.9% of patients had positive culture at day 45-60 compared with 50% in the 150/200 mg per day group. However only 1-5 colonies of cryptococci grew in most of these positive cultures in both treatment groups.

Table 4: Results of microbiological study of the cerebrospinal fluid in 52 AIDS patients with cryptococcal meningitis treated with fluconazole 150-200 mg per day or 400 mg per day

| | 150/200 mg = 17 patients | | 400 mg = 35 patients | |
	Day 30	Day 45-60	Day 30	Day 45-60
* CSF positive				
India ink	9/16	7/14 (50%)	12/32	5/31 (16%)
Culture	8/16	7/14 (50%)	14/32	4/31 (13%)
* Death	1/17	3/17	2/35	2/35
* No sample	0	0	1/35	2/35

Twenty nine patients had 47 localizations of cryptococcal infection outside the central nervous system (Table 3). In 25 of these patients, blood, urine, skin or bronchoalveolar fluid samples also became negative. Patient No. 13 had fever and positive blood cultures prior to and during fluconazole treatment at 200 mg/day. When the fluconazole dosage was raised to 400 mg on day 20 negative blood cultures and apyrexia resulted. Two patients: No. 4 and 10 failed to respond, and all cultures including CSF cultures remained positive. Patient No. 13 and 30 had several urine cultures positive until death. However in two patients (Pts. 41 and 43) who died with evolutive meningitis, were cleared from extra CNS localizations.

Examination for cryptococcal antigens was positive in the CSF of 47/51 patients and in the serum of 46/50 patients for whom this examination was carried out, with titer values between 1/10 and $1/10^6$. It was negative for the CSF of 4 patients and in the serum of 4 other patients. No patient displayed an absence of antigen in the CSF and in the serum simultaneously. In 8 patients a return to normal was observed in the detection of antigen in the CSF and in 8 patients a return to normal was observed in the serum. Among these patients, 3 were negative both in the serum and the CSF.

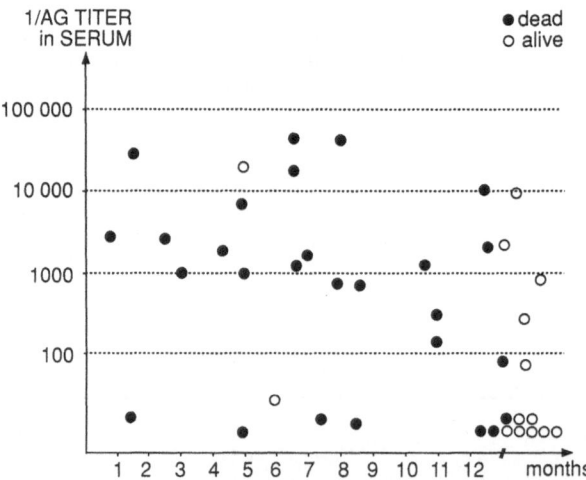

Figure 1. Correlation between the cryptococcal antigen titer in the serum and the survival of 41 patients alive (white circle) or dead (black circle).

While examination for cryptococcal antigens in the serum and the CSF
was of great diagnostic value, it would seem to us that the prognostic
value of the antigen titer was less so. Figs. 1 and 2 show that the
earliest deaths did not always correspond to the highest titer values,
certain subjects with low titer values dying early. On the other hand, a
persistence of high serum values despite therapy seemed to us a bad
prognosis, just as a marked increase in serum or CSF values generally
presaged a relapse. During the fluconazole therapy most patients showed a
low residual antigen value of between 1/4 and 1/100 despite favourable
clinical progress and negative CSF and blood cultures for several months.

MAINTENANCE THERAPY

Thirty-nine patients whose CSF became negative in culture under
initial therapy with fluconazole continued to receive maintenance
fluconazole at a dosage of 150 mg twice a week (1 Pt.), 100 mg/day (14
Pts.), or 200 mg/day (24 Pts.). The average follow-up was 7 months (range
10 days-18 months). Most patients had two successive negative lumbar
puncture culture before changing over to the maintenance therapy, although
7/39 patients who were apparently clinically cured only had one negative
lumbar puncture in culture before changing to maintenance therapy
but the following lumbar puncture(s) were also negative: 4/39 patients only

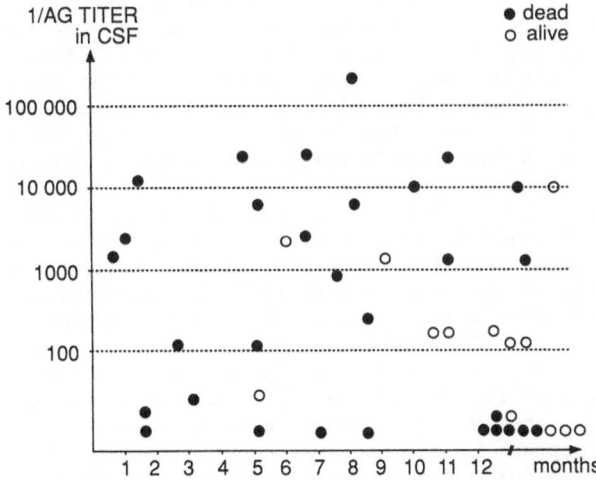

Figure 2. Correlation between the cryptococcal antigen titer in the CSF and
the survival of 43 patients alive (white circle) or dead (black
circle).

had one negative CSF culture: Pts. 36 and 50, Pt. 39 refused subsequent lumbar punctures and Pt. 48 treated with anti-vitamin K presented contraindications to lumbar punctures. The 13 other patients, not included in the maintenance therapy evaluation, were either deceased (5 Pts.), or placed on another antifungal therapy in view of the persistent positive results of the CSF cultures (5 Pts.),or on maintenance therapy although the CSF was not sterilized (2 Pts.), or refused the maintenance therapy (1 Pt.).

Of those patients who received 150 mg twice a week or 100 mg/day 4/15 relapsed, and in the group treated with 200 mg/day 5/24 relapsed; the rate of relapse was 23% (9/39) with an average period of 3.7 months (limits 0.4-9) after the start of maintenance therapy. Three patients out of 39 were considered in the evaluation of the prophylaxis of the relapses solely for the period during which the therapy at 200 mg/day was administered normally. After taking 200 mg/day for 169 days, patient 36 was mistakenly placed on 50 mg/day and relapsed one month later (not counted as a failure at 200 mg); patient 42 when one month pregnant, halted therapy for 140 days and then resumed fluconazole at the end of pregnancy after a normal lumbar puncture; patient 48 stopped maintenance at day 142 and did not relapse during the following 200 days. Patient 14 was not included in the evaluation, refused maintenance therapy and relapsed 80 days later; patient 46 stopped maintenance after 19 days and relapsed 130 days later.

The average survival time of patients treated solely with fluconazole has been calculated from the first day of therapy. Patients lost from the study after at least 6 months' therapy (Pts. 20, 24) were included in the calculations with their survival times.

The average survival time was 5.8 months (range 18 days-13 months) for 10 patients treated with (150)-200 mg per day as initial therapy. Among these patients 9 died and 1 was lost from the study after 6 months (Pt. 14). The survival time was 10.4 months (range 17 days-24 months) for 31 patients treated with 400 mg/day; 14 are still alive with 13 on maintenance therapy. The average survival time of the 17 patients treated with 400 mg/day who died was 7.6 months (range 17 days-17.5 months), the average survival time of the 14 patients treated with 400 mg/day initially and still alive at the writing of this report was 14.5 months (range 6-24 months). The overall survival time of the 31 patients treated with fluconazole (200 mg and 400 mg) was 10.4 months. This value decreases to 9.7 months if we exclude the 2 patients (Pts. 29 and 31) with AIDS but HIV negative who have a long survival time, 24 and 17 months respectively, and are still undergoing therapy. The overall survival time of 9 patients initially treated with fluconazole followed by other antifungal agents, for reasons of clinical and/or mycological inefficiency, was 12 months (range 51 days-18 months). Two patients were excluded from the survival time calculations due to insufficient information on the follow-up (Pt. 49) or errors in the protocol (Pt. 36) after the initial therapy. At the time of writing of this report the overall mortality for all therapies was 30 patients out of 52, five were lost from the study: 3 after the initial therapy, two after 6 and 11 months of apparent cure, seventeen are still alive. In 10/30 cases the cryptococcal infection contributed to or was the direct cause of death; five times due to failure of the initial therapy, five times after relapse during a course of maintenance therapy with partial or zero response to an increase in dosage. The other causes of death were miscellaneous opportunistic infections during the final phase of AIDS. Death was presumed to be not due to cryptococcus if no cryptococci could be isolated and there was no increase of antigen levels or clinical evidence of meningitis.

TOLERANCE

The clinical and biological tolerance was very good. In no case did fluconazole administration have to be stopped definitively due to toxicity, side effects were reported in 22 patients; increase in transaminases to more than three times initial values: 7 patients; increase in alkaline phosphatases and γ-GT to three times initial values: 5 patients for each of these parameters; eosinophilia: 7 patients; thrombopenia: 2 patients; digestive disorders - nausea, diarrhoea, vomiting: 10 patients; pruritus: 2 patients; skin rash: 1 patient. Biological anomalies were often present before therapy or appeared during therapy, remaining stable or returning spontaneously to normal values despite the continuation of fluconazole therapy. The abnormal clinical conditions were mild. These symptoms are not specific and could be attributed to AIDS, other intercurrent infections or other medication taken simultaneously. The gastro-intestinal disorders, eosinophilia and increase in γ-GT would appear to be related to the fluconazole in our opinion. Eight patients were treated simultaneously with rifampicin for mycobacterial infections; it was not observed clinically that this association diminished the effect of fluconazole, although the serum levels of fluconazole were not measured systematically, and one patient (Pt. 22) relapsed one month after starting rifampicin. In one patient treated with fluconazole and then an anti-vitamin K to treat a phlebitis (Pt. 48) the excessive drop in prothrombin levels justified a reduction in dosage of the anticoagulant. One patient was treated with fluconazole (200 mg) during the first month and then during the last two months of a pregnancy.

DISCUSSION

The clinical benefit obtained by 90% of patients after 60 days of treatment with fluconazole was impressive , in particular the higher dose of 400 mg per day was accompanied by a high rate (27/31) of microbiological cure. These results can be favourably compared against the relatively poor outcomes after the standard treatment of amphotericin B plus flucytosine (Kovacs et al., 1985; Zuger et al., 1988; Dupont and Drouhet, 1987). However, publications on this combination in patients with AIDS with details on survival time, drug efficacy and relapse rate are rare and include only a small number of cases.

Chuck and Sande (1989) have recently published results obtained with 40 patients treated with intravenous amphotericin B at 0.5 mg/day and with 49 patients treated with associated amphotericin B: 0.5 mg/day and 5-fluorocytosine: 5 g/day. The results for both regimens after two months' therapy were combined: 17/89 (21%) patients died and 63/89 (79%) patients were alive, the average survival rate was 4.7 months with amphotericin B and 6.1 months with associated amphotericin B and 5-fluorocytosine. The figures are better for our series, especially the regimen at 400 mg which resulted in a survival time greater than 10 months. In the same study, Chuck and Sande obtained with their patients a direct examination of positive CSF for *Cryptococcus* in 7/39 patients (17.9%) and a positive culture in 5/39 (12.8%) after 2 months of therapy; this corresponds perfectly with the results of our series under fluconazole (400 mg) with respective values of 5/31 (16.1%) and 4/31 (12.9%). Dismukes et al. (1989) have presented preliminary comparative results of a random study including 42 patients treated with 200 to 400 mg/day of fluconazole and 23 patients treated with amphotericin B with a negative CSF in culture in the first group in 19/42 (45%) and in the second group in 11/23 (48%). The respective mortality being 31% and 26%. Larsen et al. (1989) in a series of 14 patients treated with 400 mg of fluconazole per day for 2 months, found 5/14 (36%) cured, 1/14 (7%) improved and 8/14 (57%) failures while amphotericin B in 7 patients resulted in 5/7 (71%) cured and 2/7 (29%) improved. The discrepancies between this study and the preceding ones concerning amphotericin B may be due to the small number of patients or to a centre effect.

Our work confirms, in a large series of patients, the data published on oral fluconazole therapy of cryptococcosis in patients with AIDS (Stern et al., 1988; Sugar and Saunders, 1988). Stern et al. (1988) reported a clinical and mycologic response in three of four patients with active meningitis and showed that 14 of 15 patients, successfully treated with amphotericin B as initial therapy, remained free of infection while receiving fluconazole during 11 to 64 weeks as a suppressive therapy. Sugar and Saunders (1988) also found oral fluconazole useful for long-term suppressive therapy and reported two culture proven relapses in 19 treated patients, with a follow up exceeding six months in 13 of them. Our series involving 39 patients submitted to maintenance therapy showed no significant difference for dosages of 100 or 200 mg/day. However, the series was short, and furthermore, the effect of dosage observed in initial therapy and the increase in survival in patients taking 200 mg as maintenance therapy leads us to prefer this dosage.

A difficult therapeutic problem is illustrated by patients under fluconazole who are asymptomatic but still have positive CSF cultures. Naturally, the aim of the treatment is to eradicate the pathogens. This seems almost impossible in AIDS patients since their immune status is getting worst with time and the available antifungal agents are only fungistatic. Even when CSF cultures become negative, the yeasts may persist in deep tissue as evidenced by histology in autopsied patients. A more modest and realistic aim of treatment would be to clear clinical signs and symptoms allowing the patient to leave hospital. For this reason and because of the short life expectancy we think that such asymptomatic patients could continue to receive fluconazole. In the case of clinical relapse or of a net increase in serum of CSF antigen titers, increasing the dose of fluconazole or prescribing amphotericin B would be the alternative therapy.

In our opinion the efficacy of fluconazole, its very good tolerance and the excellent compliance of patients make this drug an alternative to amphotericin B in the treatment of cryptococcal meningitis in AIDS

A comparative randomized study of fluconazole versus amphotericin could indicate which drug has the best efficacy, however, such a study would take a long time especially if the difference between the drugs is small. Possibly more valuable studies would be to evaluate how high the dosage of fluconazole can be increased, and to test potential drug combinations such as fluconazole plus flucytosine or, in the future, fluconazole and itraconazole.

REFERENCES

Bennett, J.E., Dismukes, W.E., Duma, T.R. et al., 1975, A comparison of amphotericin B alone and combined with flucytosine in the treatment of cryptococcal meningitis, N. Engl. J. Med., 301:126.

Byrne, W.R., and Wajszczuk, C.P., 1988, Cryptococcal meningitis in AIDS: successful treatment with fluconazole after failure of amphotericin B, Ann. Intern. Med., 108:384.

Chuck, S.L., and Sande, M.A., 1989, Infections with Cryptococcus neoformans in the acquired immunodeficiency syndrome, N. Engl. J. Med., 21:794.

Dismukes, W.E., 1988, Cryptococcal meningitis in patients with AIDS, J. Infect. Dis., 157:624.

Dismukes, W., Cloud, G., Thompson, S., Sugar, A. et al., NIAID mycoses study group, 1989, Fluconazole versus amphotericin B therapy of acute cryptococcal meningitis, Abstract No. 1065, ICAAC.

Dupont, B., and Drouhet, E., 1987, Cryptococcal meningitis and fluconazole, Ann. Intern. Med., 106:778.

Eng, R.H.K., Bishburg, E., Smith, S.M., and Kapila, R., 1986, Cryptococcal infections in patients with acquired immune deficiency syndrome, Am. J. Med., 81:19.

Gans de ,J., Eeftinck Schattenkerk, J.K.M., and Ketel, V.R.J., 1988, Itraconazole as maintenance treatment for crytococcal meningitis in the acquired immune deficiency syndrome, Br. Med. J., 296.

Graybill, J.R., 1987, Fluconazole efficacy in animal models of mycotic diseases, in: "Recent Trends in the discovery, Development and Evaluation of Antifungal Agents", R.A. Fromtling, ed., Prous Science, Barcelona.

Katlama, C., Leport, C., Matheron, S. et al., 1984, Acquired immunodeficiency syndrome in Africans, Ann. Soc. Belge Med. Trop., 64:379.

Kovacs, J.A., Kowacs, A.A., Polis, M. et al., 1985, Cryptococcosis in the acquired immunodeficiency syndrome, Ann. Intern. Med., 103:533.

Larsen, R.A., and Leal, M.E., 1989, Fluconazole compared to amphotericin B as treatment of cryptococcal meningitis, Abstract No. 1062, ICAAC.

Perfect, J.R., Savani, D.V., and Durack, D.T., 1986, Comparison of itraconazole and fluconazole in treatment of cryptococcal meningitis and candida pyelonephritis in rabbits, Antimicrob. Agents Chemother., 29:579.

Piot, P., Taelman, H., Kapita, B.M. et al., 1984, Acquired immunodeficiency syndrome in a heterosexual population in Zaire, Lancet, 2:65.

Stern, J.J., Hartman, B.J., Sharkey, P., Rowland, V., Squires, K.E., Murray, H.W., and Graybill, R., 1988, Oral fluconazole therapy for patients with acquired immunodeficiency syndrome and cryptococcosis: experience with 22 patients, Am. J. Med., 85:477.

Sugar, A.M., and Saunders, C., 1988, Oral fluconazole as suppressive therapy of disseminated cryptococcosis in patients with acquired immunodeficiency syndrome, Am. J. Med., 85:481.

Van Cutsem, J., Van Gerven, F., and Janssen, P.A.J., 1987, The in vitro and in vivo antifungal activity of itraconazole, in: "Recent Trends in the Discovery, Development and Evaluation of Antifungal Agents", R.A. Fromtling, ed., Prous Science, Barcelona.

Van 't Wout, J.W., de Graeff-Meeder, E.R., Paul, L.C., Kuis, W., and Van Furth, R., 1988, Treatment of two cases of cryptococcal meningitis with fluconazole, Scand. J. Infect. Dis., 20:193.

Viviani, M.A., Tortorana, A.M., and Giani, P.C., 1987, Itraconazole for cryptococcal infection in the acquired immunodeficiency syndrome, Ann. Intern. Med., 106.

Zuger, A., Lonie, E., Molzmann, R.S., Simberkoff, M.S., and Rahal, J.J., 1988, Cryptococcal disease in patients with the acquired immunodeficiency syndrome, Ann. Intern. Med., 104:234.

ORAL ITRACONAZOLE THERAPY OF CRYPTOCOCCAL MENINGITIS AND CRYPTOCOCCOSIS IN

PATIENTS WITH AIDS

David W. Denning, Richard M. Tucker, John S. Hostetler,
Satpreet Gill and David A. Stevens

Div. of Infectious Diseases and Microbiology, Dept. of
Medicine and Pathology, Santa Clara Valley Medical Center,
751 South Bascom Avenue, San Jose, CA, 95128 USA
Institute for Medical Research, San Jose, CA 95128
Div. of Infectious Diseases, Dept. of Medicine, Stanford
University School of Medicine, Stanford, CA 94305

SUMMARY

We studied the efficacy of oral itraconazole in 57 AIDS patients with
cryptococcosis. Diagnoses included cryptococcal meningitis (48 patients),
cryptococcemia (28 patients), cryptococcuria (16 patients), pulmonary
cryptococcosis (12 patients), and 2 patients who were symptomatic, antigen
positive and culture negative. Patients received itraconazole 200 mg b.i.d.
orally given as sole therapy in 55 of 57 (96%); five patients with low
serum concentrations received 200 mg t.i.d. Therapy was monitored by
clinical and radiological response, culture and cryptococcal antigen
testing. Twenty three of 36 (64%) evaluable patients with cryptococcal
meningitis had complete responses (clinical resolution and negative
cultures), 8 (22%) a partial response and 5 (14%) failed therapy. In those
not previously treated, only 2 of 29 (7%) failed therapy. Meningitis
recrudesced on therapy in 13 of 31 (42%) responding patients; the median
time to recrudescence in these 13 patients was 4 months after starting
itraconazole. Cryptococcemia was abolished in 17 of 19 (89%) evaluable
patients together with clinical resolution. Among the 9 evaluable patients
with pulmonary cryptococcosis, 8 have responded to therapy (89%).
Cryptococcuria was abolished in 4 of 10 (40%) evaluable patients. Among the
whole group of 57 patients, relapse occurred in 4 of 7 (57%) patients
discontinuing therapy. Prior or concurrent therapy with rifampin (for
mycobacterial disease) was associated with one patient failing therapy and
another having to be withdrawn with no response at eight days. Our response
rates compare favorably to amphotericin B therapy with or without
flucytosine. Eighty-five of 96 (91%) isolates of C. neoformans were
susceptible in vitro to itraconazole (MIC≤3.13 mcg/ml), 6 were borderline
(MIC 6.25 mcg/ml) and 3 were resistant (≥12.5 mcg/ml). Oral itraconazole is
promising for the primary or salvage therapy of cryptococcal meningitis and
other forms of cryptococcosis in patients with AIDS.

INTRODUCTION

Over 100,000 patients with the acquired immune deficiency syndrome (AIDS) have now been reported in the U.S.A. (Centers for Disease Control, 1989). The reported incidence of cryptococcal meningitis in patients with AIDS in the U.S.A. varies between 2% (Dismukes, 1988) and 9% (Zuger et al., 1986) whereas in other areas the incidence is even higher (up to 40% in Africa). Other forms of cryptococcosis in AIDS including fungemia, pulmonary disease and prostatic disease are not infrequent, and may not always occur with meningitis. Assuming a lifetime incidence of all forms of cryptococcosis of 10%, about 10,000 individuals will already have had the disease as part of AIDS and between 8,000 and 12,000 patients will develop cryptococcosis in the USA alone over the next 3 years. Of all the opportunistic infections afflicting patients with AIDS, patients with extra-pulmonary cryptococcosis have a markedly shorter survival time, a median of 8.4 months (Horsburgh et al., Interscience Conference on Antimicrobial Agents and Chemotherapy, 1988, Abstracts). The provision of nontoxic, effective therapy for these patients, which can be administered orally, is therefore a pressing problem.

Among patients with cryptococcal meningitis in the pre-AIDS era, therapy prevented death in 41-76% of cases (Bennett et al., 1979; Hay et al., 1980; Diamond and Bennett, 1974). Many surviving patients were left with severe neurological sequelae. The clinical response to conventional therapy for cryptococcal meningitis in AIDS patients was about the same; 42% in one series (Kovacs et al., 1985), 75% in another (Zuger et al., 1986) and 79% in a third (Chuck and Sande, 1989), but relapse is more common, probably 50-65% without suppressive therapy (Dismukes, 1988). Many AIDS patients who relapse die of cryptococcal disease, despite therapy. Conventional therapy has comprised amphotericin B with flucytosine in some cases. Toxicity with both drugs is a problem; 35% of patients on flucytosine had to discontinue it in one series (Zuger et al., 1986) and most patients receiving intravenous amphotericin B experience some toxicity.

For these reasons attention has focused on the new azoles which are relatively nontoxic, orally administered alternatives to amphotericin B. The vast majority of *Cryptococcus neoformans* isolates are susceptible to itraconazole *in vitro* (Denning et al., 1989a). In a rabbit model of cryptococcal meningitis, itraconazole was effective in treating disease (Perfect et al., 1986) despite its lack of CSF penetration. Efficacy of itraconazole in other animal models has been shown (Van Cutsem et al., 1987; Graybill and Ahrens, 1984). Clinical data has suggested efficacy in cryptococcosis in patients with AIDS (Denning et al., 1989a; Viviani et al., 1987; de Gans et al., 1988). In this report we extend our observations to 57 patients with follow-up extended to a maximum of 20 months from the start of therapy.

MATERIALS AND METHODS

Patients

All patients enrolled had evidence of active cryptococcal disease (Denning et al, 1989a). All patients gave written informed consent and the study was approved by the institutional review board at Santa Clara Valley Medical Center (SCVMC). The data are summarized as of mid-October 1989. Some of these patients have been reported previously (Denning et al., 1989a, 1989b, 1990); their details are updated here.

Cerebrospinal fluid (CSF) was collected for fungal cultures and determination of cryptococcal antigen titer from all patients prior to initiation of itraconazole. Blood and urine was collected for fungal culture from most patients prior to initiation of itraconazole and respiratory cultures from some. Clinical response was assessed by a

combination of symptoms, fever, physical examination findings and
radiography if appropriate. Culture response was assessed by repeated
fungal cultures of previously positive sites. Serological response was
assessed by repeated measurements of CSF and blood cryptococcal antigen
titers. Possible toxicity of itraconazole was assessed by patient
interviews and by measurement at frequent intervals of hematologic
parameters; serum electrolytes, urea, creatinine, uric acid, calcium,
phosphate, cholesterol, triglycerides, and glucose; liver function tests
and urinalysis. Serum concentrations of itraconazole were obtained at
steady state (after 2 weeks of therapy).

Patients enrolled early in the study received oral itraconazole 200 mg
twice daily. A loading dose of 200 mg t.i.d. was given for 4 days to 11
patients. On the basis of low serum concentrations, 5 patients received
200 mg t.i.d. on a longer term basis. The treatment plan was to continue
therapy for 6 months beyond the last evidence of disease activity unless
the patient chose to discontinue the medication, died, failed to respond or
recrudescence occurred. However, duration of therapy has ranged from 1 day
to 20 months, with a median of 4.0 months thusfar in the evaluable
patients. We have defined complete response as complete resolution of all
clinical symptoms and signs of disease and negative cultures. Partial
response (only for meningitis) is defined as major symptomatic response
without CSF cultures becoming negative. Failure is defined as no parameter
improving after at least 2 weeks of itraconazole therapy. Recrudescence is
defined as emergence of cryptococcal disease on therapy with itraconazole
after either a partial or complete response, and relapse as the emergence
of cryptococcal disease after cessation of itraconazole therapy or during
therapy if serum concentrations fell from a therapeutic concentration to an
undetectable level. Any patient receiving less then two weeks of therapy is
deemed unevaluable. Follow up continued until a year after stopping
itraconazole and cure is defined as a one year disease-free interval.

Our laboratory procedures have been previously described (Denning et
al, 1989a, 1990).

Fisher's exact probability test and the Wilcoxon Rank Sum test were
used for analysis.

RESULTS

We report on 57 patients with cryptococcal disease and AIDS. Fifty-six
of the patients are male, and 55 of these probably acquired their HIV
infections through homosexual contact. One man and the one woman almost
certainly acquired their infections from blood product transfusion. The
patients' ages ranged from 21 to 73 years with a mean of 37.8 years. Among
the 46 patients with previously untreated cryptococcal infection, the
episode treated with itraconazole was the marker infection for AIDS in 18
(38%). At the time of writing 35 (61%) had had *Pneumocystis carinii*
pneumonia (PCP) and 7 (12%) had had cytomegalovirus disease requiring
therapy, preceding or during itraconazole therapy. Most patients received
zidovudine at some time but almost all had required dose modification or
discontinuation for reasons of toxicity. A small number of patients (≤15)
had taken other antiviral agents including acyclovir, AL721, compound Q and
others; the majority of patients took other herbal medicines.

Initial culture and serological results are given in Table 1. Only 2
(4%) patients were culture negative from all sites at the start of therapy.
One had been previously CSF culture positive and was clinically and
serologically failing amphotericin B therapy. The other was culture
negative from all sites despite a serum cryptococcal antigen of 1:32 and
high fever, and stopped itraconazole after 3 months; 10 days later cultures
from CSF and blood became positive. Not all patients had samples of sputum
and urine cultured, and these different denominators are reflected in the
table. Urine cultures were usually obtained without prostatic massage. Two

patients were culture positive from bone marrow, and one from a liver biopsy.

Cryptococcal Meningitis

Forty-eight patients had evidence of meningitis. Forty-seven were CSF culture positive prior to commencing itraconazole, 34 of 46 (74%) had positive India ink preparations of the CSF and 47 (98%) had positive cryptococcal antigen titers in the CSF. The median titer of the positive CSF antigens was 1:256 with a geometric mean of 1:205 and they ranged from positive undiluted to 1:16,384. Only 42 of 46 (91%) had positive serum cryptococcal antigen titers, ranging from 1:2 to 1:130,272, with a median of 1:1024 and a geometric mean of 1:981. Not all these initial sera were tested after pronase digestion. Positive blood cultures were obtained in 24 of 34 (71%) patients with meningitis.

Table 1. Enrollment characteristics

	No. Positive/No. Tested (%)
Cryptococcal Meningitis	48/57 (84)
CSF culture	47/48 (98)
India Ink	34/46 (74)
CSF cryptococcal antigen	47/48 (98)
Blood culture	24/34 (71)
Urine culture	15/24 (63)
Sputum/BAL/Pleural fluid culture	5/7 (71)
Serum cryptococcal antigen	42/46 (91)
Cryptococcemia	28/38 (74)
Without meningitis	4/28 (14)
Serum cryptococcal antigen	26/28 (93)
Pulmonary Cryptococcosis*	12#
Without meningitis or cryptococcemia	3/12 (25)
Urinary Cryptococcus	16/25 (64)
Without meningitis, pulmonary disease or cryptococcemia	1/16© (6)
Culture Negative	2/57 (4)
CSF and serum cryptococcal antigen positive (and rising) with headache and fever	1/2 (50)
Serum antigen positive (1:32) with fever	1/2 (50)

* Positive culture from respiratory site and/or radiologic abnormalities that regressed on therapy.
# Relevant samples, x-rays and data were not collected from all patients so our denominator is incomplete.
© 16 patients had positive urine cultures for *C. neoformans*.

Table 2. Cryptococcal meningitis

	No.	Response						Death on therapy[+] No. (mos.)	
		CR	CR+RC	PR	PR+RC	F	U		
Previously untreated									
No therapy	25	6*	4	2*	4	2	7	3	(0.1, 1.5, 9.0)
<7 days Amp B	12	8	2	0	1	0	1	1	(4.5)
Total	37	14	6	2	5	2	8	4	
Previously treated									
Failing prior therapy	6	1#	2	1#	0	1	1^	4	(0.1, 0.8, 1.5, 4.3)
Relapse post-therapy	5	0	0	0	0	2	3	3	(0.2, 0.5, 0.6)
Total	11	1	2	1	0	3	4	7	

CR: complete response, RC=recrudescence, PR=partial response, F=failure, U=unevaluable
#: patients given concurrent, but reduced dose amphotericin B
^: failing ketoconazole
+: deaths also included in response columns
*:1 patient relapsed in each group

Of the 48 patients with meningitis, 37 had not received an evaluable cour
of therapy and 11 had had a previous course of therapy. Twenty-five
patients had received no therapy prior to itraconazole and 12 had received
up to seven days of amphotericin B. Eight (22%) of these 37 patients were
unevaluable (Tables 2, 3). Among the 29 evaluable patients in this group c
37, 20 (69%) had a complete response and 7 (24%) a partial response (Table
3); two patients (7%) failed therapy. Four of these 37 patients patients
died on therapy at a variety of intervals. Complete response was
significantly more frequent (10/11, 91%) in evaluable patients who receive
≤7 days amphotericin immediately prior to itraconazole than those who
received no prior therapy (10/18, 56%) (p=0.048). If the 2 partial
responders with documented low serum itraconazole concentrations are
excluded from this analysis, all 10 evaluable patients who received ≤7 day:
amphotericin had complete responses (p=0.02, comparison with 10/17 who
received no prior therapy).

Table 3. Unevaluable patients

More than 13 days therapy

Pulmonary Kaposi's sarcoma and cryptococcal pneumonia Evaluation unreliable	1
Noncompliant (CNS disease not evident 4 months after a 1 month course of itraconazole)	1
Total	2

Less than 14 days therapy

Respiratory deaths (days 3 and 5 of itraconazole)	2
Respiratory decompensation	1
Pancreatitis and respiratory decompensation	1
Worsening meningitis on rifampin	1
Intractable vomiting	1
Sudden death, cause unclear (day 1)	1
Death due to gram negative sepsis and neutropenia	1
Noncompliant patient, doctor	3
Total	11

Of the eleven patients who had received previous therapy; 5 had
relapsed after a course of amphotericin B with or without flucytosine, 5
were failing therapy with amphotericin B and 1 was failing ketoconazole
when commenced on itraconazole. Among the patients started on itraconazole
because they were failing their prior therapy, 3 of 5 (60%) evaluable
patients had a complete response, one had a partial response, one failed
therapy. In contrast the patients who relapsed after a course of
amphotericin B fared poorly; 2 failed therapy and 3 were unevaluable. One
of the unevaluable patients probably responded but was noncompliant in
follow up. Three of these 5 patients died on therapy within 3 weeks of
starting itraconazole and a fourth died on amphotericin B within a month o
discontinuing itraconazole.

Overall therefore, complete responses were seen in 64% of evaluable
patients with meningitis, partial responses in 22% and failures in 14%. If
the 2 patients who received concurrent amphotericin B therapy are excluded

the respective response rates are 65%, 21% and 15%. The mean and median duration of therapy was 7.2 and 5.3 months respectively for the complete responders and 2.4 and 2.7 months respectively for the partial responders. No statistically significant association between response or lack of response and pre-treatment culture positive extrameningeal sites (n=25), CSF antigen titer (n=35) or serum sodium concentration (n=38) was seen.

We have computed the time to sterilization of the CSF in all patient groups and these are shown in Table 4. Early in the study, some lumbar punctures were not done for up to 3 months into therapy so it is likely that the actual time to sterilization is much less; the figures given are maximum times. The median time to sterilization from the day of starting itraconazole was 30 days for all patients if an assumption is made that the CSF was culture positive until the day before the first documented negative CSF sample. If one assumes the time of conversion occurred mid-way between the last culture positive CSF and the first negative CSF sample, the median for all patients is 22 days. For only complete responders (who became culture negative by definition) the median times to sterilization are 27 and 13.5 days, respectively, using these assumptions.

Table 4. Frequency over time of culture negative CSF specimens in patients with meningitis on itraconazole

Patient group	Days after starting itraconazole										
	≤8	9-15	16-22	21-30	31-45	46-60	61-75	76-90	91-120	121-150	151-200
CR, no RC	4/7	9/11	9/11	10/12	10/11	9/10	10/10	9/9	7/7	8/8	7/7
CR, RC	1/2	2/4	3/5	5/6	8/8	5/5	5/5	4/4	2/2	1/1	1/1
PR	0/8	0/8	0/8	0/7	0/5	0/5	0/3	0/2	0/2	0/1	0/1
F	0/5	0/5	0/3	0/2	0/1	0/1					
U	0/3										
Total	5/25	11/28	12/27	15/27	18/25	14/21	15/18	13/15	9/11	9/10	8/9

CR=complete response, RC=recrudescence, PR=partial response, F=failure, U=unevaluable
Numerator=culture negative CSF specimens, denominator=number of CSF specimens cultured

Various assumptions have necessarily been made in constructing this table. These are:
1) Interval between 2 sterile CSF specimens is also sterile, and likewise the interval between 2 culture positive specimens is assumed to be culture positive.
2) Interval between last CSF sampling and death (if not due to cryptococcosis) is sterile, if the last CSF was sterile.
3) Last lumbar puncture is last data point for surviving patients.
4) Interval between sterile CSF and recrudescence is negative for first half of that interval.
5) Interval between a culture positive CSF specimen and a culture negative one is assumed to be positive until the day the negative specimen is obtained.
6) The CSF sample demonstrating recrudescence (and therefore culture positive) and any subsequent CSFs are not included in above table (see Fig. 1).

Of 14 complete responders who had a positive India Ink test which was subsequently repeated, 11 (79%) converted to negative and 3 remained positive. Five of the former group (45%) and one of the latter subsequently recrudesced on therapy. Three partial responders remained India Ink positive in the CSF (60%), one became negative and one became positive having been negative. Two failing patients converted from negative to positive and another remained positive. Two unevaluable patients converted from positive to negative although both remained culture positive; one was noncompliant at 12 days after starting itraconazole because of lethargy; the other had intractable vomiting. A uniform amount of CSF was not processed for India Ink and changes in positivity of this test may be influenced by this.

In 14 of 21 evaluable complete responders, CSF cryptococcal antigen titers declined 2 dilutions or more and in 5 cases to zero. The titers remained static in 2 patients, one of whom recrudesced and the other has been on therapy only 2 months. Among the 7 evaluable partial responders, 4 showed a 2-fold or greater decline in CSF antigen titers and 3 remained unchanged. In patients failing therapy, the CSF titer rose in 2, declined in one and remained static in a fourth. The CSF was not sampled a second time from one patient who failed. The serum responses also mirrored the clinical responses in most instances. Among the complete responders, 17 of 19 evaluable serum cryptococcal antigen titers fell 2 dilutions or more, one was unchanged and one rose early in therapy despite a falling CSF antigen. Five of 7 evaluable partial responders had unchanged (by 2 dilutions or more) serum antigen titers, one had a falling titer and one a rising titer. Two of 4 paired sera among the failures showed rising titers, and two were unchanged.

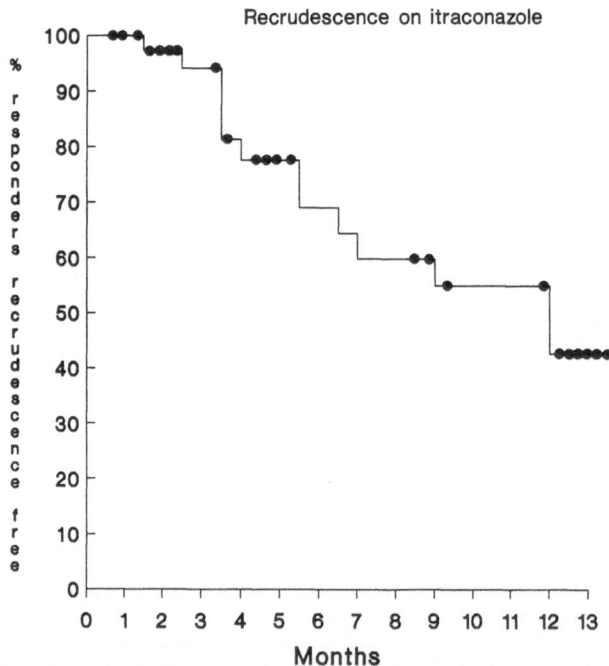

Figure 1.   Kaplan-Meier plot of proportion of meningitis responders (complete and partial) remaining recrudescence-free over time. Each dot represents the end of observation for each responding patient who did not recrudesce.

Recrudescence occurred in 13 of 31 (42%) responding patients with meningitis. This occurred at 1.5 to 12 months of therapy (median 4 and mean 5.2 months) (Fig. 1). The clinical features of recrudescence were initially a recurrence of headache (without fever) or other symptoms which were previously present, such as backache. In all cases CSF cultures became positive if they had been negative. In 2 patients conversion of CSF culture from negative to positive preceded clinical deterioration by 1 to 2 months. In most other patients, deterioration occurred rapidly and both CSF and blood cultures became positive. In patients with low CSF antigen titers a marked rise was seen at the same time or slightly preceding, clinical deterioration. Serum titers usually rose in addition, sometimes dramatically. Recrudescence occurred despite therapeutic serum concentrations of itraconazole in those in whom they were measured at the time of recrudescence and unchanged *in vitro* susceptibility of the infecting organism.

Of the 13 patients who recrudesced, 7 clinically responded to institution of amphotericin B and 2 failed amphotericin B. Two patients clinically responded to and one failed fluconazole. One patient with culture conversion without symptoms remained on itraconazole and was found dead at home 1.5 months later, having been functioning well the week before. In summary, 2 of 13 (15%) died directly as a result of recrudescence.

One patient (No. 179) who developed recurrent symptoms of meningitis after 10 months therapy refused lumbar puncture and started to take 20 garlic tablets daily. The symptoms remitted without any other change in therapy. The Chinese use garlic to treat cryptococcal meningitis (Anonymous, 1980), so this observation is of considerable interest.

Cryptococcemia

Four patients had cryptococcemia without meningitis. All four (100%) responded to itraconazole. One patient recrudesced after 12 months therapy, one is still on therapy, having been somewhat noncompliant, one relapsed 2 months after finishing 9 months therapy and one is disease free 2 months after stopping therapy given for 12 months. Of the 19 evaluable patients with cryptococcemia at the start of therapy, 17 had negative blood cultures on therapy, and 2 (11%) remained positive. Serum concentrations of itraconazole in these 2 failing patients exceeded the MIC of the organism. One of these patients failed therapy and committed suicide after 1 month of itraconazole, the other was a partial responder who recrudesced on therapy.

Urinary Cryptococcosis

At least 16 patients had positive urine cultures prior to starting therapy. Urine cultures were not done systematically at the start of the study. A few patients grew *Candida* sp. from the urine which may have obscured the growth of *C. neoformans* (the colonies usually look identical on routine fungal media). Of these 16 patients, 6 were unevaluable or did not have their urine recultured for fungi. Six of 10 (60%) patients remain culture positive on therapy. One patient with disease apparently limited to the urinary tract failed therapy probably related to undetectable serum concentrations of itraconazole associated with concurrent rifampin therapy.

Pulmonary Cryptococcosis

At least 12 patients had evidence of pulmonary disease radiologically, on culture or usually both. Three patients had only pulmonary disease. One who had concurrent pulmonary Kaposi's sarcoma is unevaluable. One patient responded and had a marked improvement in his general state of health, took itraconazole for 5.3 months and has not relapsed 6 months after stopping therapy. The third patient has failed therapy after 3.5 months taking only 100 mg b.i.d. Among the 9 pulmonary patients with other sites of disease, 2 are unevaluable with respect to their pulmonary disease, but the rest all showed radiological improvement; follow up sputum cultures were not always obtained or possible to obtain. Four patients had marked mediastinal adenopathy which improved in the 3 evaluable patients.

Relapse

Six patients have discontinued therapy (all but two of their own volition, earlier than we advised) and one patient was given phenytoin with a subsequent fall in serum itraconazole concentrations to undetectable levels. Of the 3 patients with meningitis, one is cryptococcal disease-free 4 months after finishing 13 months of therapy, and the other two have relapsed, one of whom died 2 weeks later. Two patients with cryptococcemia and the symptomatic patient with only a positive serum cryptococcal antigen stopped therapy and 2 relapsed, both with meningitis and cryptococcemia. One is cryptococcal disease free two and a half months after discontinuation of 12 months therapy. One patient with pulmonary disease remains cryptococcal disease free 6 months after completing 5.3 months of itraconazole. Overall, therefore, the relapse rate is 57%. Relapse generally occurred within 2 months of discontinuation of itraconazole.

Survival

The median survival for all AIDS patients is now about 13 months (Horsburgh et al., Interscience Conference on Antimicrobial Agents and Chemotherapy, 1988, Abstracts), however considerably shorter survival times are reported among patients with cryptococcosis, a median of 8.4 months in one large study (Horsburgh et al., 1988). We have collected detailed data on survival in the patients in this report (Table 5).

Table 5. Mortality of evaluable patients

	No.
Alive on therapy	13
Alive off therapy	13
Dead	18

Causes of Death		Patient No.
Cryptococcal meningitis on itraconazole	2	176, 199
Cryptococcal meningitis off itraconazole	5	110, 167, 180, 182, 219
Weight loss and inanition	3	101, 126, 127
AIDS dementia	1	220
Sudden death	1	192
Bradycardia followed by sudden death	1	134
Respiratory failure, ?PCP (after 4.5 mos. itraconazole)	1	234
CMV pneumonitis	1	228
Bilateral pneumothoraces, interstitial pneumonitis	1	202
CNS lymphoma	1	221
Suicide	1	192

Thirty-one patients are still alive and so the survival statistics are minimum survival times. For all 57 patients, the mean survival time was at least 12.8 months from the diagnosis of AIDS and the range was 1 to 33 months with a median of 12 months. From the day of starting itraconazole survival ranged from 1 day to 20 months with a mean of 7.1 months and a median of 6 months. Twenty-seven patients commenced itraconazole 12 months or more before this data was summarized mid-October 1989. Fourteen (52%) of these have survived for 12 months or more from the time they started itraconazole. In patients with meningitis (including unevaluable patients), survival was slightly shorter, with a mean and median exceeding 12.1 and 10 months overall and 6.5 and 6 months after starting itraconazole. If only evaluable patients with all sites of infection are considered, overall survival with AIDS exceeds a mean and median of 13.3 and 12 months respectively and 8.3 and 8.5 months respectively after starting itraconazole. Minimum survival from the diagnosis of cryptococcal infection is a mean and median of 8.7 and 8 months for all patients, is 8.0 and 7.3 respectively for patients with meningitis (including unevaluable patients) and is 12.2 and 15 months respectively for the 9 patients without meningitis.

Adverse Experiences

Three patients suffered a deterioration in pulmonary function shortly after starting itraconazole therapy. Two patients did not receive a loading dose and one did (No. 291). All 3 had extremely high serum cryptococcal antigen titers (1:16,384, 1:65,532 and 1:130,272). *In vitro* susceptibility indicated that all 3 of the organisms were highly susceptible to itraconazole (MIC/MFC = 0.8/0.8, <0.4/0.4 and 1.6/3.1). One of these patients (No. 244) died after 3 days and no autopsy was done. Another (No. 291) developed hypotension and hypoxemia 24 hours after starting itraconazole. He underwent a bronchial washing and biopsy 24 hours after deteriorating which showed the pulmonary parenchyma to be packed with cryptococcal organisms with no other pathogen. This episode resolved with continued therapy. A third patient (No. 185) was discharged from the hospital on itraconazole, but returned on day 5 with severe shortness of breath, hypoxemia, and bilateral pulmonary infiltrates (Fig. 2 and 3). He was hypoxic on 100% O_2 by mask. Bronchoscopy revealed *C. neoformans* on brush and BAL, without other pathogens; no biopsy was done. He was then treated with amphotericin B to which he responded.

One patient died suddenly within 36 hours of starting itraconazole. He did not receive a loading dose. No autopsy was done. He had had AIDS for 12 months prior to developing cryptococcal meningitis and cryptococcemia. Many cardiac abnormalities including cardiac cryptococcosis have been described in AIDS (Kinney and Monsuez, 5th International Conference on AIDS, 1989, Abstracts), as have autonomic neuropathy and cerebrovascular lesions which could account for death.

Another patient with very severe disseminated disease, a serum cryptococcal antigen titer of 1:65,536 and a CSF antigen titer of 1:4096 required high dose dexamethasone to control vomiting. Within a week he was markedly neutropenic and a bone marrow examination showed multiple cryptococcal organisms present. He was not taking zidovudine. He then developed gram negative sepsis and died. He had had AIDS for 10 months.

Toxicity

Thirty three patients (58%) exhibited no toxicity. Possible toxicity is listed in Table 6. Nausea was the commonest symptom associated with the administration of itraconazole. It usually resolved during further therapy. Lethargy was severe in both cases and caused both patients to discontinue

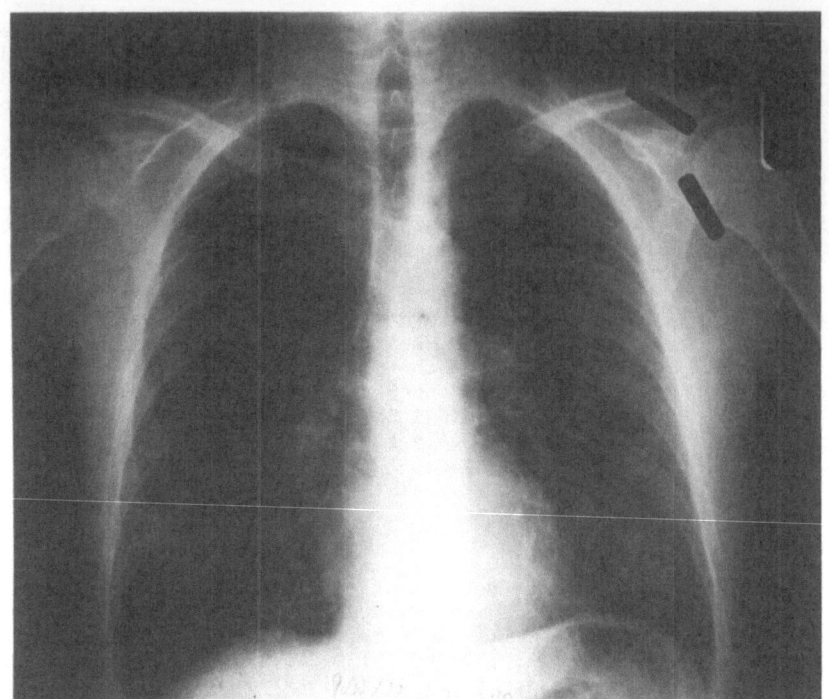

Figure 2. Chest x-ray of patient 185 at diagnosis of cryptococcosis.

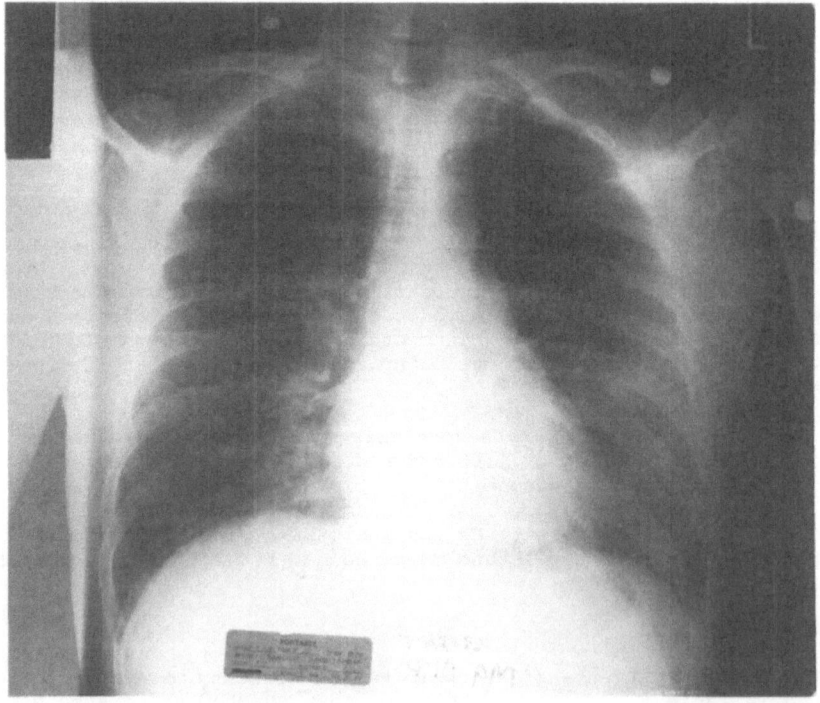

Figure 3. Chest x-ray of patient 185 five days after starting itraconazole showing bilateral interstitial infiltrates and increased right middle lobe and left mid-zone shadowing.

itraconazole. However an ACTH stimulation test was abnormal in one patient
and lethargy did not reappear on rechallenge with itraconazole after
hydrocortisone replacement therapy. Neutropenia was only seen in 2
patients, both with severe disease, one of whom was also taking zidovudine.
In the patient not on zidovudine, neutropenia led to gram negative sepsis
and death. Hypokalemia was seen in 5 patients, 2 of whom were also on
amphotericin B. This effect is occasionally seen with itraconazole in non-
AIDS patients as well (Tucker et al., 1988). Triglycerides were elevated in
5 patients, but they were not usually measured in the fasting state. In one
of these patients with very high triglycerides (>1000 mg/dl), pancreatitis
developed and resolved despite continuation of itraconazole. Another
patient developed pancreatitis within 5 days of starting itraconazole
(serum calcium was slightly elevated prior to starting). He recovered off
itraconazole. No autopsy data exist in the literature on the frequency of
pancreatic cryptococcosis and the possibility exists that this event is
akin to pulmonary decompensation as described earlier. Liver function test
abnormalities are hard to evaluate in AIDS patients but some abnormalities
were seen. It was also the case that some patients with elevated enzymes on
entry to the study had resolution of abnormalities on itraconazole.

With the possible exception of pancreatitis, no life threatening
toxicity has been observed with itraconazole in these 57 patients.

Table 6. Possible toxicity

	No. of episodes
Nausea	7
Increased liver function tests	7
Hypokalemia	5
Increased triglycerides	4
Increased LDH	3
Lethargy	2
Pancreatitis	2
Neutropenia	2
Increased alkaline phosphatase	2
Vomiting	1
Rash	1
Anemia	1

Pharmacokinetics

Serum samples were obtained from 40 patients. Itraconazole serum
concentrations were lower than we have noted previously (Tucker, et al,
1988) and varied from 0 to 12.5 mcg/ml with a 4 hour post-dose mean of
4.34 mcg/ml (36 patients) and a 12 hour post-dose mean of 4.04 mcg/ml (15
patients). Elevation of the daily dose to 200 mg t.i.d. (5 patients)
resulted in improvement in serum concentrations in 2 patients. Among the
others, 1 was taking rifampin, one had chronic persistent diarrhea and
levels were not repeated in the third. Eleven patients received loading
doses as described in Methods and comparative data in a small number of
patients indicates that steady state is achieved faster following loading
doses, probably by day 7 in most patients. In 3 patients repeated

measurements of serum concentrations over several months indicated rising concentrations with time, in some instances to twice the concentration measured at 2 weeks of therapy.

Susceptibility Studies

We have tested 94 isolates of *C. neoformans* against itraconazole. The results are displayed in Table 7. The geometric mean of the MIC and MFC values is 2.58 mcg/ml and 15.15 mcg/ml respectively. Eighty-five (91%) of the MICs fall below the mean serum concentrations of itraconazole, in contrast to the MFCs of which only 21 (26%) do. Tissue concentrations and local macrophage concentrations of itraconazole may be higher than serum concentrations. An MIC or MFC for a particular patient's isolate higher than the serum concentrations of itraconazole did not preclude a complete response.

Relationship of Response to Itraconazole Concentrations and Susceptibility Results

We used the Wilcoxon Rank Sum test to compare the patient outcome with peak and mean serum itraconazole concentrations. The analysis was constrained because there were few failures and some data was missing on these. We found no relationship between peak serum itraconazole concentration and primary response in the overall group (p=0.54) although 2 of the 6 partial responders who had serum itraconazole concentrations measured had values under 1 mcg/ml compared with 0 of 25 complete responders (p=0.04). Peak serum itraconazole concentrations were not different in complete responders who recrudesced and complete responders who did not (p=0.45). One exclusion we have made could have affected some of these non-correlations; it should be recalled that we defined relapse as reappearance of disease activity after blood levels were undetectable (due to failure to take itraconazole or drug interaction).

This analysis suggests that if the patient has a clinical response to itraconazole, a steady state serum concentration under 1 mcg/ml is more likely to be related to persistent positive CSF cultures or, if the concentration falls during therapy (to ≤1 mg/ml) to relapse. Aside from this, neither serum itraconazole concentration or susceptibility results

Table 7. Minimum inhibitory concentrations and minimum fungicidal concentrations of 94 isolates of C. neoformans tested against itraconazole

CONCENTRATION mcg/ml	MIC	MFC
≤0.39	14	1
0.78	31	5
1.56	29	9
3.13	11	6
6.25	6	17
12.5	2	22
≥12.5	1	23
Total	94	82

presently can predict clinical events except possibly the pulmonary events in patients with very low MFCs previously discussed.

DISCUSSION

We have previously documented the efficacy of itraconazole for cryptococcosis in a group of 26 patients with AIDS (Denning et al., 1989a). The present study updates our experience to a total of 57 patients and 104 disease sites in patients with AIDS. The response rate of 89% in cryptococcemia and 86% in cryptococcal meningitis compares favorably to the best results achieved by amphotericin B and 5FC in patients with and without AIDS (Bennett et al., 1989; De Wytt et al., 1982; Dismukes, 1987; Staib and Seibold, 1988; Chuck and Sande, 1989). In an early study the response rate with amphotericin B and 5FC in a series of 24 AIDS patients with cryptococcal fungemia and meningitis was only 42% (Kovacs et al., 1985) although these authors took four weeks as their assessment time. Of the 10 responders, 4 relapsed between 4 and 7 months post therapy. Mortality within 8 weeks of diagnosis was 38%. In our patients only 12 of 57 (21%) died within 8 or 10 weeks of diagnosis and 4 of diseases not related to cryptococcosis (suicide, CMV pneumonitis, PCP, CNS lymphoma). In a more recent study of cryptococcal meningitis in 89 AIDS patients, 79% were alive at 8 weeks (Chuck and Sande, 1989). In our patients with meningitis, 77% (37 of 48) were alive at 8 weeks.

Patients with AIDS and nonpulmonary cryptococcal infections have a worse prognosis than other AIDS patients with opportunistic infections. As some 50% of these cryptococcal infections are the marker infection for AIDS, this is particularly discouraging. In one study, patients with meningitis survived for 93 days (median), and those with cryptococcemia for 40 days after diagnosis of fungal infection, but these survival times included 6 patients who received no therapy (Gal et al., 1987). In a study of 10 patients treated with amphotericin B in Germany, the median survival from the diagnosis of cryptococcosis was 6 months (Staib and Seibold, 1988). In New York City, AIDS patients with cryptococcal infections as the marker infection for AIDS survived a median of 7 months from diagnosis; 30% of patients survived 12 months (Saltzman et al., 4th International Conference on AIDS, 1988, Abstracts). Among 106 AIDS patients treated in San Francisco, the median survival of patients treated with amphotericin B with or without flucytosine was 6.4 and 5.7 months after the diagnosis of cryptocococcis in those with meningitis and those without, respectively (Chuck and Sande, 1989). Another study from New York indicated an improved median survival from 3 to 6 months in the periods 1979-85 to 1986-88 but it was not stated whether this time was from the diagnosis of cryptococcal infection or from the diagnosis of AIDS (Gold et al., 5th International Conference on AIDS, 1989, Abstracts). The Centers for Disease Control, Atlanta noted a median survival among 3022 AIDS patients with extra-pulmonary cryptococcosis of 8.4 months from the diagnosis of AIDS (Horsburgh et al., Interscience Conference on Antimicrobial Agents and Chemotherapy, 1988, Abstracts).

Our data in this report are therefore remarkable. Including the unevaluable patients, we have demonstrated a median survival from the diagnosis of AIDS exceeding 12 months. In only 38% of our patients was cryptococcosis the marker infection for AIDS suggesting that our patients were on average more immunocompromised than other series where 50-100% of patients had cryptococcosis as their marker infection. In our patients with meningitis, the median survival exceeds 10 months. Fourteen of 27 (52%) patients have already survived for 12 months or greater from the time of starting itraconazole. From the diagnosis of cryptococcosis, the median survival in our patients exceeds 7.3 months in those with meningitis (48 patients) and 15 months in those without (9 patients).

The cure of cryptococcal meningitis in patients with AIDS remains elusive. Two of 3 patients relapsed after stopping therapy. One is cryptococcal disease-free 4 months after stopping itraconazole having taken it for 15 months. One patient with only pulmonary disease is cryptococcal disease-free 6 months after stopping therapy. Two of 3 patients with cryptococcemia have relapsed; the third is well, 2 months after stopping itraconazole. As fungicidal concentrations of itraconazole are achieved in serum in about 25% of patients, even with poor and declining host defenses it may be possible to cure a small proportion of patients. For patients with AIDS suppression of cryptococcosis without cure may be sufficient if the agent being used for treatment has little toxicity, as is true of itraconazole.

Following recrudescence, 9 of 12 patients responded to other therapy; 7 to amphotericin B with or without 5FC, 2 to fluconazole. Two patients failed amphotericin B and one failed fluconazole. Early studies have generally indicated that relapse after or recrudescence on conventional therapy has been unresponsive to further therapy in AIDS patients (Kovacs et al., 1985). More recent data does not support this conclusion (Chuck and Sande, 1989) as 11 of 13 patients survived long enough to receive another 1500 mg of amphotericin B. Our data indicating that itraconazole is effective primary therapy and that conventional amphotericin B therapy is effective when patients recrudesce during itraconazole therapy is strong evidence that itraconazole is a genuine alternative therapy for cryptococcal meningitis. It was also effective as salvage therapy (5 evaluable patients); 3 patients had a complete response (60%), one a partial response (20%) and 1 failed itraconazole (20%). However as therapy for patients relapsing after previous therapy, it performed poorly as 2 of 5 patients failed therapy and 3 died within 3 weeks of commencing itraconazole. It may therefore prove that it is best used for primary therapy, unless vomiting prevents administration, with amphotericin B and 5FC reserved for salvage therapy rather than the reverse.

We do not have a good explanation for the recrudescence of disease in the 13 patients while on treatment. *In vitro* development of resistance to itraconazole was not noted in 3 pairs of isolates before and after recrudescence. We can only postulate that these patients' immunity waned further allowing massive replication of organisms, despite presumably adequate tissue concentrations of itraconazole. This hypothesis is supported by 2 observations. Patient No. 182 was doing well until zidovudine had to be stopped to allow gancyclovir administration for CMV retinitis. Shortly after this, his cryptococcal disease flared. Patient No. 206 was a complete responder until he received 3 weekly injections of compound Q. After the second, cryptococcal meningitis returned. We are intrigued by the observation in one patient that garlic tablets combined with itraconazole effected a clinical response following possible recrudescence. Garlic has both anticryptococcal and immunostimulatory activity, especially on macrophage function (Anonymous, 1980).

Infection of the urinary tract, probably the prostate, is a common site of infection as part of disseminated cryptococcosis in AIDS. In our patients with meningitis 65% were urine culture positive, a figure that might have been higher had prostatic massage been routinely done. Amphotericin B therapy frequently fails to sterilize the urine of AIDS patients with disseminated cryptococcosis (Larsen et al., 1989). In the 10 patients with positive urine cultures before therapy who were recultured, 4 remain negative on therapy. Our data do not establish whether prostatic and urine infection is suppressed or eradicated.

CSF cryptococcal antigen levels and India Ink preparations proved to be a sensitive and reliable way of following therapy in our patients with meningitis. CSF cryptococcal antigen titers declined more quickly than serum titers, especially if the sera were pronase treated. It is likely the normal mechanism of removal of the capsular polysaccharide of *C. neoformans* is deranged in most, if not all, AIDS patients as it is probably dependent on macrophage function. CSF cultures were also helpful but as some did not become negative and there was often considerable delay before reports were issued, antigen titers and India Ink preparations were generally more rapid indicators of progress or otherwise during therapy. CSF cell counts, protein and glucose were generally not helpful in diagnosis or following therapy, especially as late and substantial rises in cell counts occurred in a few patients at 4-8 weeks despite complete responses and later return of the cell counts to normality.

Evaluation of failure or response in the first 2 weeks of therapy is not helped by any of these laboratory objective parameters in the illest patients in whom the clinical assessment is also difficult. Fever is often absent in these patients and its presence may be due to other coexisting pathogens. Clearing of headache and improvement (often subjective) in cognition and general well being were the most helpful early indicators of improvement. However, some patients appeared to deteriorate in these parameters, prior to improvement. In some patients, elevated CSF pressures persisted despite improvement in other parameters. Successful treatment of the patient with elevated CSF pressures may well require other therapeutic modalities, such as frequent lumbar punctures, shunting of CSF and/or corticosteroid therapy. Five patients with meningitis reported here had markedly elevated opening pressures at the onset of itraconazole therapy. Three of these failed therapy. Two received no other therapy for elevated CSF pressures and both died within 3 weeks of starting itraconazole. A third, who also failed therapy, who had cranial nerve abnormalities in addition, was treated with frequent lumbar punctures with resolution of most abnormalities until he failed to return for lumbar puncture as scheduled; after this he deteriorated. Two other patients were continued on itraconazole despite high CSF pressures (>500 mm CSF in one) and eventually both were complete responders. One had a ventriculoperitoneal shunt placed and the other received dexamethasone and underwent frequent lumbar punctures. The persistence of elevated CSF pressures on therapy may not imply failure of the antifungal agent. In patients with high CSF pressures the actual antifungal agent used may be less important than these other therapeutic endeavors in the early phases of treatment.

We found no association between response and pre-therapy CSF or serum cryptococcal antigen titers. In non-AIDS patients, CSF antigen titers did not correlate with outcome whereas serum titers did (Dismukes et al., 1987). One study showed a 100% mortality in patients with CSF antigen titers over 1:10,000 (Zuger et al., 1986) but only one of our patients had a titer this high (he clinically responded, but is unevaluable because no follow up lumbar puncture was done and he was noncompliant). We also found no association between outcome and India Ink positivity, unlike others (Zuger et al., 1986). We found no correlation between positive blood cultures and outcome in meningitis patients, in contrast to other studies (Diamond and Bennett, 1974; Dismukes et al., 1987; Chuck and Sande, 1989). We also found no association between pre-treatment serum sodium and outcome as others have (Chuck and Sande, 1989). Whether this lack of correlation represents a different population of patients, too small a group of failures to show an effect, or whether the action of itraconazole obscures otherwise poor prognostic factors if amphotericin B were used, we cannot tell.

The observation that 3 patients had severe respiratory compromise shortly after starting itraconazole is of considerable importance clinically and theoretically. We speculate, in lieu of detailed pathophysiological data, that these events are akin to the Jarisch-Herxheimer reaction of penicillin treatment for secondary syphilis. All 3 isolates from these patients were extremely susceptible to the cidal effect of itraconazole. Rapid lysis of organisms may have led to release of inflammatory mediators leading to hypoxia and hypotension. Such events are recognized in different organs for syphilis, various bacteria and mycobacteria, some protozoa (e.g., cerebral cystercercosis) but have not been described for any fungal infection. Rapid fungicidal activity in the context of extremely severe disease may have occasioned such an occurrence. Clinical management of this event may be improved by the use of corticosteroids, although very early or prophylactic therapy may be very important if blockade of inflammatory mediator production is to occur.

Detrimental drug interactions were critical in determining outcome in some patients. Both rifampin and phenytoin administration resulted in failures at the start of therapy and relapse because of undetectable serum itraconazole considerations. In other patients we have documented a negative effect of carbamazepine on serum concentrations. Others' data indicate a possible negative interaction with phenobarbital (Legendre, unpublished). As these interactions preclude the use of the 3 first line anticonvulsants in a disease in which seizures do occur, this is clearly a difficulty with the use of itraconazole. On the other hand, we have used sodium valproate in one AIDS patient without an apparent interaction (serum itraconazole concentrations 5 mcg/ml) and we would not anticipate any interaction with diazepam, clonazepam or paraldehyde. Rifampin is a first-line agent for *M. tuberculosis* infections, although successful therapy can proceed without it as we demonstrated in one patient in this series. Its necessity for *M. avium-intracellulare* infections is probably less and the use of ciprofloxacin, clofazamine, ethambutol, with or without amikacin, none of which to our knowledge interact with itraconazole, is often efficacious in our experience.

This *in vitro* and *in vivo* data document the high activity of itraconazole, against *C. neoformans*. Responses despite serum concentrations below MIC's and undetectable CSF concentrations suggest tissue concentrations of itraconazole may be critical. This may be particularly true of meningeal and brain parenchymal concentrations. Itraconazole is lipophilic, and pharmacokinetic studies indicate the drug accumulates in several tissues (Stevens, 1988). Our results are better than the early reports of itraconazole therapy using 200 mg a day possibly because we used a larger dose, 400 mg a day. The efficacy of itraconazole therapy compared to conventional amphotericin B with or without flucytosine is not known and can only be established by controlled trials. Likewise, the efficacy of itraconazole, while similar or superior to that published for fluconazole (Dupont, 1988; Stern et al., 1988; Larsen and Leal; Dismukes et al., Interscience Conference on Antimicrobial Agents and Chemotherapy, 1989, Abstracts), requires a controlled trial for comparison. Our data suggest that itraconazole is much less toxic and is better tolerated by patients than amphotericin B. In addition, itraconazole is oral rather than intravenous therapy, which has many advantages. As patients who received a few days of amphotericin prior to itraconazole were the subgroup with the very best complete response rate, a short course of induction therapy with amphotericin might improve the present itraconazole results even further. Combination therapy of itraconazole with, for example, flucytosine may yield superior results to monotherapy. We believe that itraconazole deserves further study to establish its place in the treatment of all forms of cryptococcosis.

Acknowledgements

The authors are indebted to the participating patients who courageously tried an unproven remedy. We are also grateful to many physicians who provided information about their patients: Robert Armstrong, Robert Atmar, Keith Barton, Jack Bissett, Carol Brosgart, Lisa Capaldini, Raymond Chinn, Dow Covington, Stanley Deresinski, David Drennan, Phillip Eastman, Scott Eberle, Marvin Freid, Ann Fukutome, Charles Gherman, Leonard Goldberg, Shelley Gordon, Jonathan Green, Marshall Kubota, Fred London, George Matula, Richard Hamill, Walter Krampf, Clinton Lane, Chinh Le, Louis Lehman, Brad Lewis, Elliot Liff, Myles Lippe, Larry Rumans, George Susens, John Swartzberg and others. We are grateful to Linda Hanson and Alon Perlman for performing *in vitro* susceptibility tests and John Hamilton and Anita Noble for cryptococcal antigen studies.

REFERENCES

Anonymous, 1980, Garlic in cryptococcal meningitis. A preliminary report of
    21 cases, Chin. Med. J., 93:123.
Bennett, J.E., Dismukes, W.E., Duma, R.J., Medoff, G., Sande, M.A., and
    Gallis, H., 1979, A comparison of amphotericin B alone and combined
    with flucytosine in the treatment of cryptococcal meningitis, N.
    Engl. J. Med., 301:126
Cauwenbergh, G., and De Doncker, P., 1987, The clinical use of itraconazole
    in superficial and deep mycoses, in: "Recent Trends in the
    Discovery, Development and Evaluation of Antifungal Agents", R.A.
    Fromtling, ed., J.R. Prous Science Publishers, Barcelona, Spain.
Centers for Disease Control, 1989, First 100,000 cases of acquired immune
    deficiency syndrome - United States, MMWR 38:561.
Chuck, S.L., and Sande, M.A., 1989, Infections with Cryptococcus neoformans
    in the acquired immunodeficiency syndrome, N. Engl. J. Med.,
    321:794.
De Wytt, C.N., Dickson, P.L., and Holt, G.W., 1982, Cryptococcal
    meningitis. A review of 32 years experience, J. Neurol. Sci.,
    53:283.
de Gans, J., Eeftinck Schattenkerk, J.K.M., and van Ketel, R.J., 1988,
    Itraconazole as maintenance treatment for cryptococcal meningitis in
    the acquired immunodeficiency syndrome, Br. Med. J., 296:339.
Denning, D.W., Tucker, R.M., Hanson, L.H., Hamilton, J.R., and Stevens,
    D.A., 1989a, Itraconazole therapy of cryptococcal meningitis and
    cryptocosis, Arch. Intern. Med., 149:2301.
Denning, D.W., Tucker, R.M., and Stevens, D.A., 1989b, Itraconazole therapy
    of fungal infections: endemic and opportunistic mycoses, in: "An
    Update on Diagnosis and Therapy of Systemic Mycoses", K. Holmberg,
    R.G. Meyer, eds., Raven Press, New York.
Denning, D. W., Tucker, R.M., Hanson, L.H., and Stevens, D.A., 1990,
    Itraconazole in opportunistic mycoses: cryptococcosis and
    aspergillosis, J. Am. Acad. Dermatol., in press.
Diamond, R.D., and Bennett, J.E., 1974, Prognostic factors in cryptococcal
    meningitis. A study of 111 cases, Ann. Intern. Med., 80:176.
Dismukes, W.E., 1988, Cryptococcal meningitis in patients with AIDS, J.
    Infect. Dis., 157:624
Dismukes, W.E., Cloud, G., Gallis, H.A., Kerkering, T.M., Medoff, G., and
    Craven, P.C., 1987, Treatment of cryptococcal meningitis with
    combination amphotericin B and flucytosine for four as compared with
    six weeks, N. Engl. J. Med., 317:334.
Dupont, B., 1988, Treatment of cryptococcal meningitis, in: "Proceedings of
    the X Congress of the International Society for Human and Animal
    Mycology", J.M. Torres-Rodriguez, ed., J.R. Prous Science Publisher,
    Barcelona.
Gal, A.A., Evans, S., and Meyer, P.R., 1987, The clinical and laboratory
    evaluation of cryptococcal infections in the acquired
    immunodeficiency syndrome, Diagn. Microbiol. Infect. Dis., 7:249.
Graybill, J.R., and Ahrens, J., 1984, R51211 (itraconazole) therapy of
    murine cryptococcosis, Sabouraudia, 22:445.
Hay, R.J., Mackenzie, D.W.R., Campbell, C.K., and Philpot, C.M., 1980,
    Cryptococcosis in the United Kingdom and the Irish Republic: an
    analysis of 69 cases, J. Infect., 2:13.
Kovacs, J.A., Kovacs, A.A., Polis, M., Wright, W.C., Gill, V.J., Tuazon,
    C.U., Gelmann, E.P., Lane, H.C., Longfield, R., Overturf, G.,
    Macher, A.M., Fauci, A.S., Parillo, J.E., Bennett, J.E., and Masur,
    M., 1985, Cryptococcosis in the acquired immune deficiency syndrome,
    Ann. Intern. Med., 103:533.
Larsen, R.A., Bozzette, S., McCutchan, A., Chiu, J., Leal, M.A., Richman,
    D.D., and The California Collaborative Treatment Group, 1989,
    Persistent Cryptococcus neoformans infection of the prostate after
    successful treatment of meningitis, Ann. Intern. Med., 111:125.

Perfect, J.R., Savani, D.V., and Durack, D.T., 1986, Comparison of itraconazole and fluconazole in treatment of cryptococcal meningitis and candida pyelonephritis in rabbits, Antimicrob. Agents Chemother., 29:579.

Polak, A., 1987, Combination therapy of experimental candidiasis, cryptococcosis, aspergillosis and wangiellosis in mice, Chemotherapy, 33:381.

Staib, F., and Seibold, M., 1988, Mycological-diagnostic assessment of the efficacy of amphotericin B + flucytosine to control *Cryptococcus neoformans* in AIDS patients, Mykosen, 31:175.

Stern, J.J., Hartman, B.J., Sharkey, P., Rowland, V., Squires, K.E., and Murray, H.W., 1988, Oral fluconazole therapy for patients with acquired immunodeficiency syndrome and cryptococcosis: experience with 22 patients, Am. J. Med., 85:477.

Stevens, D.A., 1988, The new generation of antifungal drugs, Eur. J. Clin. Microbiol. Infect. Dis., 7:732.

Sugar, A.M. ,and Saunders, C., 1988, Oral fluconazole as suppressive therapy of disseminated cryptococcosis in patients with acquired immunodeficiency syndrome, Am. J. Med., 85:481.

Tucker, R.M., Williams, P.L., Arathoon, E.G., and Stevens, D.A., 1988, Treatment of mycoses with itraconazole, Ann. NY Acad. Sci., 544:451.

Van Cutsem, J., Van Gerven, F., and Janssen, P.A.J., 1987, Activity of orally, topically, and parenterally administered itraconazole in the treatment of superficial and deep mycoses: animal models, Rev. Infect. Dis., 9:S15.

Viviani, M.A., Tortorano, A.M., Woestenborghs, R., and Cauwenbergh, G., 1987, Experience with itraconazole in deep mycoses in Northern Italy, Mykosen, 30:233.

Viviani, M.A., Tortorano, A.M., Giani, P.C., Arici, C., Goglio, A., and Crocchiolo, P., 1987, Itraconazole for cryptococcal infection in the acquired immunodeficiency syndrome, Ann. Intern. Med., 106:166.

Zuger, A., Louie, E., Holzman, R.S., Simberkoff, M.S., and Rahal, J.J., 1986, Cryptococcal disease in patient with the acquired immunodeficiency syndrome. Diagnostic features and outcome of treatment, Ann. Intern. Med., 104:234.

TREATMENT OF DERMATOMYCOSIS IN AIDS PATIENTS

H. Degreef

Dept. of Dermatology, U.Z. Sint Rafaël
B3000 Leuven, Belgium

The acquired immunodeficiency syndrome (AIDS) is a highly lethal new epidemic that predominantly affects homosexual and bisexual men, intravenous drug abusers, Haïtans, Africans, hemophiliacs, infants born to mothers in high risk groups for this syndrome, and female sexual contacts of male intravenous drug abusers (Farthing et al., 1988). Significant progress has been made concerning the epidemiology, aetiology, and clinical behaviour of this complex disease and its sequelae since the first cases were reported in 1981.

AIDS has many faces, some of which have important dermatologic implications such as Kaposi's sarcoma lesions, seborrheic dermatitis, papular and follicular eruptions, vasculitis, purpura, oral "hairy"-leukoplakia and the yellow nail syndrome (Farthing et al., 1988).

The most common superficial mycotic infection in AIDS is certainly *Candida* stomatitis and esophagitis. Its treatment has already been discussed. However, there is not yet agreement about whether seborrheic dermatitis must be considered a pityrosporosis or a disease in which *Pityrosporum ovale* plays an important role; nor is there agreement on the role of this microorganism in the very common cases of dermatitis in AIDS patients that resemble seborrheic dermatitis. We do know that this disease reacts favorably to a peroral treatment with ketoconazole and that seborrheic dermatitis occurs less frequently in AIDS patients and in ARC patients who are treated with ketoconazole for oral candidiasis (Goodman et al., 1987). Because of the lack of insight into the pathogenesis of this disease, one cannot go further into its treatment in a discussion of the treatment of dermatomycoses.

We will thus concentrate primarily on the treatment of intertriginous candidiasis, dermatophytoses and onychomycosis, pityriasis versicolor, and pityrosporon folliculitis.
Dermatophytoses are among the most common skin diseases and thus can frequently be met in AIDS patients, generally presenting with the well-known clinical phenomena on the feet and in the groin. Although one may assume that AIDS patients are given a very thorough skin examination, there is no agreement about whether these infections occur more frequently than in people without AIDS. According to Pierard (1986); they do not occur more frequently in AIDS patients; according to others (Muhlemann et al.,

1986; Goodman et al., 1987; Matis, 1987), they occur frequently, up to 30%, in AIDS patients. In general, these infections occur four times more often in AIDS and ARC patients than in a control population (Torssander et al., 1985), but there remains a clear controversy.

There is agreement, however, that dermatophytoses in these patients occur in a more extensive, more serious form and particularly in deviant clinical forms in which the lesions often take on an eczematous or psoriasiform aspect. The number of infection sites is also sometimes greater. *T. rubrum* is the most common causative agent.

Onychomycosis occurs frequently not only in ARC-AIDS patients but also in homosexuals in general (Matis, 1987). One wonders if the life style (e.g., frequent sauna baths) might not be a contributing factor for the more frequent cases of onychomycosis among homosexuals.

Proximal white subungual onychomycosis caused by *T. rubrum* and onychomycosis caused by dermatophytes together affecting the paronychium should suggest symptomatic AIDS (Noppakun and Head, 1986; Kaplan et al., 1987; Weismann et al., 1988). Dermatophytoses have a chronic evolution in AIDS patients. They are characterized by therapy resistance and their tendency to recur. No comparative studies between healthy people and AIDS patients are available to enable us to judge the value of the existing antimycotics. Such studies, however, will be confronted with major problems, such as the death of the patients, drug absorption, and patient compliance.

The newer antimycotics such as itraconazole or terbinafine have also not been studied systematically with these patients.

HIV-negative patients also can have chronic dermatophyte infections. A clear cause cannot always be determined, although an underlying deficiency of the cellular immunity is often assumed. Therefore, it is not surprising that the treatment of dermatomycoses in AIDS patients will be based on the way these chronic infections are treated in HIV-negative patients.

However, one will approach the treatment in a practical way, which often means the use of broad-spectrum antifungal drugs whereby the mycotic skin and nail infections as well as the oropharyngeal lesions and seborrheic dermatitis, which may be present, can be treated. Finally, one must also take into account drug absorption because of the potential gastrointestinal disturbances and drug-interaction, as is the case with rifampicin (Engelhard et al., 1984).

SYSTEMIC THERAPY

For these reasons, a broad-spectrum antimycotic such as ketoconazole is preferred to griseofulvin.

Long-term treatment with 200 mg of ketoconazole per day will thus be necessary in most cases. Upon resistance, the dosage can be increased to 400 and even to 600 mg per day, as is the case with other immunodeficiency syndromes such as chronic mucocutaneous candidosis. It is evident that close follow-up of liver function is necessary with these patients. The numerous other medications, with their potential hepatotoxic activity, probably increase the risk of this undesirable side effect.

As already noted, interaction with other medications must be kept in mind. We mention here (Jones, 1987):

- Drugs that should not be used concurrently with ketoconazole, or the combination should be used judiciously: cyclosporin, warfarin and other coumarin anticoagulants, rifampicin, and hepatotoxic drugs.

- Certain drugs may affect the absorption of ketoconazole: H2-receptor antagonists, antimuscarinic agents, antisecretory drugs (omeprazole) and antacids.
- Others may potentially interact with ketoconazole, but the clinical significance of the interaction remains to be qualified: phenytoin, chlordiazepoxide, methylprednisolone and other corticosteroids, ethanol and other drugs metabolized by hepatic mixed-function oxidases.

In this way, one may expect favorable results with dermatophytoses, onychomycoses caused by dermatophytes, pityriasis versicolor, and intertriginous candidiasis.

Griseofulvin has the advantage of being active only against dermatophytes. When no yeast-cell infections are present, this antimycotic can be considered (Muhlemann et al., 1984) but here, too, a higher dosage is often necessary to achieve the desired result.

Alternative newer antimycotic, which can be considered if the results are unsatisfactory, are fluconazole for yeast-cell infections and terbinafine for dermatophytoses. Finally itraconazole (100 to 200 mg per day) is a broad-spectrum antimycotic that can be considered an alternative therapy for both yeast-cell and dermatophyte infections. The anticipated therapeutic activity of this new antimycotic on *Pityrosporum ovale*, however, would be less than that of ketoconazole (Faergemann, 1984). Finally, we would argue for the simultaneous use of local therapy.

LOCAL THERAPY

Favorable results have been described for exclusively local therapy in cases of superficial cutaneous mycotic infections treated, for example, with nystatin, miconazole, ketoconazole, clotrimazole, or bifonazole (Goodman et al., 1987; Kaplan et al., 1987). The medication must often be given continuously (Kaplan et al., 1987). These means can certainly be used as adjuvant therapy. An exclusively local therapy, although sometimes active, will generally be insufficient to treat dermatomycotic infections. A combination of local and systemic therapy will produce a higer percentage of cures, but even with higher dosages one still has to take therapy resistance into account. After the cessation of successful therapy, recurrences will have to be dealt with, so that intermittent long-term systemic prophylaxis with broad-spectrum antimycotics will be required.

CONCLUSION

We may thus conclude that there is no specific treatment for dermatomycoses in ARC/AIDS patients. There have been no controlled studies on the value of the various antimycotics available, and such studies will probably be very difficult to conduct. From a practical standpoint, intensive treatment with broad-spectrum antimycotics will generally be the indicated approach, and intermittent long-term prophylaxis will very often be desirable. Local therapy can be given as support.

REFERENCES

Engelhard, D., Stuttman, H.R., and Marks, M.I., 1984, Interaction of ketoconazole with rifampicin and isoniazid, N. Engl. J. Med., 311:1681.

Faergemann, J., 1984, *In vitro* and *in vivo* activities of ketoconazole and itraconazole against *Pityrosporon orbiculare*, <u>Antimicrob. Agents Chemother.</u>, 26:773.

Farthing, C.F., Brown, S.E., and Staughton, R.C.D., 1988, A colour atlas of AIDS and HIV disease, 2nd edition, Wolfe Medical Publications Ltd., London, England.

Goodman, D.S., Teplitz, E.D., Wisher, A., Klein, R.S., Burk, P.G. and Hershenbaum, E., 1987, Prevalence of cutaneous disease in patients with acquired immunodeficiency syndrome (AIDS) or AIDS-related complex, <u>J. Am. Acad. Dermatol.</u>, 17:210.

Hulsebosch, H.J., and Brakman, M., 1989, Aids en de huid, <u>Ned. Tijdschr. Geneeskd.</u>, 133:873.

Jones, H.E., 1987, Drug interaction, <u>in:</u> "Ketoconazole today", Adis Press Ltd., Manchester.

Kaplan, M.H., Sadick, N., McNutt, N.S., Meltzer, M., Sarngadharan, M.G., and Pahwa, S., 1987, Dermatologic findings and manifestations of acquired immunodeficiency syndrome (AIDS), <u>J. Am. Acad. Dermatol.</u>, 16:485.

Matis, W.L., 1987, Dermatologic findings associated with human immunodeficiency virus infection, <u>J. Am. Acad. Dermatol.</u>, 17:746.

Muhlemann, M.F., Anderson, M.G., Paradinas, F.J., Key, P.R., Dawson, S.G., Evans, B.A., Murray-Lyon, I.M., and Cream, J.J., 1984, Early warning skin signs in AIDS and persistent generalized lymphadenopathy, <u>Br. J. Dermatol.</u>, 114:419.

Noppakun, N., and Head, S.H., 1986, Proximal white subungual onychomycosis in a patient with acquired immune deficiency syndrome, <u>Int. J. Dermatol.</u>, 25:586.

Pierard, G.E., Pierard-Franchimont, C., and Lapière, Ch.M., 1986, Signes cutanés du SIDA, <u>Rev. Med. Liège</u>, 41:189.

Torssander, J., Wasserman, J., Morfeldt-Mansson, L., Betrini, B., and Von Steding;, L.V., 1985, Persistent generalized lymphadenopathy: Immunological and mycological investigations, <u>Acta Derm. Venereol.</u>, 65:515.

Weismann, K., Knudsen, E.A., and Pedersen, C., 1988, White nails in AIDS/ARC due to *Trichophyton rubrum* infection, <u>Clin. Exp. Dermatol.</u>, 13:24.

Absidia corymbifera, 9
Acquired immunodeficiency syndrome,
     see AIDS
ACTH stimulation, test 180, 185, 317
Acyclovir, 307
Adrenal, 200
     cryptococcosis, 31
     histoplasmosis, 180, 185
     itraconazole, 228
     ketoconazole, 228, 267
African histoplasmosis, 164
AIDS
     Africa, 6, 9
     alternariosis, 31
     Asia, 6
     aspergillosis, 30-31, 155-156,
          158, 200, 265
     Belgium, 116
     blacks, 5
     blastomycosis, 30, 179
     Brazil, 4, 6
     Burundi, 6
     Canada, 6
     candidemia, 75, 96
     candidosis, 29, 31, 33, 35, 67,
          83, 93, 143
     coccidioidomycosis, 8, 27, 29,
          171, 179
     Congo, 6
     cryptococcal meningitis, 287, 306
     cryptococcosis, 8, 27, 29, 31, 35,
          39, 103, 115, 123
     cryptosporidiosis, 4, 7-8
     Denmark, 6
     dermatophytes, 35, 135, 141, 325
     epidemiology, 3, 4, 5, 9
     Europe, 9
     Federal Republic of Germany, 5,
          6
     France, 4-6
     Haiti, 6
     hispanics, 5
     histoplasmosis, 8, 27, 29, 31, 35,
          163, 179
     indicative diseases, 7
     Italy, 5, 6
     Kenya, 6
     Malawi, 6
     Mexico, 6

     mucormycosis, 30
     nocardiosis, 30
     opportunistic infections 3, 4, 7,
          29
     paracoccidioidomycosis, 37, 148,
          179
     pathogenesis, 17
     PCP 37, 55
     penicilliosis marneffei, 30, 147
     seborrheic dermatitis, 144
     Spain, 6
     sporotrichosis, 30, 147
     Switzerland, 6
     Tanzania, 6
     The Caribbean, 6
     thrush, 27, 280
     toxoplasmosis, 4, 7-8, 39, 76, 297
     Uganda, 4, 6
     United Kingdom, 6
     USA, 4-7, 9
     Western Europe, 6
     whites, 5
     Zaire, 6
     Zambia, 6
AIDS related complex, 19, 28, 141,
          180, 255
AL721, 307
Aldosterone, 202
Allergic bronchopulmonary
     aspergillosis, 156
Alloxan-diabetic, 211
Allylamines, 223, 225, 271-272, 326
Alternaria, 29, 31, 44, 144, 151,
     156
Alveolar macrophages, 131
Alveolitis, 156
American histoplasmosis, 164
Amorolfine
     mode of action, 224
Amphotericin B, 45, 76, 79-80, 150,
          171-172, 184-185, 187, 246,
          256, 266, 280, 288, 302, 307,
          313
     aspergillosis, 213
     Candida albicans
          IC30 value 256
     coccidioidomycosis, 174, 176, 266
     cryptococcal meningitis, 202, 287,
          306, 319

cryptococcosis, 218, 266
histoplasmosis, 183, 187
liposomes, 204, 251, 271
mode of action, 223
Penicillium marneffei, 148
Pneumocystis carinii, 60
Rhizopus, 218
sporotrichosis, 151
systemic candidosis, 213
urine, 249
Anemia, 18
Animal models, 207-208, 211, 213,
    218
Anti-*Candida albicans*
    IgA, 84
    IgG, 86
Anti-cryptococcal antibodies, 126
Anti-Leu3a, 14
Anti-P25 antibodies, 18
ARC <u>see</u> AIDS related complex
Aromatase
    ketoconazole, 228
Aspartyl-protease, 15
Aspergilloma, 156
Aspergillosis, 30, 31, 151, 155-156,
    158, 200, 265
    amphotericin B, 213
    animal models, 213
    itraconazole, 202, 268-269
    ketoconazole, 202, 268-269
Aspergillus fumigatus, 8, 60, 155,
    213
2-Aza-2,3-dihydrosqualene, 225
Azidothymidine, 137-138, 185
Azole antifungals
    ergosterol biosynthesis, 223-224,
        226, 228-239
    pharmacokinetics, 245, 246, 247
AZT, 137, 138, 185, 307, 320

Bacillus megaterium
    cytochrome P450, 225
Bacterial infections, 4, 8
Bifonazole, 327
    selectivity, 227
Blastomyces dermatitidis, 8, 42
Blastomycosis, 147
    itraconazole, 202, 268
B-lymphocytes, 32, 36, 96
Bone marrow precursors, 21

Candida
    diarrhea, 99
    DNA types, 67
    germ tube formation, 95
Candida albicans, 7, 8, 28, 30, 59,
        75, 80, 83, 141, 155, 157,
        211, 255
    47kD antigen, 87
    cytochrome P450, 226
    DNA, 70, 71
    epidemiology, 67-69
    gastrointestinal tract, 83
    growth
        effects of

fluconazole, 228
itraconazole, 228
mannan, 88
P45014DM, 226
serotype A, 33-35
serotype B, 33-35, 69-70, 72
Southern blot hybridization
    patterns, 72
Candida endocarditis, 204
Candida esophagitis, 4, 7, 27, 76
Candida glabrata, 9, 141, 224, 228-
    229, 265
    cytochrome P450, 232
    ergosterol synthesis, 229-233
    fluconazole, 228, 283
    growth
        effects of
            fluconazole, 228
            itraconazole, 211, 228
            ketoconazole, 267, 281
            miconazole, 228
Candida inconspicua
    fluconazole, 283
Candida krusei, 9, 75-76
    fluconazole, 283
Candida leukoplakia, 280
Candida mucositis, 95, 98
Candida parapsilosis, 9, 75, 76, 79-
    80
Candida peritonitis, 97
Candida pneumonia, 96
Candida pseudotropicalis, 9
Candida sepsis, 96
Candida stomatitis, 325
Candida tropicalis, 9
    fluconazole, 283
    P45014DM, 226
Candida vaginitis, 97, 200
Candidal oropharyngitis, 7
Candidemia, 75, 76, 79, 80, 96, 200,
    265
Candidiasis, <u>see</u> candidosis
Candidosis, 3, 8, 27, 31, 35-37, 41,
    75, 83, 86, 143, 202, 213, 265
    immunological aspects, 83, 85-87
    immunosuppressive effects, 35
CD4+ 14, 16-22, 32, 37, 142, 191-194
CD8+, 18, 142, 191-193
CD16+, 19
CDC, 3, 4, 5, 7, 27, 155
    classification system, 27
Cell-mediated immunity, 3, 131, 183
Cellular tropism, 14
Chloramphenicol, 208, 212
Cholesterol side-chain cleavage
    ketoconazole, 228
Cholesterol-7α-hydroxylase
    ketoconazole, 228
Chromoblastomycosis
    itraconazole, 268, 269
Chromomycosis, 158
Chronic mucocutaneous candidosis,
    35, 143
Cimetidine, 201
Ciprofloxacin

itraconazole, 322
Cladosporium
  itraconazole, 269
Clofazamine
  itraconazole, 322
Clotrimazole, 96, 144, 280, 327
  enzyme induction, 251
  metabolic clearance, 250
  selectivity, 227
Coccidioidal meningitis, 200
Coccidioidal skin test, 173
Coccidioides immitis, 9, 41, 155,
    171, 173, 191
Coccidioidomycosis, 3, 7, 27, 37,
    45, 171, 175, 193, 202, 266
  amphotericin B, 176, 266
  fluconazole, 176
  itraconazole, 176, 202, 269
  ketoconazole, 266
Collagen disease, 115
Compound Q, 307, 320
Concanavalin A, 60
Core proteins, 13-14
Corticosteroids, 37, 76, 115, 144,
    157-158, 185, 201, 207, 327
Cortisol, 201
  histoplasmosis, 180
Cryptococcaemia, 118
Cryptococcal meningoencephalitis,
  Burundi, 28
  Zaire, 28
Cryptococcal antigens, 36, 126, 131
Cryptococcal meningitis, 4, 7, 80,
    115, 117-118, 139, 141, 200,
    287
  5-FC, 287, 306, 319
  amphotericin B, 319
  diagnosis, 119
  fluconazole, 287-289, 296-299,
    301-302
  itraconazole, 305-306, 308, 310,
    311-312, 319-321
  rabbit
    itraconazole, 306
Cryptococcemia
  itraconazole 305, 313
Cryptococcosis, 7-8, 36, 40-41, 45,
    103-104, 112, 115-116, 118,
    124, 144, 155, 158, 180, 203,
    220
  5-FC, 218, 266
  amphotericin B, 266
  animal models, 218
  Burundi, 39
  central nervous system, 117
  clinical aspects, 116
  diagnosis, 288
  epidemiology, 116
  France, 38-39, 287
  immunological aspects, 123-125,
    130-131
  immunosuppressive effects, 35
  itraconazole, 202, 305-306, 319,
    321-322
  prevalence, 115-116

  Belgium, 116
  France, 116
  Germany, 116
  The Netherlands, 116
  U.K., 39, 116
  U.S.A., 39, 116, 287, 306
Cryptococcuria
  itraconazole, 305
Cryptococcus albidus, 104
Cryptococcus diffluens, 59
Cryptococcus neoformans, 7, 9, 30,
    93, 103, 115, 123, 129, 155,
    287-302, 305-322
  amphotericin B, 266, 302
  birds, 103
  capsular polysaccharide, 36
  capsule, 124-126
  cell wall, 125
  cell-mediated immunity, 129
  Central Africa, 105
  DNA, 107
    polymorphism, 107
  ecology, 103-105
  fluconazole, 270, 287-302
  itraconazole, 213, 302, 305-322
  mating type α, 104
  peach juice, 103
  pigeon droppings, 103
  prevalence, 105
  serotype A, 104, 107
  Zaire, 105
Cryptococcus neoformans var *gattii*,
    103, 105, 111,
  serotypes, 106
Cryptosporidiosis, 4, 7-8
  enteritis, 4
Cunninghamella bertholletiae, 9
Curvularia, 151
Cyclodextrins, 187, 213, 251
  itraconazole, 268, 270
Cyclophosphamide, 208
Cyclosporins, 201, 326
  cytochrome P450, 226, 251
  itraconazole, 268
  ketoconazole, 251, 267
Cytochrome P450, 201, 224-228, 232,
    250-251
  14α-demethylase, 224-227
  adrenal, 228
  Bacillus megaterium, 225
  bifonazole, 227
  Candida albicans, 226-228, 232
  Candida glabrata, 232
  Candida tropicalis, 226
  clotrimazole, 227, 251
  fluconazole, 226, 232
  itraconazole, 226-228, 232
  ketoconazole, 226-228, 232
  kidney, 228
  liver, 228
  miconazole, 227, 232
  nor-ketoconazole, 227
  placenta, 228
  Pseudomonas putida, 225
  Saccharomyces cerevisiae, 226

saperconazole, 232
    skin, 228, 251
    testis, 227-228
Cytokines, 18, 20, 22
Cytomegalovirus, 4, 8, 9, 94, 307
Cytosine deaminase, 223
Cytosine permease, 223

Dandruff
    Pityrosporum, 43
Darling's disease, 164
Deferoxamine, 218
Dementia, 8
14α-Demethylase, 224
    Aspergillus spp., 228
    Candida albicans, 226-228
    Candida glabrata, 229-232, 239
    econazole, 225
    fluconazole, 231-235
    Histoplasma capsulatum, 233-239
    imazalil, 227
    itraconazole, 201, 226, 228, 231-
        236, 239
    ketoconazole, 225, 228, 232
    Leishmania mexicana, 228
    miconazole, 230-233, 239
    Pityrosporum ovale, 228
    Saccharomyces cerevisiae, 226,
        229, 239
Dendritic cells, 14, 20
Dermatophytes, 135, 141, 142
    bifonazole, 327
    clotrimazole, 327
    fluconazole, 327
    griseofulvin, 143
    itraconazole, 143, 327
    ketoconazole, 267, 327
    miconazole, 327
    nystatin, 327
    tolnaftate, 225
Dermatophytosis, 31, 135-136, 142,
        208, 220, 326
Dexamethasone, 315
11-Deoxycortisol, 228
Diagnosis, 119, 280, 288
Diarrhea, 7, 8, 28, 94, 99
Dihydrofolate reductase
    Pneumocystis carinii, 61
1,25-Dihydroxyvitamin D3, 228
Dilantin
    ketoconazole, 267
4,4-Dimethyl-Δ8,14,24-
        cholestatrienol, 226
4,4-Dimethyl-Δ8,14,24(28)-
        ergostatrienol, 226
4,14-Dimethylzymosterol, 229, 231
Dimorphic fungi
    immunological aspects, 191, 193
Dixon's medium
    malt, 137
DNA, 14, 107, 223
    polymerase, 15
    polymorphism, 107, 111

Econazole

14α-demethylase, 225
Endonuclease, 15
Env gene, 13
Envelope glycoproteins, 14
Epidermophyton floccosum, 31, 43
Ergosterol, 223-224, 228, 266
    Aspergillus, 228
    azole antifungals,223-239
    biosynthesis
        and fluconazole, 228
        and inhibitors, 223-239
        and itraconazole, 226-228,
            232, 239
        and ketoconazole, 226-228,
            232, 239
        and miconazole, 230-233, 239
    Candida albicans, 226, 228, 239
    Candida glabrata, 228-233, 239
    Histoplasma capsulatum, 233-239
    Leishmania mexicana, 228
    Pityrosporum ovale, 228
Erianthus giganteus, 104
Esophageal candidosis, 28, 32, 94
    Zaire, 28
Ethambutol
    itraconazole, 322
External envelope protein, 14

Fatty acids, 141, 233
5-FC, 34, 79, 148, 150, 288, 319
    Candida albicans
        IC30 value, 256
    cryptococcal meningitis, 287, 306
    cryptococcosis, 218, 266
    mode of action, 223
    Penicillium marneffei, 148
    pharmacokinetics, 245
    resistance, 33, 34
    urine, 249
Fc receptors, 15
Ferritin, 125, 128
Filobasidiaceae, 103
Filobasidiella, 103, 123
Filobasidiella neoformans,104
Filobasidium floriforme, 104
Fluconazole, 45, 96, 200, 203, 220,
        225
    adverse reactions, 201, 301
    alkaline phosphatases, 301
    azotemia, 200
    bioavailability, 246
    Candida albicans
        cytochrome P450, 232
    Candida glabrata, 283
        cytochrome P450, 232
    Candida inconspicia, 283
    Candida krusei, 283
    Candida tropicalis, 283
    candidemia, 270
    candidosis, 144
    coccidioidal meningitis, 271
    coccidioidomycosis, 171, 176
    cryptococcal meningitis, 45, 203,
        287-289, 296-299, 301-302
    cryptococcosis, 218

Cryptococcus neoformans, 270, 287–302, 320
cyclosporins, 200-201
eosinophilia, 301
epidermal necrolysis 202
ergosterol biosynthesis, 226, 228, 231, 236, 239
γ-GT, 301
gastro-intestinal disorders, 301
glipizide, 201
haem iron complex, 226
hemodialysis, 200
histoplasmosis, 186
MIC90, 246
oral candidosis, 281-283
P45014DM, 226
pharmacokinetics, 246,
phenytoin, 201
prothrombin levels, 301
    times, 201
rifampin, 201
sporotrichosis, 269
Stevens Johnson syndrome, 202
sulfonylureas, 200
systemic candidosis, 213
thrombopenia, 301
tissue distribution, 247
tolbutamide, 201
transaminases, 301
urine, 249
warfarin, 200-201
Fluorocytosine see 5-FC
5-Fluorodeoxyuridylate, 223
5-Fluorouracil, 223
5-Fluorouridine triphosphate, 223
Fonsecaea pedrosoi
    fluconazole, 268
    itraconazole, 268, 269
    ketoconazole, 268
    SCH 39304, 268
Fusariosis, 151
Fusarium, 156, 158, 265
    proliferatum, 8

Gag gene, 13
Galactoxylomannan, 124
Gancyclovir, 185, 320
Garlic, 320
Gastrointestinal candidosis, 212
    animal models, 212
Geotrichum, 150
Glucuronoxylomannan, 36, 124
GP41, 13, 15, 19
GP120, 13, 14, 15, 19
Griseofulvin, 266
    dermatomycosis, 326
    dermatophytes, 143
    pharmacokinetics, 245
    resistance, 200
GTPase, 15

Helminthic infections, 4
Helper T-cell, 85, 88
Herpes simplex, 4, 8-9, 94, 143-144

Hisoplasma capsulatum var. *farciminosum*, 163
Histoplasma capsulatum, 9, 32, 40, 155, 180, 186, 191-193, 224, 233
    Africa, 165
    antigens, 194
    14α-demethylase, 228, 233-239
    epidemiology, 166, 167
    ergosterol synthesis, 233-239
        and fluconazole, 236, 239
        and itraconazole, 234-239
        and ketoconazole, 234-236
    growth
        and fluconazole, 233
        and itraconazole, 233
    South America, 164
    U.S.A, 164
Histoplasma capsulatum var. *capsulatum*, 7, 148, 163
    epidemiology, 164-165
Histoplasma capsulatum var. *duboisii*, 9., 147, 163
Histoplasmin skin test, 164
Histoplasmosis, 7, 9, 27, 37, 40-41, 45, 76, 144, 163, 165, 167, 181, 187, 193, 233, 266
    amphotericin B, 183
    Belgium, 41
    clinical aspects, 166, 180-182,
    diagnosis, 181
    epidemiology, 179
    fluconazole, 186
    France, 40-41
    immunosuppressive effects, 35
    itraconazole, 186-188, 202, 268
    ketoconazole, 184-185, 266
    Sweden, 40-41
    Switzerland, 40-41
    USA, 40
    Zaire, 41
HIV, 3, 4, 13-14, 17-18, 27, 75, 83, 85
    biological cycle, 14
    cellular tropism, 14-15
    genome, 15
    structure, 13
    wasting syndrome, 8
HIV-1, 6, 15, 68
    antigen, 288
HIV-2, 6, 15
    antigen, 288
H2 Receptor antagonists, 200-201, 327
Human immunodeficiency virus see HIV
Hydrocortisone, 208
11ß-hydroxylase
    ketoconazole, 228
17α-Hydroxylase
    itraconazole, 228
    ketoconazole, 228
21-Hydroxylase
    keoconazole, 228
4-Hydroxyretinoic acid, 228
Hypergammaglobulinemia, 18, 20

Hyperglobulinaemia, 142
Hyperplastic candidosis, 84
Hyphomycetales, 163
Hyphomycetes, 163
Hypokalemia, 202

IgA, 84, 85
IgE, 156
IgG, 86, 87, 126
IgM, 36, 87, 126
Imazalil
  Penicillium italicum, 237
  selectivity, 227
Immunotherapy, 194
Integrase, 15
Interferon, 45, 95, 142
Interleukins, 18, 21, 142
Intestinal candidosis, 213
Invasive pulmonary aspergillosis,
    157
Δ8-Δ7 Isomerase, 224
Isospora belli, 7
Isosporiasis, 3, 8
Itraconazole, 45, 96, 150, 200, 203,
    212, 220, 250, 326
  ACTH stimulation test, 315
  adverse reactions, 187, 315, 317
  alkaline phosphatase, 317
  amikacin, 322
  aromatase, 228
  aspergillosis, 202, 213, 268-269
  bioavailability, 201, 246
  brain levels, 247
  Candida albicans, 211
      cytochrome P450, 232
      IC30 value, 256
  Candida glabrata
      cytochrome P450, 232
  candidosis, 144
  cerebrospinal fluid, 247
  cholesterol, 201
  chromoblastomycosis, 268
  ciprofloxacin, 322
  clofazamine, 322
  coccidioidomycosis, 171, 176, 268
  cryptococcal meningitis, 45, 202,
      203, 305, 308, 310-312, 319
  cryptococcemia, 313
  cryptococcosis, 218, 305-307, 319,
      321-322
  Cryptococcus neoformans, 305-322
      MFC, 318
      MIC, 318
  cyclodextrins, 251
  cyclosporins, 200-201
  14α-demethylase, 225-239
  dermatophytes, 143, 204
  elimination rate, 249
  ergosterol biosynthesis, 226, 228,
      231, 236, 239
  ergosterol depletion, 234, 236
  esophageal candidosis, 96
  ethambutol, 322
  and 5-fluorocytosine, 220
  Fonsecaea pedrosoi, 268-269

formulation, 251, 270
gastrointestinal candidosis, 212
haem iron complex, 226
histoplasmosis, 45, 186, 187-188,
    236
hypokalemia, 317
intestinal candidosis, 213
intravenous 251
3-Ketosteroid reductase, 234-238
mineralocorticoids, 268
neutropenia, 317
oral candidosis, 96, 251, 281
oropharyngeal candidosis, 251
P45014DM, 225-228
paracoccidioidomycosis, 269
Penicillium marneffei, 148, 219
pharmacokinetics, 246, 261, 317
phenytoin, 200, 322
Pityrosporum ovale, 327
plasma levels, 248, 317
plasma protein binding, 250
pulmonary cryptococcosis, 313
rifampin, 201, 313, 322
selectivity, 227
skin, 249, 261
skin candidosis, 211
sporotrichosis, 151, 269
sulfonylureas, 200
systemic candidosis, 213
tinea versicolor, 202
tissue distribution, 247
tissue levels, 248, 261
triglycerides, 317
urinary cryptococcosis, 313
vaginal candidosis, 211-212
vaginal epithelium, 249
vaginal tissue, 249
warfarin, 200
IV-drug users, 5

JC virus, 4, 9

Kaposi's sarcoma, 3, 4, 8, 19, 22,
    28, 76, 310
Ketoconazole, 80, 150, 185, 186,
    199, 325, 327
  Aspergillus, 268
  bioavailability, 246
  blastomycosis, 267
  Candida albicans, 211
      colony-forming units, 262
      cytochrome P450, 232
      IC30 value, 256
  Candida glabrata, 267, 281
      cytochrome P450, 232
  candidemia, 270
  candidosis, 144
  cholesterol, 201
  coccidioidomycosis, 171, 173, 266
  cyclosporins, 201, 251, 267
  14α-demethylase, 225, 228
  dermatophytes, 267, 326
  dilantin, 267
  ergosterol biosynthesis, 226, 236,
      239

ergosterol depletion, 234, 236
Fonsecaea pedrosoi, 268
formulation, 251
gastrointestinal candidosis, 212
haem iron complex, 227
hepatotoxicity, 202, 267
histoplasmosis, 184, 266, 267
hormonal effects, 201
intestinal candidosis, 213
3-Ketosteroid reductase
    Histoplasma capsulatum, 236
liver
      14α-demethylase, 228
mucosal candidosis, 267
paracoccidioidomycosis, 267, 269
Penicillium marneffei, 148
pharmacokinetics, 246, 259
Pityrosporum, 144
plasma protein binding, 250
Pseudallescheria boydii, 267
resistance, 144
seborrheic dermatitis, 43
selectivity, 227
skin, 249, 259
skin candidosis, 211
steroid synthesis, 267
suspension
    oral candidosis, 280-281
thrush, 267
tissue levels, 261
vaginal candidosis, 211-212
vaginal epithelium, 249
zidovudine, 262
3-Ketosteroid reductase
  Histoplasma capsulatum
    itraconazole, 235-239
    ketoconazole, 235-239
  Penicillium italicum
    imazalil, 237

Langerhans cells, 14, 21, 32, 86
Lanosterol, 226, 229, 231
Leishmania mexicana mexicana
  ergosterol, 228
  itraconazole, 228
Leukoencephalopathy, 4, 8-9
Leukopenia, 18, 181
Leukotrienes, 142
Liposomes, 204, 246, 271-272
Listeria monocytogenes, 220
Liver
  P45014DM, 225
17,20-Lyase, 228
Lymphadenopathy, 28, 41, 94, 143
Lymphoblastic leukemia, 44, 156
Lymphoblastic lymphoma, 76
Lymphokines, 45, 130, 193
Lymphoma, 4, 8

Macrophages, 15, 17, 27, 32, 142
Maduromycosis, 158
Malassezia furfur, 8, 31, 32, 43, 158
Mannan, 35, 87, 124
  immunosuppressive effect, 35, 88

Mechlorethamine hydrochloride, 208
Melanin, 269
24-Methylenedihydrolanosterol, 225-226, 229, 235-236
14α-Methyl-ergosta-Δ8,24(28)-dien-3ß,6α-diol, 230, 235
14α-Methylfecosterol, 231
14α-Methylfecosterone, 235-239
Metronidazole, 208
Miconazole, 150, 280, 327
  Candida glabrata, 228
    cytochrome P450, 232
    ergosterol biosynthesis, 228, 230-233, 239
  14α-demethylase, 225
  metabolic clearance, 250
  Penicillium marneffei, 148
  selectivity 227
ß-Microglobulinemia, 18
Microsporum canis, 208
Mode of action
  5-FC, 223
  allylamines, 225
  amorolfine, 224
  amphotericin B, 223
  azole antifungals, 225-239
  morpholines, 224
  thiocarbamates, 225
Moniliaceae, 163
Monocytes, 193
Morpholines, 223, 224
mRNA, 16
mtDNA, 107
  hybridization patterns, 108, 109, 110-111
Mucocutaneous candidosis, 88
Mucorales, 30, 158
Mucormycosis, 29
Mycobacterium, 192
Mycobacterium avium, 4, 8, 76, 116, 186
Mycobacterium kansasii, 4
Mycobacterium tuberculosis, 8
Mycotic indicators, 7
Myristylated protein, 15

Naftifine
  squalene epoxidase, 225
NANRU mouse strain, 208
Nef gene, 16
Nef protein, 15
Neoplasms, 4
Neopterine, 18
Neutropenia, 80, 208
Neutrophilic granulocytes, 124
Neutrophils, 86, 126, 131, 157, 193
Nibiscus siriaca, 104
Nocardia, 29-31
Non-Hodgkin's lymphomas, 3
Nor-ketoconazole, 227
Nystatin, 96, 99, 256, 280, 327

Obtusifoliol, 231
Obtusifolione, 235-239

Oesophageal candidosis, 67, 68, 83–84, 139, 143
OKT4a, 14
Onychomycosis, 136, 143, 272, 325, 326
Oral candidosis, 67, 84 88, 143, 255, 279
  diagnosis, 279
  fluconazole, 282–283
  itraconazole, 281
  ketoconazole, 280–281, 325
Oropharyngeal candidosis, 7, 94, 141

P13, 13–14
P18, 13
P25, 13
P450-haem iron complex
  azole antifungals, 226–227
Papovavirus, 4, 9
Paracoccidioides brasiliensis, 9, 30, 191
Paracoccidioidomycosis, 147
  Brasil, 42
  itraconazole, 269
  ketoconazole, 269
Paronychia, 143
PCP, see *Pneumocystis carinii* pneumonia
Penicilliosis
  animal models, 219
Penicillium italicum, 237
Penicillium marneffei, 9, 29, 41, 147–150, 152, 219
  5-FC ,148
  amphotericin B, 148
  clinical aspects, 148
  itraconazole, 148, 219
  ketoconazole, 148
  miconazole, 148
Pentamidine, 172
  Pneumocystis carinii, 55
Pericarditis, 41
Perlèche, 280
Peroxidase, 124
Phaeohyphomycetes, 151, 207
Phagocytosis, 131
Phenytoin
  fluconazole, 201
  itraconazole, 322
Pityriasis capitis, 43
Pityriasis versicolor, 138, 144, 325
Pityrosporosis, 135, 325
Pityrosporum, 42, 135, 141, 144
  antigen, 144
  ketoconazole, 144
Pityrosporum folliculitis, 144
Pityrosporum ovale, 31–32, 43, 137–139, 325, 327
  ecological aspects, 136
    itraconazole, 327
    ketoconazole, 327
    pathology, 138
Pneumocystis carinii, 7, 27, 37, 55, 58, 94, 116, 155, 192
  ß-1,3 glucan, 60

  chromosomes, 55
  cysts, 57
  dihydrofolate reductase, 61
  DNA content, 60
  Golgi apparatus, 58
  life cycle, 55, 56
  mitochondria, 55
  N-acetylglucosamine, 60
  nucleus, 55
  phylogenetic tree, 59
  pneumonia,3, 4, 8, 29, 55, 76, 98, 141, 172, 307
  rRNA, 61
    sequences, 59
  sporozoites, 55
  5s rRNA, 59
  taxonomy, 55, 61
  thymidylate synthase, 61
  trimethoprim-sulphamethoxazole, 172
  trophozoites, 55–57
Pneumocystis carinii pneumonia, 3, 4, 8, 29, 76, 98, 141, 158, 172, 307
Polymorphonuclear leukocytes, 142
Polymorphonuclear phagocytes, 157
Polyproteins, 15
Prednisolone, 208
Prostaglandin E2, 86
Protease, 15
Proviral DNA, 14, 22
Pseudallescheria boydii, 29, 156
  ketoconazole, 267
Pseudomonas putida
  cytochrome P450, 225
Pulmonary candidosis, 98
Pulmonary cryptococcosis
  itraconazole, 305, 313

Ranitidine, 187
Ras oncogen, 15
14-Reductase, 224
Regulator of viral expression, 15
Reticuloendotheliosis, 164
Retinoic acid, 228
Rev gene, 16
Reverse transcriptase, 15
Rhinocladiella atrovirens, 9
Rhizopus spp.
  amphotericin B, 218–219
Rhodotorula, 30
Rifampicin, 201
  fluconazole, 200–201, 301
  itraconazole, 200, 313, 322
  ketoconazole, 251, 267

Saccharomyces cerevisiae, 59, 60
  cytochrome oxidase, 111
  DNA, 111
  P45014DM, 225, 226
Saccharomycosis, 151
Salmonella, 8
Saperconazole, 220
  Candida
    cytochrome P450, 232

Sarcoidosis, 115
SCH 39304, 187, 204, 267, 271
  Aspergillus, 268
  coccidioidomycosis, 171
  cryptococcosis, 271
  Fonsecaea pedrosoi, 268
  histoplasmosis, 187, 271
  in vitro potency, 187
  Sporothrix schenckii, 268
  pharmacokinetics, 268
Schizosaccharomyces octosporus, 61
Seborrheic dermatitis, 30, 42, 138-
      139, 144, 325
  Pityrisporum, 42
Simian Immunodeficiency Virus (SIV),
      13, 15
Skin, 136, 138, 143, 144, 150, 325
  blistering, 257-261
  candidosis
      animal models, 211
        azole antifungals, 211
  cytochrome P450, 228, 251
  retoic acid metabolism, 228
Sodium valproate
  itraconazole, 322
Sporothrix schenckii, 42, 150, 158,
      191
Sporotrichosis, 29, 147, 150, 152,
      156
  fluconazole, 268
  itraconazole, 268
  ketoconazole, 268
  SCH 39304, 268
Squalene epoxidase, 223
  Candida albicans, 225
  inhibitors, 224, 225
  liver
      guinea pig, 225
      rat, 225
  naftifine, 225
  Trichophyton mentagrophytes, 225
  thiocarbamates, 225
  terbinafine, 225
Starling birds, 164
Stevens Johnson syndrome, 202
Streptozotocin, 208, 211
Strongyloidiasis, 4
Suppressor T-cells, 85, 88

Tat gene, 16
T-cells, 19, 28, 41, 84-88, 115,
      155, 193
  proliferation, 86, 87
T-helper cells, 27-28, 95, 126, 130,
      158
Terbinafine, 326
  dermatophytes, 272
  onychomycosis, 271
  squalene epoxidase, 225
Terconazole
  14α-demethylase, 225
Testosterone, 201
Tetanus toxoid, 192
Thiocarbamates, 225
Thrombocytopenia, 18, 181

Thrush, 7, 37, 76, 94-95, 97, 280
Thymidylate synthase, 60, 223
Thymocyte activating factor, 18
Tinea capitis, 32, 43
Tinea corporis, 43, 143
Tinea cruris, 43, 143
Tinea pedis, 32, 43, 136, 143
Tingo Maria fever, 164
Tolciclate
  squalene epoxidase, 225
Tolnaftate
  squalene epoxidase, 225
Torulopsis candida, 9
  fluconazole, 283
Torulopsis glabrata, 9, 141, 211
      (see also *Candida glabrata*)
Toxoplasma encephalitis, 297
Toxoplasma gondii, 4, 7, 39, 76
Transferrin, 141
Tremella, 104
Trichophytin, 192
Trichophyton, 28
  allylamines, 225
Trichophyton beigelii, 265
Trichophyton interdigitale, 31
Trichophyton mentagrophytes, 208,
      225
Trichophyton rubrum, 9, 31, 43, 136,
      142
Trichosporon, 30
Trichosporon beigelii, 31, 32, 43,
      151
Trichosporonosis, 151
Trimethoprim-sulphamethoxazole, 172
  Pneumocystis carinii, 55
tRNA-lysine, 14
T-suppressor cells, 36
Tuberculosis, 8, 148
Tuberculous meningitis, 117
Tumor necrosis factor, 18

Urinary cryptococcosis
  itraconazole, 313

Vaginal candidosis, 67, 83, 86, 95,
      97, 211-212, 220, 265
  animal models, 211
Vaginal *Candida glabrata*, 212, 220
Viral genome, 13-14
  protein x, 15
  reverse transcriptase, 14
  RNA, 14
Virion infectively factor 15
Vulvovaginal candidosis, 97

Warfarin, 200-201
World Health Organization, 4

Yersinia enterolytica, 220

Zidovudine, 185, 307, 315, 320
  ketoconazole, 262
Zygomycosis, 151
  animal models, 218